Lecture Notes in Computer Science 15610

Founding Editors

Gerhard Goos
Juris Hartmanis

Editorial Board Members

Elisa Bertino, *Purdue University, West Lafayette, IN, USA*
Wen Gao, *Peking University, Beijing, China*
Bernhard Steffen ⓘ, *TU Dortmund University, Dortmund, Germany*
Moti Yung ⓘ, *Columbia University, New York, NY, USA*

The series Lecture Notes in Computer Science (LNCS), including its subseries Lecture Notes in Artificial Intelligence (LNAI) and Lecture Notes in Bioinformatics (LNBI), has established itself as a medium for the publication of new developments in computer science and information technology research, teaching, and education.

LNCS enjoys close cooperation with the computer science R & D community, the series counts many renowned academics among its volume editors and paper authors, and collaborates with prestigious societies. Its mission is to serve this international community by providing an invaluable service, mainly focused on the publication of conference and workshop proceedings and postproceedings. LNCS commenced publication in 1973.

Martin S. Krejca · Markus Wagner
Editors

Evolutionary Computation in Combinatorial Optimization

25th European Conference, EvoCOP 2025
Held as Part of EvoStar 2025, Trieste, Italy, April 23–25, 2025
Proceedings

 Springer

Editors
Martin S. Krejca
Institut Polytechnique de Paris
Ecole Polytechnique
Palaiseau, France

Markus Wagner
Monash University
Clayton, VIC, Australia

ISSN 0302-9743　　　　　　　　ISSN 1611-3349 (electronic)
Lecture Notes in Computer Science
ISBN 978-3-031-86848-1　　　　ISBN 978-3-031-86849-8 (eBook)
https://doi.org/10.1007/978-3-031-86849-8

© The Editor(s) (if applicable) and The Author(s), under exclusive license to Springer Nature Switzerland AG 2025

This work is subject to copyright. All rights are solely and exclusively licensed by the Publisher, whether the whole or part of the material is concerned, specifically the rights of translation, reprinting, reuse of illustrations, recitation, broadcasting, reproduction on microfilms or in any other physical way, and transmission or information storage and retrieval, electronic adaptation, computer software, or by similar or dissimilar methodology now known or hereafter developed.
The use of general descriptive names, registered names, trademarks, service marks, etc. in this publication does not imply, even in the absence of a specific statement, that such names are exempt from the relevant protective laws and regulations and therefore free for general use.
The publisher, the authors and the editors are safe to assume that the advice and information in this book are believed to be true and accurate at the date of publication. Neither the publisher nor the authors or the editors give a warranty, expressed or implied, with respect to the material contained herein or for any errors or omissions that may have been made. The publisher remains neutral with regard to jurisdictional claims in published maps and institutional affiliations.

This Springer imprint is published by the registered company Springer Nature Switzerland AG
The registered company address is: Gewerbestrasse 11, 6330 Cham, Switzerland

If disposing of this product, please recycle the paper.

Preface

A plethora of real-world optimization problems owe their hardness to the combinatorial explosion resulting from the sheer amount of the problems' parameters and their range. Methods from evolutionary computation lend themselves naturally to solving such problems, as their iterative and randomized nature allows for great flexibility as well as problem-specific adjustments, enabling users to focus on the most important problem aspects. The success of such an approach is proven each year anew, with improved algorithms, analysis methods, as well as rigorous guarantees, going beyond the state of the art. By now, the domain of evolutionary computation spans a wide spectrum of diverse algorithmic paradigms, problems, and topics, ranging from mathematical guarantees about the algorithms' performance and behavior over the design of better algorithms and harder and more informative benchmarks to the large-scale application of these algorithms to problems that challenge the limits of what is currently feasible. We are very happy to contribute, as every year, to this exciting and important field with a great selection of top papers that cover a broad range of the defining aspects of evolutionary computation. The articles in this volume showcase recent theoretical and experimental advances in combinatorial optimization, evolutionary algorithms, metaheuristics, and related research fields.

This volume contains the proceedings of EvoCOP 2025, the *25th European Conference on Evolutionary Computation in Combinatorial Optimisation*. The conference was held in the lovely city of Trieste, Italy, during April 23–25, 2025. The EvoCOP conference series started in 2001, with the first workshop specifically devoted to evolutionary computation in combinatorial optimization, and it became an annual conference in 2004. EvoCOP 2025 was organized together with EuroGP (the 28th European Conference on Genetic Programming), EvoMUSART (the 14th International Conference on Artificial Intelligence in Music, Sound, Art and Design), and EvoApplications (the 28th European Conference on the Applications of Evolutionary Computation, formerly known as EvoWorkshops), in a joint event collectively known as EvoStar 2025. Previous EvoCOP proceedings were published by Springer in the *Lecture Notes in Computer Science* series (LNCS volumes 2037, 2279, 2611, 3004, 3448, 3906, 4446, 4972, 5482, 6022, 6622, 7245, 7832, 8600, 9026, 9595, 10197, 10782, 11452, 12102, 12692, 13222, 13987, and 14632). The table on the next page reports the statistics for each of the previous conferences.

This year, 16 out of 43 papers were accepted after a rigorous double-blind process in which submissions received on average 3.9 reviews each, resulting in a 37% acceptance rate. We would like to acknowledge the quality and timeliness of our high-quality and diverse program committee members' work. Each year, the members give freely of their time and expertise in order to maintain the high standards in EvoCOP and to provide constructive feedback to help authors improve their papers. Decisions considered both the reviewers' reports and the evaluation of the program chairs. The 16 accepted papers cover a variety of topics, ranging from benchmark creation, over genetic programming, heuristics for real-world and NP-hard problems, as well as the foundations of evolutionary computation algorithms and other search heuristics, to both mixed-binary and multi-objective optimization. Fundamental and methodological aspects deal with the solution quality aspects and run time analysis of evolutionary algorithms and estimation-of-distribution algorithms as well as with the construction of good benchmarks and visualizing them. Applications cover problem domains such as routing, scheduling, and packing problems. We believe that the range of topics covered in this volume reflects the current state of research in the fields of metaheuristics and combinatorial optimization.

EvoCOP	LNCS vol.	Submitted	Accepted	Acceptance (%)
2025	15610	43	16	37.2
2024	14632	28	12	42.9
2023	13987	32	15	46.8
2022	13222	28	13	46.4
2021	12692	42	14	33.3
2020	12102	37	14	37.8
2019	11452	37	14	37.8
2018	10782	37	12	32.4
2017	10197	39	16	41.0
2016	9595	44	17	38.6
2015	9026	46	19	41.3
2014	8600	42	20	47.6
2013	7832	50	23	46.0
2012	7245	48	22	45.8
2011	6622	42	22	52.4
2010	6022	69	24	34.8
2009	5482	53	21	39.6
2008	4972	69	24	34.8
2007	4446	81	21	25.9
2006	3906	77	24	31.2
2005	3448	66	24	36.4
2004	3004	86	23	26.7
2003	2611	39	19	48.7
2002	2279	32	18	56.3
2001	2037	31	23	74.2

We would like to express our appreciation to the various people and institutions making EvoCOP 2025 a successful event. First, we thank the local organization team, led by Luca Manzoni, Eric Medvet, Giorgia Nadizar, and Gloria Pietropolli from the

University of Trieste in Italy. Our acknowledgments also go to SPECIES, the Society for the Promotion of Evolutionary Computation in Europe and its Surroundings. We extend our acknowledgments to Nuno Lourenço from the University of Coimbra, Portugal, for his dedicated work with the submission and registration system, to João Correia from the University of Coimbra, Portugal, for the EvoStar publicity and social media service, to Francisco Chicano from the University of Málaga, Spain, for managing the EvoStar website, and to Sérgio Rebelo and Jessica Parente from the University of Coimbra, Portugal, for their important graphic design work. We wish to thank our prominent keynote speakers, Daniela Besozzi and Tea Tušar. Finally, we express our appreciation to Anna I. Esparcia-Alcázar from SPECIES, Europe, whose considerable efforts in managing and coordinating EvoStar helped toward building a unique, vibrant, and friendly atmosphere.

Special thanks also to Christian Blum, Francisco Chicano, Carlos Cotta, Peter Cowling, Jens Gottlieb, Jin-Kao Hao, Jano van Hemert, Bin Hu, Arnaud Liefooghe, Manuel Lopéz-Ibáñez, Peter Merz, Martin Middendorf, Gabriela Ochoa, Luís Paquete, Günther R. Raidl, Thomas Stützle, Sébastien Verel, and Christine Zarges for their hard work and dedication at past editions of EvoCOP, making this one of the reference international events in evolutionary computation and metaheuristics.

April 2025

Martin S. Krejca
Markus Wagner

Organization

Organizing Committee

Conference Chairs

Martin S. Krejca	École Polytechnique, IP Paris, France
Markus Wagner	Monash University, Australia

Local Organization

Luca Manzoni	University of Trieste, Italy
Eric Medvet	University of Trieste, Italy
Giorgia Nadizar	University of Trieste, Italy
Gloria Pietropolli	University of Trieste, Italy

Publicity Chair

João Correia	University of Coimbra, Portugal

EvoStar Coordinator

Anna Esparcia-Alcázar	Universitat Politècnica de València, Spain

EvoCOP Steering Committee

Christian Blum	Artificial Intelligence Research Institute (IIIA-CSIC), Spain
Francisco Chicano	University of Málaga, Spain
Peter Cowling	Queen Mary University of London, UK
Jens Gottlieb	SAP AG, Germany
Jin-Kao Hao	University of Angers, France
Bin Hu	AIT Austrian Institute of Technology, Austria
Arnaud Liefooghe	University of Lille, France
Manuel Lopéz-Ibáññez	University of Manchester, UK
Martin Middendorf	University of Leipzig, Germany
Gabriela Ochoa	University of Stirling, UK
Luís Paquete	University of Coimbra, Portugal

Günther Raidl   Vienna University of Technology, Austria
Jano van Hemert   Optos, UK
Sébastin Verel   University of the Littoral Opal Coast, France
Christine Zarges   Aberystwyth University, UK

Society for the Promotion of Evolutionary Computation in Europe and its Surroundings (SPECIES)

Penousal Machado (President)
Mario Giacobini (Secretary)
Francisco Chicano (Treasurer)

Program Committee

Richard Allmendinger   University of Manchester, UK
Maria João Alves   University of Coimbra/INESC Coimbra, Portugal
Denis Antipov   Sorbonne University, France
Matthieu Basseur   University of the Littoral Opal Coast, France
Christian Blum   Spanish National Research Council (CSIC), Spain
Jakob Bossek   Paderborn University, Germany
Alexander Brownlee   University of Stirling, UK
Maxim Buzdalov   Aberystwyth University, UK
Josu Ceberio   University of the Basque Country, Spain
Francisco Chicano   University of Málaga, Spain
Carlos Cotta   University of Málaga, Spain
Nguyen Dang   St. Andrews University, UK
Bilel Derbel   University of Lille, France
Karl Doerner   University of Vienna, Austria
Benjamin Doerr   École Polytechnique, IP Paris, France
Mehdi El Krari   British Antarctic Survey, UK
Mohamed El Yafrani   Aalborg University, Denmark
Talbi El-Ghazali   University of Lille, France
Jonathan Fieldsend   University of Exeter, UK
Carlos M. Fonseca   University of Coimbra, Portugal
Carlos García-Martínez   University of Córdoba, Spain
Adrien Goeffon   University of Angers, France
Andreia Guerreiro   University of Coimbra, Portugal
Jin-Kao Hao   University of Angers, France
Geir Hasle   SINTEF Digital, Norway
Mario Hevia Fajardo   University of Birmingham, UK
Ekhine Irurozki   Télécom Paris, France

Anja Jankovic	RWTH Aachen University, Germany
Thomas Jansen	Aberystwyth University, UK
Andrzej Jaszkiewicz	Poznań University of Technology, Poland
Jiby Mariya Jose	Indian Institute of Information Technology Kottayam, India
Marie-Eleonore Kessaci	University of Lille, France
Ahmed Kheiri	University of Manchester, UK
Timo Kötzing	Hasso Plattner Institute, University of Potsdam, Germany
Frederic Lardeux	University of Angers, France
Per Kristian Lehre	University of Birmingham, UK
Johannes Lengler	ETH Zurich, Switzerland
Rhydian Lewis	Cardiff University, UK
Jose A. Lozano	University of the Basque Country, Spain
Gabriel Luque	University of Málaga, Spain
Manuel López-Ibáñez	University of Manchester, UK
Krzysztof Michalak	Wroclaw University of Economics and Business, Poland
Nysret Musliu	Vienna University of Technology, Austria
Frank Neumann	University of Adelaide, Australia
Gabriela Ochoa	University of Stirling, UK
Pietro S. Oliveto	Southern University of Science and Technology, China
Beatrice Ombuki-Berman	Brock University, Canada
Andre Opris	University of Passau, Germany
Luis Paquete	University of Coimbra, Portugal
Mario Pavone	University of Catania, Italy
Paola Pellegrini	IFSTTAR, France
Francisco Pereira	Instituto Superior de Engenharia de Coimbra, Portugal
Daniel Porumbel	CEDRIC, CNAM, France
Abraham Punnen	Simon Fraser University, Canada
Chao Qian	Nanjing University, China
Wu Qinghua	Huazhong University of Science and Technology, China
Günther Raidl	Vienna University of Technology, Austria
Elena Raponi	Leiden University, Netherlands
María Cristina Riff	Federico Santa María Technical University, Chile
Marcus Ritt	Federal University of Rio Grande do Sul, Brazil
Eduardo Arturo Rodriguez Tello	Cinvestav Tamaulipas, Mexico
Andrea Roli	University of Bologna, Italy
Jonathan Rowe	University of Birmingham, UK

Günter Rudolph	TU Dortmund University, Germany
Valentino Santucci	University for Foreigners of Perugia, Italy
Frédéric Saubion	University of Angers, France
Marcella Scoczynski Ribeiro Martins	Federal University of Technology – Paraná, Brazil
Kevin Sim	Edinburgh Napier University, UK
Thomas Stützle	Université Libre de Bruxelles, Belgium
Dirk Sudholt	University of Passau, Germany
Andrew M. Sutton	University of Minnesota, USA
Sara Tari	University of the Littoral Opal Coast, France
Renato Tinós	University of São Paulo, Brazil
Niki van Stein	Leiden University, Netherlands
Nadarajen Veerapen	University of Lille, France
Sébastien Verel	University of the Littoral Opal Coast, France
Hao Wang	Leiden University, Netherlands
Carsten Witt	Technical University of Denmark, Denmark
Furong Ye	Leiden University, Netherlands
Christine Zarges	Aberystwyth University, UK
Fangfang Zhang	Victoria University of Wellington, New Zealand
Weijie Zheng	Harbin Institute of Technology, China

Contents

A Runtime Analysis of the Multi-valued Compact Genetic Algorithm
on Generalized LEADINGONES 1
  Sumit Adak and Carsten Witt

Evolutionary Anytime Algorithms 18
  Aishwaryaprajna and Jonathan E. Rowe

Studies on Survival Strategies to Protect Expert Knowledge in Evolutionary
Algorithms for Interactive Role Mining 33
  Simon Anderer, Nicolas Justen, Bernd Scheuermann,
  and Sanaz Mostaghim

Diversification Through Candidate Sampling for a Non-iterated
Lin-Kernighan-Helsgaun Algorithm 50
  Tarek Boufar, Omar Rifki, and Matthieu Basseur

Instance Space Analysis and Algorithm Selection for a Parallel Batch
Scheduling Problem 66
  Francesca Da Ros, Luca Di Gaspero, Marie-Louise Lackner,
  and Nysret Musliu

Meta-learning of Univariate Estimation-of-Distribution Algorithms
for Pseudo-Boolean Problems 84
  Olivier Goudet, Adrien Goëffon, Frédéric Saubion, and Sébastien Verel

A Selective Vehicle Routing Problem for the Bloodmobile System 101
  Aldy Gunawan, Samuel Alan Darmasaputra, Sy Hoang Do,
  and Vincent F. Yu

A Genetic Approach to the Operational Freight-on-Transit Problem 116
  Corentin Juvigny, Diego Delle Donne, and Laurent Alfandari

LON/D — Sub-problem Landscape Analysis in Decomposition-Based
Multi-objective Optimization 133
  Arnaud Liefooghe, Gabriela Ochoa, and Sébastien Verel

Visualizing Pseudo-Boolean Functions: Feature Selection
and Regularization for Machine Learning 150
  Corentin Masson, Xavier F. C. Sánchez-Díaz, and Ole Jakob Mengshoel

Mixed-Binary Problems Optimized with Fast Discrete Solver 167
   Timo Kötzing and Aishwarya Radhakrishnan

Feature-Based Evolutionary Diversity Optimization of Discriminating
Instances for Chance-Constrained Optimization Problems 184
   *Saba Sadeghi Ahouei, Denis Antipov, Aneta Neumann,
   and Frank Neumann*

Adaptive Neighborhood Search Based on Landscape Learning: A TSP
Study 200
   Malek Sarhani and Stefan Voß

Healthcare Facility Location Problem and Fitness Landscape Analysis 217
   *Justin Scouarnec, Corinne Lucet, Sara Tari,
   and Laure Brisoux Devendeville*

Generating (Semi-)active Schedules for Dynamic Multi-mode Project
Scheduling Using Genetic Programming Hyper-heuristics 232
   Yuan Tian, Yi Mei, and Mengjie Zhang

Price-and-Branch Heuristic for Vector Bin Packing 249
   Ze Wang, Tim Süß, Nikolay Popov, and Lars Nagel

Author Index 267

A Runtime Analysis of the Multi-valued Compact Genetic Algorithm on Generalized LEADINGONES

Sumit Adak(✉) and Carsten Witt

DTU Compute, Technical University of Denmark, Kgs. Lyngby, Denmark
{suad,cawi}@dtu.dk

Abstract. In the literature on runtime analyses of estimation of distribution algorithms (EDAs), researchers have recently explored univariate EDAs for multi-valued decision variables. Particularly, Jedidia et al. gave the first runtime analysis of the multi-valued UMDA on the r-valued LEADINGONES (r-LEADINGONES) functions and Adak and Witt gave the first runtime analysis of the multi-valued cGA (r-cGA) on the r-valued OneMax function. We utilize their framework to conduct an analysis of the multi-valued cGA on the r-valued LEADINGONES function. Even for the binary case, a runtime analysis of the classical cGA on LEADINGONES was not yet available. In this work, we show that the runtime of the r-cGA on r-LEADINGONES is $O(n^2 r^2 \log^3 n \log^2 r)$ with high probability.

Keywords: Estimation of distribution algorithms · multi-valued compact genetic algorithm · genetic drift · LEADINGONES

1 Introduction

An optimization technique known as *estimation of distribution algorithms* (EDAs) builds a probabilistic model that is subsequently used to generate new search points based on earlier searches. Three main phases are involved when creating an EDA; by using the existing probabilistic model, a population of individuals is first sampled; next, the population's fitness is ascertained; and then, a new probabilistic model is generated based on the population's fitness. The main difference between them and evolutionary algorithms (EAs) is that the latter evolve a population, whilst the former evolve a probabilistic model. A number of studies [6, 10, 42] have shown that EDAs can outperform EAs.

Different probabilistic models and update techniques give rise to distinct algorithms within EDAs. Each approaches offers unique advantages and challenges, making it suitable for different types of optimization problems. According to the strength of the underlying probabilistic model, EDAs can be categorized as *univariate* and *multivariate* algorithms. Only one variable is used in the model of each problem variable by univariate algorithms, and, on the other hand, multivariate algorithms employ multiple variables to model a problem variable. A more detailed classification of EDAs is provided by Pelikan et al. [32]. A couple of popular examples of multivariate EDAs include the *factorized distribution algorithm* (FDA) [29], the *extended compact genetic algorithm* (ecGA) [19], the *mutual-information-maximization input clustering* (MIMIC) [9], the *bivariate marginal distribution algorithm* (BMDA) [33], and the *Bayesian optimization algorithm* (BOA) [31]. Univariate EDAs examples include the *univariate marginal distribution algorithm* (UMDA) [30], the *population-based incremental learning* (PBIL), [4], the *compact genetic algorithm* (cGA) [18], and many more. This manuscript is devoted to the theoretical investigations of univariate EDAs, particularly to the multi-valued compact genetic algorithm introduced in [5].

OneMax [28] and LEADINGONES [34] are the pseudo-Boolean functions for EDAs that are most frequently examined theoretically. Other functions have been examined as well, the probably most well-known of which is BINVAL [29]. Traditional evolutionary algorithms are frequently utilized for diverse search spaces, while EDAs are typically applied for problems that involve binary decision variables. Moreover, researchers made the first moves toward utilizing EDAs in scenarios where decision variables have more than two values [1,5,35]. Specially, Jedidia et al. [5] and Adak and Witt [1] explore univariate EDAs for multi-valued decision variables. By adding r probability values for each variable, they address a multi-valued problem. This article addresses multi-valued LEADINGONES.

In the literature on runtime analyses of EDAs, there are several analyses of the UMDA on LEADINGONES and variants [7,8]. But a runtime analysis of even the simple binary cGA on LEADINGONES was to date missing. Hence, one of our goals is to provide a first runtime analysis of the cGA on LEADINGONES. In fact, in this article, we address a more general case by providing the first runtime analysis of the multi-valued cGA (r-cGA) on the r-valued LEADINGONES (r-LEADINGONES) function, giving additional insight on the performance of multi-valued EDAs. Specifically, we bound the runtime of the r-cGA on the r-LEADINGONES problem by $O(n^2 r^2 \log^3 n \log^2 r)$ (Theorem 2) with high probability (defined as probability $1 - o(1)$). For $r = O(1)$, this bound is close to the typical runtime around $\Theta(n^2)$ that simple EAs [16] and other EDAs with optimal parameter settings [12] have on this problem.

The manuscript is organized as follows: Sect. 2 defines the multi-valued LEADINGONES function and summarizes the earlier work on our technical domains. Section 3 elaborates on the multi-valued EDA framework for the multi-valued cGA. The main technical results including genetic drift analysis and the runtime execution of the r-cGA on r-valued LEADINGONES function are presented in Sect. 4 and 5. For the *hypothetical population size* K, which is the key parameter of the r-cGA, the experiments in Sect. 6 present the empirical runtime across the parameter. Finally, the manuscript concludes with a brief summary. Due to space restrictions, some of our proofs are not included in this paper. They can be found in the preprint [2].

2 Background

2.1 Preliminaries

We focus on the maximization of functions of the form $f: \{0,1,\ldots,r-1\}^n \to \mathbb{R}$, also called r-valued (or multi-valued) fitness functions. For an individual $x \in \{0,1,\ldots,r-1\}^n$, we call $f(x)$ the *fitness* of x.

Let $n \in \mathbb{N}_{\geq 1}$ and $r \in \mathbb{N}_{\geq 2}$. We state the definition of r-LEADINGONES as already defined in [5]. For all $x = (x_1,\ldots,x_n) \in \{0,1,\ldots,r-1\}^n$, we have

$$r\text{-LEADINGONES}(x) := \sum_{i=1}^{n} \prod_{j=1}^{i} \mathbb{1}\{x_j = 0\}$$

and the function returns the number of consecutive 0s starting from the leftmost position. Note that the unique maximum is the all-0s string in r-LEADINGONES function. However, a more general version can be defined by choosing an arbitrary optimum $a \in \{0,\ldots,r-1\}^n$, and defining, for all $b \in \{0,\ldots,r-1\}^n$, $r\text{-LEADINGONES}_{a,\sigma}(b) = \sum_{i=1}^{n} \prod_{j=1}^{i} \mathbb{1}\{b_\sigma(j) = a_\sigma(j)\}$, where σ is a permutation of $\{1,\ldots,n\}$ [3]. Note that, for r-LEADINGONES, the maximum fitness value is n.

2.2 Related Work

In this work, we concentrate on the runtime evaluation of r-valued compact genetic algorithm (r-cGA) on the multi-valued LEADINGONES function. There are many theoretical articles starting from classical evolutionary algorithms to EDAs. EDAs are commonly utilized to address a wide range of complex problems, as highlighted in recent studies [15,39]. Droste presented the first runtime analysis of the cGA on linear pseudo-Boolean functions [15]. It was also shown that the expected runtime for any function has a lower bound of $\Omega(K\sqrt{n})$ and an upper bound of $O(Kn)$ for every linear function. Furthermore, it was noted that the difference in runtime between two linear functions suggests that EDAs optimize problems within this class in distinct ways. Most theoretical research on EDAs has focused on pseudo-Boolean optimization [22]. Among the most commonly studied pseudo-Boolean functions for EDAs are OneMax and LEADINGONES [27,28]. In addition to these, BINVAL is another widely recognized function that has been explored [7], although other functions have also been analyzed [15,29]. A framework for EDAs for optimizing problems with more than two choice variables from the multi-valued domain was recently introduced by Jedidia et al. [5]. They demonstrate how the multi-valued UMDA effectively solves the r-valued LEADINGONES problem. Subsequently, Adak and Witt [1] provide the first runtime analysis of a r-valued OneMax function using the multi-valued cGA within their framework. Together, their work demonstrates how EDAs can be tailored for multi-valued problems and used to define their parameters.

The analysis of EDAs for complex problems is a particularly active field of research nowadays [22]. Very recently, Hamano et al. [17] explored a probabilistic model-based technique with a sample size of two and an underlying distribution derived from the family of categorical distributions, which they termed categorical compact genetic algorithm (ccGA). It turns out that this algorithm is equivalent to the r-cGA investigated in this paper. Theoretically, they have investigated the dependency of the number of dimensions, the number of possible categories, and the learning rate on the runtime. In the categorical domain, they have explored the tail bound of the runtime on two linear functions: categorical OneMax (COM), which is equivalent to the r-OneMax function mentioned earlier, and KVAL, an extension of the BINVAL function. Furthermore, more information regarding the theory and application of EDAs can be found in [22,23,32].

3 The Multi-valued cGA

The compact genetic algorithm (cGA) [18] is one of the most popular univariate EDAs. It has only one parameter $K \in \mathbb{R}_{>0}$, which is called hypothetical population size [10] and it maintains a vector of probabilities (called frequencies). In each iteration of the cGA, it creates two solutions independently. Further, by comparing the fitness values of the two solutions, each frequency is updated (increases or decreases) by $1/K$ in the direction of the better offspring.

An extended version of cGA is the r-cGA where it supports the multi-valued variables instead of binary only [5]. Algorithm 1 defines the r-cGA. It uses marginal probabilities (denoted as frequencies) $p_{i,j}^{(t)}$ corresponding to the probability at time t of position i and value j. Further, in each iteration it creates two solutions x and y independently. After comparing the fitness values of x and y, it updates the frequencies by $\pm 1/K$ in the direction of the better offspring. Note that K indicates the strength of the update of the probabilistic model.

More precisely, the probabilistic model of the r-cGA is defined by an $n \times r$ matrix (the frequency matrix), where each row $i \in \{1, \ldots, n\}$ forms a vector $p_i := (p_{i,j}^{(t)})_{j \in \{0,\ldots,r-1\}}$ (the frequency vector at position i). In the frequency matrix, initially each frequency is set to $1/r$, leading to a uniform distribution when sampling the first individuals. We create two individuals $x, y \in \{0, \ldots, r-1\}^n$. Then, for all $i \in \{1, \ldots, n\}$ and all $j \in \{0, \ldots, r-1\}$, the probability that x_i and y_i has value j is $p_{i,j}^{(t)}$. By comparing the fitness values of x and y, we update the frequency by $1/K$. After updating the frequencies and before restricting them to an interval (see next paragraph), each frequency vector sums to 1 in this model, because exactly one frequency is increased by $1/K$ and exactly one frequency is decreased by this same amount.

In order to avoid the fixation at 0 or 1, the framework introduced in [5] restricts all frequencies to the interval $[1/((r-1)n), 1-1/n]$; see the paper for details. Note that the restriction procedure may also update frequencies belonging to values in $\{0, \ldots, r-1\}$ that were not sampled in any of the two individuals. The restriction ensures that there is always a positive probability to sample an

---

Algorithm 1: r-valued Compact Genetic Algorithm (r-cGA) for the maximization of $f : \{0, \ldots, r-1\}^n \to \mathbb{R}$

Initialization : $t \leftarrow 0$
$$p_{i,0}^{(t)} \leftarrow p_{i,1}^{(t)} \leftarrow p_{i,2}^{(t)} \cdots \leftarrow p_{i,r-1}^{(t)} \leftarrow \tfrac{1}{r} \text{ where } i \in \{1, 2, \ldots, n\}$$

1 **while** *termination criterion not met* **do**
2   **for** $i \in \{1, 2, \ldots, n\}$ independently **do**
3        $x_i \leftarrow j$ with probability $p_{i,j}^{(t)}$ w.r.t. $j = 0, \ldots, r-1$
4        $y_i \leftarrow j$ with probability $p_{i,j}^{(t)}$ w.r.t. $j = 0, \ldots, r-1$
5   **if** $f(x) < f(y)$ **then**
6   swap x and y
7   **for** $i \in \{1, 2, \ldots, n\}$ **do**
8   **for** $j \in \{0, 1, \ldots, r-1\}$ **do**
9           $\overline{p}_{i,j}^{(t+1)} \leftarrow p_{i,j}^{(t)} + \frac{1}{K}(\mathbb{1}\{x_i = j\} - \mathbb{1}\{y_i = j\})$
10          $p_{i,j}^{(t+1)} \leftarrow$ restrict $\overline{p}_{i,j}^{(t+1)}$ to be within $[\frac{1}{(r-1)n}, 1 - \frac{1}{n}]$ (see [5])
11     $t \leftarrow t + 1$

---

individual of optimum value; on the negative side, the more complicated update mechanism for $r \geq 3$ rules out the so-called well-behaved frequency assumption [39] that has been useful for the binary cGA. In the rest of the paper, we denote $1/((r-1)n)$ as lower border and $1 - 1/n$ as upper border. We are interested in the number of function evaluations that are needed to sample a solution of optimum value. This is proportional to the value of t in the algorithm. Further, this number is referred to as *runtime* or *optimization time*.

To analyze the runtime, we define the concept of *critical position* (introduced in [12] for the binary cGA) according to r-LEADINGONES. Informally, a position is called critical if all the frequencies for value 0 of lower position (left of the current position) have gained the maximum (upper border) probability. Formally, a position $i \in \{1, \ldots, n\}$ is called critical if and only if the frequencies $p_{j,0}^{(t)}$ have never been greater than $1 - 1/n$ at any point in the past where $j \in \{1, \ldots, i-1\}$, and the frequency $p_{i,0}^{(t)}$ is less than $1 - 1/n$. In this paper, we define the index of critical position at time t by m_t ($t \geq 0$). Obviously, m_t is non-decreasing over time. A major part of our analysis will deal with bounding the time until m_t increases by at least 1.

4 Genetic Drift

In EDAs, genetic drift is the result of random fluctuations brought on by the process's stochasticity rather than a clear signal from the goal function that would cause a frequency to reach the extreme values. Researchers have examined genetic drift in EDAs in detail in a number of runtime analyses [11,14,25,39–41], as well as in the works of Shapiro [36–38]. Given the significance of having a solid grasp of genetic drift, we now apply the framework from [13] and build on the insights from [1,5] to study genetic drift specifically for the r-cGA.

In this section, we will prove an upper bound on the effect of genetic drift for r-cGA in a similar fashion as Ref. [1,5,13]. This allows us to determine the parameter values for EDAs that avoid the usually unwanted effect of genetic drift. In the following section, we discuss genetic drift and prove a concentration result for *neutral positions*. An upper bound for positions with *weak preference* is also included. Next, we first describe the stochastic processes underlying the probabilistic model in the algorithm.

4.1 Behavior of the Probabilistic Model

We look in detail into how the r-cGA optimized r-LEADINGONES and define the change in frequency in one step as $\Delta_{i,j} := \Delta_{i,j}^t := p_{i,j}^{(t+1)} - p_{i,j}^{(t)}$, where $i \in \{1, \ldots, n\}$ and $j \in \{0, \ldots, r-1\}$. Particularly, we are interested in $\Delta_{i,0}$ which is crucial to find the optimum on r-LEADINGONES. The decision to update the frequency in the next step depends on the strings x and y sampled at current time. Specially, we inspect the effect of a particular position in the r-LEADINGONES values. To find this, we calculate the fitness of strings x and y up to position i ($i > 1$) as $r\text{-LO}_i(x) := \sum_{k=1}^{i} \prod_{j=1}^{k} \mathbb{1}\{x_j = 0\}$ and $r\text{-LO}_i(y) := \sum_{k=1}^{i} \prod_{j=1}^{k} \mathbb{1}\{y_j = 0\}$. Then, r-LEADINGONES experiences two kinds of steps which are called *random-walk step (rw-step)* and *biased step (b-step)* [39] depending on the value of position i and lead to an increase or decrease of frequencies. In the rest of this whole section, we temporarily ignore the corrections made to frequencies after clamping them to the interval $[\frac{1}{(r-1)n}, 1 - \frac{1}{n}]$. A closer analysis of the update scheme of the r-cGA, as defined in [5], reveals that this correction will decrease the value of a frequency by no more than $\frac{1}{(n-1)(r-1)}$ compared to the case without borders. Also, for such a correction to happen, a value whose frequency equals the lower border $\frac{1}{(r-1)n}$ has to be sampled. By a union bound over at most $r-1$ such values, this happens with probability at most $1/n$, so the expected negative correction of frequencies after clamping is at most $\frac{1}{(r-1)(n-1)^2}$. This term will be considered in our drift analyses in Sect. 5.

Starting from the left-most position (position 1), if the lowest position which is sampled differently in the two individuals x and y is less than i ($i > 1$), then position i is not relevant for the ranking of the samples and it performs a random-walk step. Formally, if the position $i > \min\{r\text{-LO}_i(x), r\text{-LO}_i(y)\}$, then the value of position i has no impact on the decision to update with respect to string x or y. Hence, the frequency will be increased or decreased by $1/K$ with identical probability, which means $p_{i,0}^{(t+1)} = p_{i,0}^{(t)} \pm 1/K$ with probability $p_{i,0}^{(t)}(1 - p_{i,0}^{(t)})$. Otherwise, it keeps the same value as previously $p_{i,0}^{(t+1)} = p_{i,0}^{(t)}$ with the remaining probability.

A biased step at position i occurs under the following condition. The positions left of i are sampled as all 0 in the two individuals x and y. That means the fitness of x and y, restricted to the first $i-1$ positions, is $r\text{-LO}_{i-1}(x) = r\text{-LO}_{i-1}(y) =$

$i-1$. In that case, if $x_i = 0$ and $y_i \neq 0$ (or $x_i \neq 0$ and $y_i = 0$), then position i determines the decision whether to update with respect to string x or y. Hence, both the events of sampling the position i increase the frequency of value 0. This scenario is called a biased step where the selection between x and y yields a bias towards increasing the frequency of $p_{i,0}^{(t)}$. So, the frequency is updated by $p_{i,0}^{(t+1)} = p_{i,0}^{(t)} + 1/K$ with probability $2p_{i,0}^{(t)}(1 - p_{i,0}^{(t)})$. Otherwise, the frequency keeps the same value as $p_{i,0}^{(t+1)} = p_{i,0}^{(t)}$ with the remaining probability.

To start our runtime analysis, we will analyze the stochastic process of the frequency for value 0 at the first position. We analyze the growth of the frequency $p_{1,0}^{(t)}$ from its starting value of $1/r$ to its maximum of $1 - 1/n$. The value of $p_{1,0}^{(t)}$ changes if the two samples of r-cGA differ in the first position in the following sense: one individual samples as 0 and the another one samples as non-zero. Since we are considering only the first position on r-LEADINGONES, this means that the individual sampling the 0 has the higher fitness value and the frequency of $p_{1,0}^{(t)}$ increases. Note that position 1 is the only position where only biased steps occur, and for all following positions, rw-steps can occur.

4.2 Analysis of Genetic Drift for the r-cGA

Genetic drift is usually studied according to the behavior of a *neutral* position of a fitness function. Let f be an r-valued fitness function. We call a position $i \in \{1, \ldots, n\}$ *neutral* (w.r.t. to f) if and only if, for all $x \in \{0, \ldots, r-1\}^n$, and the value of x_i has no influence on the value of f. More formally, a position i is neutral if, for all individuals $x, x' \in \{0, \ldots, r-1\}^n$, whenever $x_j = x'_j$ for all $j \in \{1, \ldots, n\} \setminus i$, it holds that $f(x) = f(x')$. A greater portion of this section follows closely [1] and is further adjusted to the present paper.

In the analysis of genetic drift, an important property of neutral variables is that their frequencies in typical EDAs without margins form martingales [13]. This observation applies to EDAs for the binary representations. Further, this statement extends to the r-cGA [1, Lemma 1], where they proved that all the frequencies belonging to neutral positions are martingales.

In [1,5], all frequencies of an EDA start at a value $1/r$. They analyze the progress of the expected value of the frequency and tolerate smaller deviations of the actual frequency value from this expected value up to $1/(2r)$ in either direction. In this article, for the r-cGA we follow the same frequencies setting starting from $1/r$ and tolerate a deviation up to $1/(2r)$ in either direction.

Further, we are applying a martingale concentration result [26, Theorem 3.15] to analyze the behavior of frequencies in r-cGA at neutral positions. In [1, Theorem 1], we provide a concentration bound for a martingale difference sequence. Then, in [1, Theorem 2], we show how the frequencies of neutral positions remain concentrated around their initial value, with a probability bound that depends on population size. In many situations, positions are not neutral for a given fitness function. However, we prove that the results on neutral positions translate to positions where one value is better compared to all other values. This is referred to as *weak preference* [13]. Formally, we can say that an r-valued fitness function f has a weak preference for a value $j \in \{0, \ldots, r-1\}$ at a position $i \in \{1, \ldots, n\}$, if and only if, for all $x_1, \ldots, x_n \in \{0, \ldots, r-1\}$, it holds that

$$f(x_1, \ldots, x_{i-1}, x_i, x_{i+1}, \ldots, x_n) \le f(x_1, \ldots, x_{i-1}, j, x_{i+1}, \ldots, x_n).$$

By applying [1, Theorem 3], we can extend the results of [1, Theorem 2] to positions with weak preference.

Theorem 1. *Let f be an r-valued fitness function with a weak preference for 0 at position $i \in \{1, \ldots, n\}$. Consider the r-cGA optimizing f with parameter K. Let $T \in \mathbb{N}$. Then we have*

$$\Pr\left[\min_{t \in \{0, \ldots, T\}} p_{i,0}^{(t)} \le p_{i,0}^{(0)} - \frac{1}{2r}\right] \le 2\exp\left(-\frac{K^2}{8Tr^2}\right).$$

5 Runtime Analysis

In this section, we present the runtime results of the r-cGA (Algorithm 1) on r-LEADINGONES. To prove our results, we have used the methods of *occupation probabilities* as given in Lemma 2 [21]. The concept is that the frequency (for sampling a zero) stays close to a so-called target state 0 after having been there once, where state 0 corresponds to frequency value $1 - 1/n$. To prove this, we will relate the frequency value to a Markov process X_t ($t \ge 0$) on $\mathbb{R}$ with a drift towards 0. In particular, the analysis will exploit that the occupation probability of state exactly 0 can only be reduced slightly in one iteration of the r-cGA.

Similar to previous works, we will derive bounds on the runtime of the r-cGA that hold with high probability (i.e., with probability $1 - o(1)$). Bounding the expected runtime presents additional challenges, which could be addressed in future work. The runtime analysis crucially depends on the event that no frequency drops below $1/(2r)$ by genetic drift. Our main runtime result is formulated in the following theorem.

Theorem 2. *With high probability, the runtime of the r-cGA on the function r-LEADINGONES with $K \ge cnr^2 \log^2 n \log r$ for a sufficiently large $c > 0$ and $K = o(n^2)$, $r = \text{poly}(n)$ is $O(nK \log r \log K)$. For $K = cnr^2 \log^2 n \log r$, the bound is $O(n^2 r^2 \log^3 n \log^2 r)$.*

To prove the above theorem, we need the following lemmas dealing with the above-mentioned occupation probabilities. The following lemma bounds the probability of a frequency being at least $1 - 2/n$ under the assumption that all frequencies left of it satisfy this bound. While analyzing how a frequency approaches its maximum, we stop our analysis at the first point in time where it has become at least $1 - 1/n - 1/K$. This ensures that the drift bounds derived in the following are not affected by capping a frequency at its upper border. By our assumption on K from Theorem 2, we have $1 - 1/n - 1/K = 1 - 1/n - o(1/n)$, which does not change the asymptotic result.

Lemma 1. *Let $K \geq cn \ln n$ for a sufficiently large constant $c > 0$ and $K = o(n^2)$. Consider an index $i \in \{2, \ldots, n\}$ and a time $t^* \geq 0$ such that $p_{i,0}^{(t^*)} = 1 - 1/n$. For a period of length $T > 0$, assume that $p_{j,0}^{(t)} \geq 1 - 2/n$ for all $t \in [t^*, t^*+T]$ and all $j < i$. Then for all $t \in [t^*, t^*+T]$, it holds that $p_{i,0}^{(t)} \geq 1 - 2/n$ with probability at least $1 - 2Te^{1-c/(6e^4) \ln n}$.*

This lemma implies the following corollary via a straightforward union bound over at most n positions.

Corollary 1. *Let $K \geq 6e^4 cn \ln n$ for a constant $c > 0$, and $K = o(n^2)$ and $T \in \mathbf{N}^+$. Then, for any point in time $t \in [0, T]$, all frequencies for value 0 left of the current critical position are bounded from below by $1 - 2/n$ with probability at least $1 - O(Tn^{-c+1})$.*

To show our lemma, we will need the following helper result from [21]. Therein, "additive drift at least d towards 0" means that $\mathrm{E}[X_t - X_{t+1} \mid X_t] \geq d$ for all $X_t > 0$, "step size at most c" means $|X_t - X_{t+1}| \leq c$ with probability 1 for all $t \geq 0$, and "self-loop probability at least p_0" means $\mathrm{P}[X_{t+1} = X_t \mid X_t] \geq p_0$ for all $t \geq 0$.

Lemma 2 (Theorem 7 in [21]). *Let a Markov process X_t, $t \geq 0$, on $\mathbf{R}_0^+$ with additive drift at least d towards 0 be given, starting at 0 (i.e., $X_0 = 0$), with step size at most c and self-loop probability at least p_0. Then we have for all $t \in \mathbf{N}$ and $b \in \mathbf{R}_0^+$ that*

$$\mathrm{P}[X_t \geq b] \leq 2e^{\frac{2d}{3c(1-p_0)}(1-b/c)}.$$

Proof of Lemma 1: This proof is crucially based on an application of Lemma 2. We consider the stochastic process on $p_{i,0}^{(t)}$, i.e., the frequency of position i at value 0, and measure its distance from the upper border in units of $1/K$-steps; more formally, let $X_t = K(1 - 1/n - p_{i,0}^{(t)})$. For notational convenience, we assume $t^* = 0$ and obtain $X_0 = 0$ from our assumption that $p_{i,0}^{(t^*)} = 1 - 1/n$. We will set $b = K/n$, corresponding to a frequency of $1 - 2/n$, and analyze the probability of the event $E_t := X_t \geq b$ for $t \geq 0$. As long as $X_t \leq b$, the probability of updating the frequency is bounded from above by $(4/n)(1 - 2/n)$ using a union bound since in one of the two individuals, position i must be sampled as 0 and differently in the other individual. Hence, we work with a self-loop probability of at least $p_0 \geq 1 - 4/n$ before the first occurrence of E_t. Clearly, we have $c = 1$ as a bound on the step size of the scaled process.

To analyze the drift of the X_t-process, we again distinguish between random-walk and biased steps. Let B_t be the event that all positions left of i are sampled as 0 in both individuals sampled by the r-cGA at time t. Under B_t, a biased step occurs, so frequency i cannot decrease and it increases with probability $q_t := 2p_{i,0}^{(t)}(1 - p_{i,0}^{(t)})$ unless it is at the upper border already. If B_t does not occur, a random-walk step occurs and each with probability q_t, the frequency increases and decreases (again up to hitting a border). Hence, if $X_t > 0$, then an increasing step happens with probability at least $(1+\mathrm{P}[B_t])q_t/2$ and a decreasing step with probability $(1-\mathrm{P}[B_t])q_t/2$. Note that by our assumption of identifying all frequencies in $[1-1/n-1/K, 1/n]$ with the upper border, the increasing steps are not cut at the upper border. We are left with bounding the probability of B_t.

By our assumptions on the frequencies of lower index, we have that for all $t \leq T$ that $\mathrm{P}[B_t] \geq ((1-2/n)^{i-1})^2 \geq 1/e^4$ since it is sufficient to sample all positions of index less than i as 0 in both individuals. Hence, together with the probabilities of increasing and decreasing steps, we have for $X_t > 0$ a drift of at least

$$\mathrm{E}[X_t - X_{t+1} \mid X_t] \geq \left(\frac{1}{2} + \frac{1}{2e^4}\right)q_t - \left(\frac{1}{2} - \frac{1}{2e^4}\right)q_t \geq \frac{q_t}{e^4}, \tag{1}$$

which, using that $q_t \geq (2/n)(1-1/n) \geq 1/n$, gives $\mathrm{E}[X_t - X_{t+1} \mid X_t] \geq \frac{1}{e^4 n}$. Adjusting by the possible negative effects of the frequency clamping mentioned in Sect. 4, the bound is still $\frac{1}{e^4 n} - K\frac{1}{(r-1)(n-1)^2} \geq \frac{1}{2e^4 n} =: d$ since $K = o(n^2)$.

This drift bound holds at any time before E_t happens. We will use a union bound to show that the probability of ever observing E_t in T steps is small enough. Plugging in our parameters in Lemma 2, we now have for all $t \in [0, T]$ that

$$\mathrm{P}[X_t \geq b] \leq 2T e^{\frac{1}{3e^4 n(4/n)}(1-K/n)} \leq 2T e^{1-\frac{1}{6e^4}c \ln n}.$$

The lemma now follows, noting that the actual starting time is t^*. □

The primary concept for the proof of Theorem 2 is that the frequencies are likely to increase from their initial values of $1/r$. The effect of genetic drift is bounded if the update strength is selected small enough, which means that all frequencies corresponding to value 0 never fall below $1/(2r)$ with high probability. In this scenario, we demonstrate how the marginal probabilities have a tendency to shift toward their upper border, which increases the likelihood of finding the optimum. The following lemma, which also utilizes Lemma 1 in its proof, establishes a positive trend towards optimal values for the r-cGA.

Lemma 3. *If $p_{j,0}^{(t)} \geq 1 - 2/n$ for all $j < i$ and $p_{i,0}^{(t)} \leq 1 - 1/n - 1/K$, then*

$$\mathrm{E}[\Delta_{i,0} \mid p_{i,0}^{(t)}] \geq \frac{p_{i,0}^{(t)}(1 - p_{i,0}^{(t)})}{2e^4} \cdot \frac{1}{K}.$$

We are now ready to prove our main result.

Proof of Theorem 2: Recall that $p_{i,j}^{(t)}$ denote the marginal probabilities where $(i,j) \in \{1, \ldots, n\} \times \{0, \ldots, r-1\}$ at time t. We already defined the change of frequency in one step as $\Delta_{i,j} := \Delta_{i,j}^t := p_{i,j}^{(t+1)} - p_{i,j}^{(t)}$. We show that, starting with a setting where all frequencies are at least $1/r$, after $O(nK \log n \log r \log K)$ iterations with probability $1 - o(1)$ the global optimum has been found and no frequency corresponding to value 0 has dropped below $1/(2r)$. The basic idea is to use additive drift analysis with tail bounds in a series of phases to bound the expected optimization under the premise of low genetic drift, and then multiplicative drift analysis with tail bounds, which includes the above-mentioned concept of critical position and the analysis of the time until the critical position increases by 1.

To select a K that makes genetic drift unlikely, we use Theorem 1. We apply the theorem to show that in $T = c'nK \log n \log r \log K$ iterations no frequency corresponding to value 0 dropped below $1/(2r)$ with high probability where c' is a given constant. For a single frequency, the probability is bounded by $O(1/n^2)$ and by the union bound over all frequencies, the probability of at least one frequency dropping below $1/(2r)$ is still bounded by $O(1/n)$:

$$2\exp\left(-\frac{K^2}{8Tr^2}\right) \leq \frac{1}{n^2} \Leftrightarrow 2\exp\left(-\frac{K^2}{8c'nr^2 K \log n \log r \log K}\right) \leq \frac{1}{n^2}$$

$$\Leftrightarrow \left(-\frac{K}{8c'nr^2 \log n \log r \log K}\right) \leq -2\ln n.$$

Hence, we choose $K \geq cnr^2 \log^2 n \log r$ where c is a constant. The rest of the proof shows that the optimum is sampled in T iterations with high probability.

Let m_t be the index of critical position at time t, where we often drop the time index for notational convenience. Then, the main idea is to bound the time to increase m by at least 1. So, this only bounds the time for a single frequency to reach its upper bound or, more precisely, for the frequency at the critical position. To bound the time for the frequency to reach its upper border (more precisely, recall that we identify this border with the interval $[1-1/n-1/K, 1/n]$) with high probability, at first, we bound the time for $p_{m,0}^{(t)}$ to increase from $1/(2r)$ to at least $1/2$ and, after that, bound the time for $p_{m,0}^{(t)}$ to increase from $1/2$ to $1 - 1/n$.

The aim is to demonstrate that $O(K \log r \log K)$ is an upper bound on the time needed for the critical position to increase. Using Lemma 1 and Corollary 1, we will analyze how the index m_t of the critical position increases. We note that the assumption of Lemma 1 holds for T iterations with probability $1 - o(1)$ if the constant c from K is chosen large enough. Finally, the total time will be bounded by multiplying the time to increase the critical positions with the number of positions n.

Note that, although the frequency is $1/r$ after initialization of the algorithm, we pessimistically assume that it has dropped to $1/(2r)$ by the time that the index of the frequency is the critical position. Also, we have assumed above that genetic drift leads to a deviation of at most $1/(2r)$ from the expected value, so every frequency that corresponds to a neutral position does not drop below $1/r - 1/(2r)$ in the number T of steps considered above. Now, we split the time for the frequency to increase from $1/(2r)$ to $1/2$ into phases. To reach the value $1/2$ starting from $1/(2r)$, we need $r/2$ phases with an increase of $1/r$ per phase, starting from phase 2, and for phase 1, we need an increase of $3/(2r)$. We consider phase indices $k = 1, 2, \ldots, r/2$.

Further, we assume that for the starting time T_k of phase k it holds $p_{m,0}^{(T_k)} \geq k/r$ for each $m \in \{1, \ldots, n\}$. After that, by applying the additive drift theorem with tail bounds [20, Theorem 2], we bound with high probability the time T_k to conclude phase k by at most

$$\mathrm{P}[T_k \geq s_k] \leq \exp\left(-\frac{s_k \varepsilon^2}{8c^2}\right)$$

where c is a bound on the step size, ε is a bound on the drift, and s_k must satisfy $s_k \geq 2D/\varepsilon$ where D is the distance that the process should bridge. Now, in our case, $c = 1/K$ since that is the maximum change of a frequency at a time, $\varepsilon = (k/r) \cdot (1/K) \cdot (1/(2e^4))$ as already derived in Lemma 3. In general, the distance to be bridged $D \leq (k+1)/r + 1/K - k/r \leq 2/r$ for $k \geq 2$ and for $k = 1$, we have $D \leq 2/r + 1/K - 1/(2r) \leq 3/r$. So, we select $D = 3/r$ in every phase and use $s_k = 2D/\varepsilon$. (Since the frequency value k/r may not be achievable exactly, we add up to $1/K$ to hit the smallest possible frequency above k/r.)

Putting all together, we get

$$\mathrm{P}[T_k \geq s_k] \leq \exp\left(-\frac{2}{\varepsilon} \cdot \frac{3}{r} \cdot \frac{K^2 \varepsilon^2}{8}\right) \leq \exp\left(-\frac{3K^2}{8r} \cdot \frac{k}{rKe^4}\right) \leq \exp\left(-\frac{3Kk}{8r^2 e^4}\right).$$

Now plugging in our assumption $K \geq cnr^2 \log^2 n \log r$, we have

$$\mathrm{P}[T_k \geq s_k] \leq \exp\left(-\frac{3cnkr^2 \log^2 n \log r}{8r^2 e^4}\right) \leq \exp\left(-\frac{3cnk \log^2 n \log r}{8e^4}\right) = n^{-\omega(1)}.$$

Now, we take a union bound over all $r/2$ phases and still have high probability of every phase finishing within at most $2D/\varepsilon = 6e^4 K/k$ steps. Further, we sum up all the s_k to bound the total time that we consider for all phases $1, \ldots, r/2$ is $O(K \log r)$ with probability at least $1 - rn^{-\omega(1)}$. Since $r = poly(n)$, this failure probability is still $o(1)$.

After the frequency has reached at least $1/2$, we need to analyze the remaining time until $p_{m,0}^{(t)}$ attains its maximum at $1 - 1/n$. Here, we assume (according to Corollary 1) all frequencies that have reached the upper border $1 - 1/n$ before to be bounded from below by $1 - 2/n$. In this phase, we apply the multiplicative drift with tail bounds [24, Theorem 2.4.5]. We have $p_{m,0}^{(t)} \geq 1/2$ as starting point. Let $q_{i,j}^{(t)} := 1 - 1/n - p_{i,j}^{(t)}$ where $(i,j) \in \{1, \ldots, n\} \times \{0, \ldots, r-1\}$ at time t. Further, we bound $p_{m,0}^{(t)}(1 - p_{m,0}^{(t)})$ by using $p_{m,0}^{(t)} \geq 1/2$ and $1 - p_{m,0}^{(t)} = q_{m,0}^{(t)} + 1/n$, then by Lemma 3, we get

$$\mathrm{E}[q_{m,0}^{(t)} - q_{m,0}^{(t+1)}] \geq \frac{2 p_{m,0}^{(t)} (1 - p_{m,0}^{(t)})}{2Ke^4} \geq \frac{2 \cdot \frac{1}{2} \cdot (q_{m,0}^{(t)} + \frac{1}{n})}{2Ke^4} \geq \frac{(q_{m,0}^{(t)} + \frac{1}{n})}{2Ke^4} \geq \frac{q_{m,0}^{(t)}}{2K}.$$

According to our model that identifies the upper frequency border and all values at least $1 - 1/n - 1/K$, the smallest possible state is $1/K$. So, here we apply the multiplicative drift theorem [24, Theorem 2.4.5] with $x_{\min} = 1/K$, $X_t = q_{m,0}^{(t)}$ and already derived $\delta = 1/(2K)$. Then, choosing $r' > 0$, it holds for $T := \min\{t \mid X_t = 0\}$ that

$$\mathrm{P}[T > (r' + \ln(X_0/x_{\min}))/\delta] \leq e^{-r'}.$$

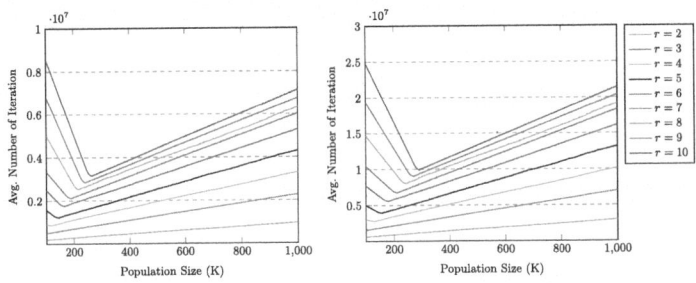

Fig. 1. Empirical runtime of the r-cGA on r-LEADINGONES; for $n = 500$ (left-hand side) and $n = 1000$ (right-hand side), $K \in \{100, \ldots, 1000\}$ and averaged over 1000 runs.

By selecting $r' = c \ln K$, where c is a constant, we get $P[T > (2c+1)K \ln K] \leq e^{-c \ln K}$. So, the remaining time for the frequency of the critical position to reach at least $1 - 1/n - 1/K$ is $O(K \log K)$ with probability $1 - o(1)$.

By adding the two stages, we have that the total time spent for a position is $O(K \log r \log K)$. Further, summing over all positions $m \in \{1, \ldots, n\}$, we obtain the total time until all frequencies for value 0 have been raised to at least $1 - 1/n - 1/K$ at least once is $O(nK \log r \log K)$ with high probability.

Note that, the r-cGA creates the optimum of r-LEADINGONES during the subsequent iteration with probability at least $(1 - \frac{2}{n})^n \geq \frac{1}{e^2}$ once for every $i \in \{1, \ldots, n\}$ it holds that $p_{i,0}^{(t)} \geq 1 - 2/n$. The probability of not creating the optimum within the next $\log n$ iterations is at most $(1 - \frac{1}{e^2})^{\log n} = o(1)$. Now, the total time to sample the optimum is $O(nK \log r \log K + \log n) = O(nK \log r \log K)$ with high probability.

Using the above choice $K = cnr^2 \log^2 n \log r$ that prevents any frequency dropping below $1/(2r)$ with high probability, the runtime of the r-cGA on r-LEADINGONES is $O(n^2 r^2 \log^2 n \log^2 r(\log n + \log r + \log \log n + \log \log r)) = O(n^2 r^2 \log^2 n \log^2 r(\log n + \log r)) = O(n^2 r^2 \log^3 n \log^2 r)$, noting that $r = poly(n)$ implies $\log r = O(\log n)$, and this holds with high probability. □

6 Experiments

In this section, we present the results of the experiments we performed to evaluate the performance of the proposed algorithm with border restrictions. We theoretically prove the expected runtime for the r-cGA on r-LEADINGONES. We implemented the algorithm using the C programming language, using the WELL1024a random number generator.

We conducted the r-cGA on r-LEADINGONES using two distinct aspects in our experiment. First, as presented in Fig. 1, we provide the average number of iterations for a variety of hypothetical population sizes ($K \in \{100, \ldots, 1000\}$). We next compare the outcomes for various $r \in \{2, \ldots, 10\}$. This empirical study

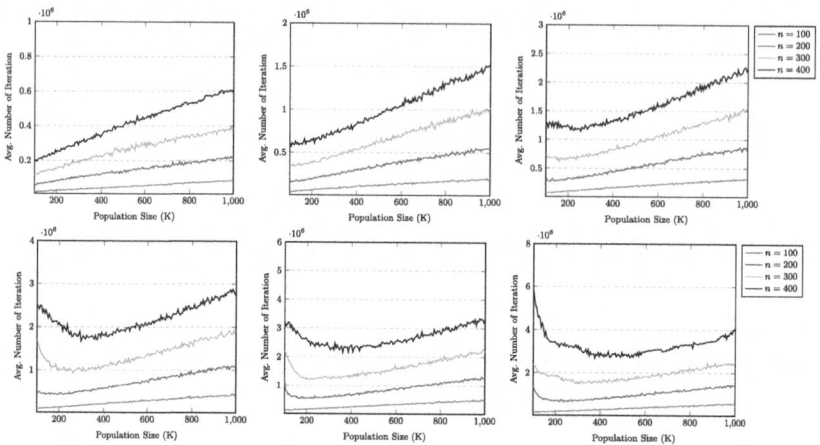

Fig. 2. Empirical runtime of the r-cGA on r-LEADINGONES; for $r = 2$ (top-left), $r = 3$ (top-middle), $r = 4$ (top-right), $r = 5$ (bottom-left), $r = 6$ (bottom-middle) and $r = 7$ (bottom-right); $n \in \{100, \ldots, 400\}$, $K \in \{100, \ldots, 1000\}$ and averaged over 200 runs.

allows us to clearly see how the runtime depends on r by comparing the findings for various r. Furthermore, we can figure out the value of K at which the minimum of runtime is met. In every instance, we note that it begins at a high value, drops to a minimum, and then rises once more throughout the remainder of K. For instance, we find that the minimum K is approximately 168 (Fig. 1: $n = 500$ and $r = 6$).

In the other aspects, we compare the results for different $n \in \{100, \ldots, 400\}$ and offer the average number of iterations for a range of hypothetical population sizes ($K \in \{100, \ldots, 1000\}$). We showed the various plots for $r \in \{2, \ldots, 7\}$ in Fig. 2. Here, we see an identical situation as in the preceding one. The empirical runtime begins at a very high number in each case, decreases to a minimum, and then increases once again for the remaining K. By comparing the outcomes, this empirical study makes it evident how the runtime depends on r. We can conclude that the theoretical analysis's bound is probably not tight based on the experimental configuration.

7 Conclusion

We have conducted a runtime analysis of a multi-valued cGA on a generalized LEADINGONES function with border restrictions and bounded its runtime with high probability. For constant r, our bound is only by polylogarithmic factors larger than the typical $\Theta(n^2)$ runtime that many randomized search heuristics exhibit on this problem. To prove the main results, we applied additive drift analysis with tail bounds and then multiplicative drift with tail bounds to analyze the growth of relevant frequencies and overall progress in the so-called critical positions. Additionally, we used occupation probabilities methods. We believe that our runtime bounds for the r-cGA on r-LEADINGONES can be improved based on the experimental results.

In this work, we have used the function r-LEADINGONES, which represents categorical values and indicates that only the appropriate value for a position can add to the fitness. The next challenge is to examine the r-cGA on multi-valued functions where each position can contribute more than two values to the fitness.

Acknowledgments. This work has been supported by the Danish Council for Independent Research through grant 10.46540/2032-00101B.

References

1. Adak, S., Witt, C.: Runtime analysis of a multi-valued compact genetic algorithm on generalized OneMax. In: Parallel Problem Solving from Nature – PPSN XVIII, pp. 53–69. Springer, Cham (2024)
2. Adak, S., Witt, C.: A runtime analysis of the multi-valued compact genetic algorithm on generalized LeadingOnes (2025). https://arxiv.org/abs/2501.09514
3. Afshani, P., Agrawal, M., Doerr, B., Doerr, C., Larsen, K.G., Mehlhorn, K.: The query complexity of finding a hidden permutation. Space-efficient data structures, streams, and algorithms: Papers in honor of J. Ian Munro on the occasion of his 66th birthday, pp. 1–11 (2013)
4. Baluja, S.: Population-based incremental learning: a method for integrating genetic search based function optimization and competitive learning. Carnegie Mellon University Pittsburgh, PA, School of Computer Science (1994)
5. Ben Jedidia, F., Doerr, B., Krejca, M.S.: Estimation-of-distribution algorithms for multi-valued decision variables. Theor. Comput. Sci. **1003**, 114622 (2024). Preliminary version in GECCO '23
6. Benbaki, R., Benomar, Z., Doerr, B.: A rigorous runtime analysis of the 2-MMASib on jump functions: ant colony optimizers can cope well with local optima. In: Proceedings of the Genetic and Evolutionary Computation Conference, pp. 4–13 (2021)
7. Chen, T., Tang, K., Chen, G., Yao, X.: Rigorous time complexity analysis of univariate marginal distribution algorithm with margins. In: Proceedings of the Eleventh Conference on Congress on Evolutionary Computation, CEC 2009, pp. 2157–2164. IEEE Press (2009)
8. Chen, T., Tang, K., Chen, G., Yao, X.: Analysis of computational time of simple estimation of distribution algorithms. IEEE Trans. Evol. Comput. **14**(1), 1–22 (2010)
9. De Bonet, J., Isbell, C., Viola, P.: Mimic: finding optima by estimating probability densities. In: Advances in Neural Information Processing Systems, vol. 9 (1996)
10. Doerr, B.: The runtime of the compact genetic algorithm on jump functions. Algorithmica **83**, 3059–3107 (2021)
11. Doerr, B., Krejca, M.S.: The univariate marginal distribution algorithm copes well with deception and epistasis. In: Proceedings of the 2020 Genetic and Evolutionary Computation Conference, pp. 17–18 (2020)
12. Doerr, B., Krejca, M.S.: A simplified run time analysis of the univariate marginal distribution algorithm on LeadingOnes. Theoret. Comput. Sci. **851**, 121–128 (2021)

13. Doerr, B., Zheng, W.: Sharp bounds for genetic drift in estimation of distribution algorithms. IEEE Trans. Evol. Comput. **24**(6), 1140–1149 (2020)
14. Droste, S.: Not all linear functions are equally difficult for the compact genetic algorithm. In: Proceedings of the 7th Annual Conference on Genetic and Evolutionary Computation, pp. 679–686 (2005)
15. Droste, S.: A rigorous analysis of the compact genetic algorithm for linear functions. Nat. Comput. **5**, 257–283 (2006)
16. Droste, S., Jansen, T., Wegener, I.: On the analysis of the (1+1) evolutionary algorithm. Theoret. Comput. Sci. **276**(1), 51–81 (2002)
17. Hamano, R., Uchida, K., Shirakawa, S., Morinaga, D., Akimoto, Y.: Tail bounds on the runtime of categorical compact genetic algorithm. Evol. Comput. 1–52 (2024)
18. Harik, G.R., Lobo, F.G., Goldberg, D.E.: The compact genetic algorithm. IEEE Trans. Evol. Comput. **3**(4), 287–297 (1999)
19. Harik, G.R., Lobo, F.G., Sastry, K.: Linkage learning via probabilistic modeling in the extended compact genetic algorithm (ECGA). In: Scalable Optimization via Probabilistic Modeling, pp. 39–61. Springer (2006)
20. Kötzing, T.: Concentration of first hitting times under additive drift. Algorithmica **75**, 490–506 (2016)
21. Kötzing, T., Lissovoi, A., Witt, C.: (1+1) EA on generalized dynamic OneMax. In: Proceedings of FOGA 2015, pp. 40–51. ACM Press (2015)
22. Krejca, M.S., Witt, C.: Theory of estimation-of-distribution algorithms. In: Doerr, B., Neumann, F. (eds.) Theory of Evolutionary Computation: Recent Developments in Discrete Optimization, pp. 405–442. Springer, Cham (2020)
23. Larrañaga, P., Lozano, J.A. (eds.): Estimation of Distribution Algorithms: A New Tool for Evolutionary Computation, vol. 2. Springer (2001)
24. Lengler, J.: Drift analysis. In: Doerr, B., Neumann, F. (eds.) Theory of Evolutionary Computation: Recent Developments in Discrete Optimization, pp. 89–131. Springer, Cham (2020)
25. Lengler, J., Sudholt, D., Witt, C.: The complex parameter landscape of the compact genetic algorithm. Algorithmica **83**, 1096–1137 (2021)
26. McDiarmid, C.: Concentration. In: Probabilistic Methods for Algorithmic Discrete Mathematics, pp. 195–248. Springer (1998)
27. Motwani, R., Raghavan, P.: Randomized Algorithms. Cambridge University Press (1995)
28. Mühlenbein, H.: How genetic algorithms really work: mutation and hillclimbing. In: Männer, R., Manderick, B. (eds.) Parallel Problem Solving from Nature, PPSN-II, pp. 15–26. Elsevier (1992)
29. Mühlenbein, H., Mahnig, T.: FDA - a scalable evolutionary algorithm for the optimization of additively decomposed functions. Evol. Comput. **7**(4), 353–376 (1999)
30. Mühlenbein, H., Paass, G.: From recombination of genes to the estimation of distributions i. binary parameters. In: International Conference on Parallel Problem Solving from Nature, pp. 178–187. Springer (1996)
31. Pelikan, M., Goldberg, D.E., Cantú-Paz, E.: BOA: the Bayesian optimization algorithm. In: Proceedings of the 1st Annual Conference on Genetic and Evolutionary Computation, pp. 525–532 (1999)
32. Pelikan, M., Hauschild, M.W., Lobo, F.G.: Estimation of distribution algorithms. In: Springer Handbook of Computational Intelligence, pp. 899–928 (2015)
33. Pelikan, M., Muehlenbein, H.: The bivariate marginal distribution algorithm. In: Roy, R., Furuhashi, T., Chawdhry, P.K. (eds.) Advances in Soft Computing, pp. 521–535. Springer, London (1999)

34. Rudolph, G.: Convergence properties of evolutionary algorithms. Verlag Dr, Kovač (1997)
35. Santana, R., Larrañaga, P., Lozano, J.A.: Protein folding in simplified models with estimation of distribution algorithms. IEEE Trans. Evol. Comput. **12**(4), 418–438 (2008)
36. Shapiro, J.L.: The sensitivity of PBIL to its learning rate, and how detailed balance can remove it. In: FOGA, pp. 115–132 (2002)
37. Shapiro, J.L.: Drift and scaling in estimation of distribution algorithms. Evol. Comput. **13**(1), 99–123 (2005)
38. Shapiro, J.L.: Diversity loss in general estimation of distribution algorithms. In: International Conference on Parallel Problem Solving from Nature, pp. 92–101. Springer (2006)
39. Sudholt, D., Witt, C.: On the choice of the update strength in estimation-of-distribution algorithms and ant colony optimization. Algorithmica **81**, 1450–1489 (2019)
40. Witt, C.: Domino convergence: why one should hill-climb on linear functions. In: Proceedings of the Genetic and Evolutionary Computation Conference, pp. 1539–1546 (2018)
41. Witt, C.: Upper bounds on the running time of the univariate marginal distribution algorithm on OneMax. Algorithmica **81**, 632–667 (2019)
42. Witt, C.: How majority-vote crossover and estimation-of-distribution algorithms cope with fitness valleys. Theoret. Comput. Sci. **940**, 18–42 (2023)

Evolutionary Anytime Algorithms

Aishwaryaprajna[1] and Jonathan E. Rowe[2(✉)]

[1] Department of Computer Science, University of Exeter, Exeter EX4 4QF, UK
aishwaryaprajna@exeter.ac.uk
[2] School of Computer Science, University of Birmingham, Birmingham B15 2TT, UK
j.e.rowe@cs.bham.ac.uk

Abstract. Evolutionary algorithms are *anytime* algorithms, as they can produce a viable solution at any time and, moreover, the quality of the solution improves over time. However, it may be hoped that more information can be gained when a solution is produced, even if the algorithm is stopped before the optimum has been found. For example, one might want to know which of the bit values in the current best solution are necessary for a good result, and which are still uncertain. We propose heuristics for efficiently gaining such information about bit values, in the context of the $(1+1)$EA. Along the way, we prove two useful general results. The first bounds the runtime for m parallel copies of a process. The second bounds the probability of having a non-optimal bit value at any given position when optimising a weakly monotonic function. Using these results, we prove bounds for the time it takes for our proposed heuristics to correctly identify bit values when the $(1+1)$EA runs on monotonic and hidden subset problems.

1 Introduction

An *anytime* algorithm is one which can produce a viable (but not necessarily optimal) solution to a problem at any time during its execution. The quality of the solution produced monotonically increases with respect to the execution time. The concept was first introduced in the context of planning problems [5] where it is essential that a decision-making algorithm produce some decision any time it is required. It was soon realised that evolutionary approaches can be naturally cast as anytime algorithms, since they maintain a population of solutions that improve over time [4]. There has been previous research on what the expected fitness will be, when stopping an evolutionary algorithm after a fixed budget of evaluations [13,14]. However, in certain contexts, one might hope for more information than simply the solution with the best so far fitness. One might also wish for details about that solution, for example which parts we are certain of, and which we are less sure about. For example, consider the well-known LEADINGONES problem, defined on the set of binary strings with length n. For a given target string $z \in \{0,1\}^n$ and bijection $\pi : \{1,\ldots,n\} \to \{1,\ldots,n\}$ we have:

$$\text{LEADINGONES}(x) = \sum_{i=1}^{n} \prod_{j=1}^{i} [x_{\pi(j)} = z_{\pi(j)}]$$

(where $[expr]$ equals 1 if $expr$ is true and otherwise 0). It is known [13] that if we run the $(1+1)$EA with mutation probability $1/n$ on LEADINGONES for, say b iterations (with some mild constraints on the value of b), then in expectation, the solution found will have $1 + 2b/n$ correct bits (up to lower order terms). However, it would also be very useful to know which bits in the given solution are the correct bits, and what the values of these are, and which bits have values that are as yet undecided.

Let $x \in \{0,1\}^n$ uniformly at random ;
while *Termination criterion not satisfied* **do**
    Let $y = x$;
    for $k = 1$ *to* n **do**
        | With probability c/n let $y_k = 1 - y_k$
    end
    if $f(y) \geq f(x)$ **then**
        | Let $x = y$;
    end
end

Algorithm 1: $(1+1)$EA with mutation probability c/n, where c is a positive constant.

We could ask similar questions of the $(1+1)$EA when run on strictly monotonic functions [8]. These fitness functions have a unique optimum $z \in \{0,1\}^n$ and satisfy the property that for any string $x \in \{0,1\}^n$, if $x_i \neq z_i$ for some index i, then changing x_i to z_i increases the fitness. Linear functions with non-negative weights are good examples. Again, if we stop before the optimal solution is found, we would like to know which bit values we are sure of, and which are still to be decided. When we are sure what a bit's value should be (according to the global optimum), we will say that it has been *confirmed*. In this paper, we will look at heuristics for determining which bits in the current solution have been confirmed. We will be interested in questions such as:

1. Will bit values always be correctly confirmed?
2. For a particular bit, what is the expected time until its value is confirmed?
3. What is the expected time until all bit values are confirmed?

It may seem intuitive that we can only confirm all the bit values once the optimal solution has been discovered by the algorithm. However, we will see that while this is the case for some classes of function (e.g. LEADINGONES), it is not true in general.

We will begin by looking at the $(1+1)$EA, with mutation probability c/n (see Algorithm 1), on strictly monotonic problems, and then extend these results to other functions, such as LEADINGONES. We will show that our proposed heuristic is not efficient for *hidden subset* problems, and we will give an alternative method for such problems, based on taking a *majority vote* in a population.

The concept of identifying a correct bit value makes sense in the context of monotonic functions with a unique optimum—the value we are looking for is the one in the optimal string. We consider whether it might also make sense for any non-monotonic functions with unique optimum. The issue here is that early stages of the search might not give us good information about the optimal bit values (consider, for example, a function where fitness increases along an arbitrary path in the hypercube). However, in special cases, such as the JUMP function, we can discover such information efficiently.

2 Strictly Monotonic Functions

Definition 1. *A function $f : \{0,1\}^n \to \mathbb{R}$ is* strictly monotonic *if there exists a string $z \in \{0,1\}^n$ such that for all $x \in \{0,1\}^n$ and all $i \in \{1,\ldots,n\}$:*

$$x_i \neq z_i \implies f(x \oplus e_i) > f(x)$$

where e_i is the string with a 1 at index i and zeros elsewhere, and $\oplus$ is exclusive-or. The string z is necessarily the unique global optimum of f.

For strictly monotonic functions, if a string contains the correct bit value at a given index, then a mutation which flips that bit and leaves the remainder untouched will produce an offspring that has lower fitness, and hence would be rejected by the $(1+1)$EA. We say that a mutation event that flips exactly one bit is *focussed* at the position of that bit. Our criterion for confirming the value of a particular bit position, then, is when a mutation focussed at that position is rejected. Acceptance of a mutated bit would not be sufficient, since the bit might be neutral with regards fitness. It is clear from the definition of strictly monotonic functions that this criterion correctly confirms bit values for this problem class. Note that the probability of a mutation focussed at a particular bit is $p = (1 - c/n)^{n-1} c/n$.

Consider a particular bit position, and how it evolves over time until its value is confirmed. We will model this as a three-state Markov chain with states 0, 1 and $\star$. State 0 corresponds to the bit having the incorrect value and being unconfirmed. State 1 corresponds to the correct value, but still unconfirmed. State $\star$ corresponds to the bit's value being confirmed. Since we are interested in the time it takes to become confirmed, we will consider this to be an absorbing state of our Markov chain. The probability that a string with the incorrect value is replaced by one with the correct value is at least p. The probability that a string with the correct value has this confirmed is exactly p. The probability that a string with the correct value is replaced by one with the incorrect value is at most

$$\left(1 - \left(1 - \frac{c}{n}\right)^{n-1}\right) \frac{c}{n} = \frac{c}{n} - p$$

since both this bit and at least one other will need to be flipped for the replacement to take place. We therefore define our Markov chain according to the state diagram in Fig. 1.

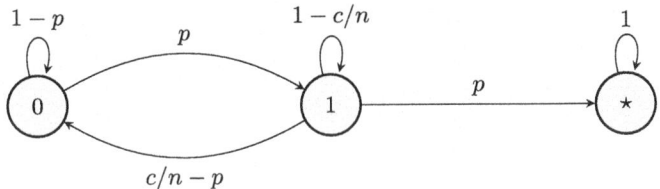

Fig. 1. Markov chain modelling the process by which a bit value gets confirmed.

By construction, the expected time it takes for a bit in a particular position to have its value confirmed is no greater than the time to absorption of this Markov chain. The transition matrix is given by

$$\begin{pmatrix} 1-p & p & 0 \\ c/n-p & 1-c/n & p \\ 0 & 0 & 1 \end{pmatrix}$$

which has fundamental matrix:

$$\frac{1}{p^2} \begin{pmatrix} c/n & p \\ c/n-p & p \end{pmatrix}.$$

We therefore see that the time to absorption[1] (and hence the time for a particular bit value to be confirmed) is at most

$$\frac{c}{np^2} + \frac{1}{p}.$$

Now, using the fact that $1 - x/2 \geq \exp(-x)$ for all $0 \leq x \leq 1$ (see, for example, [7]) we have:

$$p = \left(1 - \frac{c}{n}\right)^{n-1} \frac{c}{n} \geq \frac{e^{-2c}c}{n}$$

(for sufficiently large n) and so we have proved:

Theorem 1. *Consider the $(1+1)EA$, with mutation probability c/n, running on a strictly monotonic problem. We say that the value at any bit position is confirmed if a focussed mutation at that position is not accepted by the algorithm. Then the expected time for a particular bit value to be confirmed is at most $(e^{4c} + e^{2c})n/c$.*

In order to investigate the expected time until all the bits have their values confirmed, we use the following lemma[2]:

Lemma 1. *Let $X^{(1)}(t), \ldots, X^{(m)}(t)$, for $t \geq 0$, be m random processes (not necessarily independent) that run synchronously in parallel. For each $i = 1, \ldots, m$,*

---
[1] See any standard text on Markov chains.
[2] A similar result, but with stronger pre-conditions, is Lemma 3 of [15].

let there be an event of interest U_i, and let T_i denote the first hitting time of event U_i by process $X^{(i)}(t)$. That is,

$$T_i = \inf\{t \mid X^{(i)}(t) \in U_i\}.$$

Let τ be an upper bound for the expected time each process takes to reach its respective event of interest, regardless of the starting point. That is, for any time t and any state s, if the process $X^{(i)}$ is in state s at time t, the expected time for event U_i to occur is at most $t + \tau$. Let $T^* = \max_i T_i$. Then for any $\kappa > 0$,

$$\Pr(T^* \geq e(\kappa + 1)\tau \ln m) \leq \frac{1}{m^\kappa}.$$

Proof. From Markov's inequality:

$$\Pr(T_i \geq e\tau) \leq \frac{E[T_i]}{e\tau} \leq \frac{1}{e}.$$

Now consider running the parallel processes in phases of length $e\tau$. Since the upper bound τ applies regardless of the starting point, we can consider these phases to be independent[3]. Thus, after s such phases, the probability that a process has not achieved its event of interest is

$$\Pr(T_i \geq se\tau) \leq \frac{1}{e^s}.$$

Choosing $s = (\kappa + 1)\ln m$ we have

$$\Pr(T_i \geq e(\kappa + 1)\tau \ln m) \leq \frac{1}{m^{\kappa+1}}$$

for all $i = 1, \ldots, m$. Therefore, by the union bound,

$$\Pr(T^* \geq e(\kappa + 1)\tau \ln m) \leq \frac{1}{m^\kappa}.$$

□

A direct application of this lemma, taking the m random processes to be the values of each of the n bit positions, and whether or not they are confirmed, yields:

Theorem 2. *Consider the $(1+1)EA$, with mutation probability c/n, running on a strictly monotonic problem. We say that the value at any bit position is confirmed if a focussed mutation at that position is not accepted by the algorithm. Then for any $\kappa > 1$, all n bit values are correctly confirmed in at most $2(e^{4c+1}/c)(\kappa + 1)n \ln n$ iterations, with probability at least $1 - 1/n^\kappa$.*

---

[3] This "independent phases" argument appears to be well-known, but it is not clear where it originated. See [1] for a previous example of its use.

It is known [8,16] that for $c < 1$, the $(1+1)$EA finds the optimal solution to any strictly monotonic problem in $O(n \ln n)$ time, and our upper bound for all bit values to be confirmed is asymptotically the same. However, when $c > 2.2$, there are strictly monotonic functions for which the runtime is exponential in n. This leads us to a paradoxical situation, in which the time it takes for the algorithm, using our criterion, to know for certain the correct value for each bit, is far less than the time it takes the algorithm to actually produce the optimal string!

3 Weakly Monotonic Functions

Definition 2. *A function $f : \{0,1\}^n \to \mathbb{R}$ is weakly monotonic if there exists a target string $z \in \{0,1\}^n$ such that for all $x \in \{0,1\}^n$ and all $i \in \{1,\ldots,n\}$:*

$$x_i \neq z_i \implies f(x_i \oplus e_i) \geq f(x)$$

where e_i is the string with a 1 at index i and zeros elsewhere. The string z is necessarily a global optimum of f (though not necessarily a unique one). Note that strictly monotonic functions are also weakly monotonic functions, taking the target string to be the optimum.

Our criterion which confirms a bit value following the rejection of a focussed mutation does not necessarily give the correct results on all weakly monotonic functions. Consider the function PLATEAU_k for some fixed $1 \leq k \leq n$ given by

$$\text{PLATEAU}_k(x) = \begin{cases} \sum_i x_i & \text{if } \sum_i x_i \leq k \\ k & \text{otherwise.} \end{cases}$$

This function is weakly monotonic (taking $z = 1^n$), but if k is large enough (e.g. $k = 3n/4$) there will be several bits that have value 1 "confirmed" even though none of these are necessary to optimise the problem. The difficulty, of course, is that there is a neutral plateau of global optima. This can be alleviated if we consider only weakly monotonic functions with a unique global optimum. Our criterion will then correctly confirm the values for all the bits in this optimum string.

A simple example of a weakly monotonic function with unique global optimum is the NEEDLE function which, for a fixed target string z is defined as $\text{NEEDLE}(x) = [x = z]$. That is, it is zero on all strings except z, for which $\text{NEEDLE}(z) = 1$. Our criterion correctly confirms that each bit has to take the values given by z, but this only happens once the optimum is found, since only then will we have focussed mutations that are rejected. The $(1+1)$EA, with mutation probability c/n, takes an average of $2^n/(1-e^{-c})$ iterations to find z (see [11]). The probability of having a focussed mutation (on any bit position) is $c(1-c/n)^{n-1} \geq c\exp(-2c)$, so the time it takes for all n bits to be subject to a focussed mutation is the same as the coupon collectors problem, slowed down by a factor of at most $\exp(2c)/c$. Therefore the expected time to confirm all the bits is less than $2^n/(1-e^{-c}) + nH_n e^{2c}/c$, where H_n is the nth harmonic number.

LEADINGONES is also an example of a weakly monotonic function with unique global optimum. The focussed mutation of a correct bit value will only be rejected if all the previous bits (according to the given permutation) are already correct. The expected time for the $(1+1)$EA to correctly identify the first k bits is $\Theta(nk)$, (see, for example, [2]), and so the time taken for the kth bit value to be confirmed is $\Theta(n(k+1))$ since the expected time for a focussed mutation on a particular bit is $\Theta(n)$. We see that we can only confirm the values of all the bits once the final bit has been correctly identified. Therefore the time to confirm all the bit values is $\Theta(n^2)$. We summarise our results as follows:

Theorem 3. *Using the focussed mutation method for the $(1+1)EA$ with mutation probability c/n (where c is a constant), we have:*

	Time to confirm one bit	Time to confirm all bits
Strictly monotonic	$O(n)$	$O(n \log n)$
NEEDLE	$O(2^n)$	$O(2^n)$
LEADINGONES	$\Theta(nk)$ for bit k	$\Theta(n^2)$

4 Hidden Subset Functions

A *hidden subset* problem is defined by taking a function $f : \{0,1\}^n \to \mathbb{R}$ and embedding it within a much larger number of bits, N, with the extra bits contributing nothing to the fitness however they are set [3,6]. If the embedded function is weakly monotonic, then so is the hidden subset version. We will consider the variant in which the number of hidden bits, n, and their placement within the larger string of N bits, is unknown. For example, consider the hidden ONEMAX function, defined on a subset $S \subset \{0,1\}^N$ of size n: $\text{ONEMAX}_S(x) = \sum_{i \in S} x_i$.

If we run the standard $(1+1)$EA with mutation probability $1/N$, then the running time will be $\Theta(N \log n)$. Since we consider N to be much larger than n, the challenge is to come up with an evolutionary algorithm that reduces this runtime, and has less dependency on N. Remarkably, in [6], a mutation scheme is proposed for the $(1+1)$EA which gives a runtime of $O(n \log^{2+\epsilon} n)$ for this problem (for any fixed $\epsilon > 0$). So the runtime depends only on n and not on N. The idea is, in each iteration, to pick a random integer i with probability p_i which is inversely proportional to $i \log^{1+\epsilon} i$ and then use a mutation probability of $1/i$ for that iteration (Algorithm 2). This scheme certainly meets the challenge of producing an algorithm that solves the problem in a time that does not depend on N. However, it creates some further issues. First, given n is unknown, we do not know when to halt the algorithm. That is, we have no idea when we have found all the relevant bits. Second, even if we run the algorithm for long enough to solve the problem, we are given an optimal string of length N, but have no clue as to which of the bits in this string are the relevant ones, and which are neutral.

Let $x \in \{0,1\}^N$ uniformly at random ;
while *Termination criterion not satisfied* **do**
$\quad$ Let $y = x$;
$\quad$ Pick $i \in \mathbb{N}$ with probability inversely proportional to $i \log^{1+\epsilon} i$;
$\quad$ **for** $k = 1$ *to* N **do**
$\quad\quad$ | With probability $1/i$ let $y_k = 1 - y_k$
$\quad$ **end**
$\quad$ **if** $f(y) \geq f(x)$ **then**
$\quad\quad$ | Let $x = y$;
$\quad$ **end**
end

Algorithm 2: $(1+1)$EA with randomly chosen mutation probability [6].

Our focussed mutation criterion for confirming bit values will work on this problem, since focussed mutations on neutral bits will always be accepted, and thus neutral bits will never be confirmed. We could therefore use this approach to identifying the relevant bits and their values. However, the probability to flip one particular bit and no others is:

$$\sum_{i=2}^{\infty} \frac{p_i}{i} \left(1 - \frac{1}{i}\right)^{N-1} \leq \sum_{i=2}^{\infty} \frac{1}{i^2} \left(1 - \frac{1}{i}\right)^{N-1}.$$

Now the function

$$h(x) = \frac{1}{x^2} \left(1 - \frac{1}{x}\right)^{N-1}$$

increases from $x = 1$ to $x = (N+1)/2$ and then decreases monotonically. So

$$\sum_{i=2}^{\infty} \frac{1}{i^2} \left(1 - \frac{1}{i}\right)^{N-1} \leq \int_1^{\lfloor (N+1)/2 \rfloor} h(x) dx + h(\lfloor (N+1)/2 \rfloor) + \int_{\lfloor (N+1)/2 \rfloor}^{\infty} h(x) dx$$

$$= \int_1^{\infty} h(x) dx + h(\lfloor (N+1)/2 \rfloor)$$

$$\leq 1/N + O(1/N^2).$$

Thus, the expected waiting time for a focussed mutation on a particular bit is $\Omega(N)$, which defeats the point of the algorithm.

We can do better using the following scheme. Run $\lambda = c \ln N$ copies of the algorithm in parallel (for some constant c). For any particular bit position, if it ever happens that more than 3/4 of the population have the same value at that bit position, then confirm this to be the correct value. To explore the effectiveness of this approach, we make use of the following:

Definition 3. *A reasonable mutation scheme is one in which the probability of a bit being flipped does not depend on the value of that bit.*

Lemma 2. *Let $X(t)$, for $t \geq 0$, represent the random sequence of strings produced by a $(1+1)EA$, with any reasonable mutation scheme. Suppose the fitness function $f : \{0,1\}^n \to \mathbb{R}$ is weakly monotonic with target string z. Then for any $1 \leq k \leq n$, and for all $t \geq 0$ it holds that*

$$\Pr(X(t)_k \neq z_k) \leq 1/2.$$

Proof. Assume, without loss of generality, that the target string is $z = 1^n$. To simplify notation we will write, for fixed $1 \leq k \leq n$, $t \geq 0$ and $a, b \in \{0,1\}$

$$\Pr(a \mapsto b) := \Pr(X(t+1)_k = b \,|\, X(t)_k = a).$$

We first show that the following condition is sufficient to prove the Lemma:

$$\Pr(0 \mapsto 1) \geq \Pr(1 \mapsto 0)$$

for all $1 \leq k \leq n$, $t \geq 0$. We fix k and proceed by induction on t. The base case $t = 0$ is clear, as $X(0)$ is generated uniformly at random, so $\Pr(X(0)_k = 0) = 1/2$. Now suppose $\Pr(X(t)_k = 0) \leq 1/2$ for some $t \geq 0$. Then we have

$$\Pr(X(t+1)_k = 0) = \Pr(0 \mapsto 0)\Pr(X(t)_k = 0) + \Pr(1 \mapsto 0)\Pr(X(t)_k = 1)$$
$$= (1 - \Pr(0 \mapsto 1))\Pr(X(t)_k = 0) + \Pr(1 \mapsto 0)(1 - \Pr(X(t)_k = 0))$$
$$= \Pr(X(t)_k = 0)(1 - \Pr(0 \mapsto 1) - \Pr(1 \mapsto 0)) + \Pr(1 \mapsto 0)$$
$$\leq (1 - \Pr(0 \mapsto 1) - \Pr(1 \mapsto 0))/2 + \Pr(1 \mapsto 0)$$

(where we have used the induction hypothesis)

$$= 1/2(1 - \Pr(0 \mapsto 1) + \Pr(1 \mapsto 0))$$
$$\leq 1/2$$

by our assumed condition. To prove this condition, let $M(t)$ be the random string which indicates which bits are mutated at time step t (before a decision is taken about whether to accept the new offspring). That is, it contains a 1 in exactly those bit positions that are mutated at time t. Then, fixing bit position k and time t:

$$\Pr(0 \mapsto 1) = \sum_{m \in \{0,1\}^n} \Pr(0 \mapsto 1 \,|\, M(t) = m)\Pr(M(t) = m)$$

noting that the choice of $M(t)$ is independent of $X(t)$. Similarly, we have:

$$\Pr(1 \mapsto 0) = \sum_{m \in \{0,1\}^n} \Pr(1 \mapsto 0 \,|\, M(t) = m)\Pr(M(t) = m).$$

For any m such that $m_k = 0$ we have

$$\Pr(0 \mapsto 1 \,|\, M(t) = m) = \Pr(1 \mapsto 0 \,|\, M(t) = m) = 0$$

so we now consider cases where $m_k = 1$. Define a function:

$$\pi : \{0,1\}^n \to \{0,1\}^{n-1}$$

such that $\pi(x)$ is the string x with bit k removed. We also define a function:

$$\rho : \{0,1\}^{n-1} \times \{0,1\} \to \{0,1\}^n$$

such that $\rho(x,b)$ is the string x with bit b inserted at position k. Then

$\Pr(0 \mapsto 1 \mid M(t) = m)$

$= \sum_{x \in \{0,1\}^n} \Pr(X(t) = x) \Pr(X(t+1)_k = 1 \mid X(t) = x, X(t)_k = 0, M(t) = m)$

$= \sum_{x \in \{0,1\}^{n-1}} \Pr(\pi(X(t)) = x) \Pr(X(t+1)_k = 1 \mid X(t) = \rho(x,0), M(t) = m).$

Similarly,

$\Pr(1 \mapsto 0 \mid M(t) = m)$

$= \sum_{x \in \{0,1\}^{n-1}} \Pr(\pi(X(t)) = x) \Pr(X(t+1)_k = 0 \mid X(t) = \rho(x,1), M(t) = m).$

Notice that, for a given x and m, the second probability in each of these sums is either 0 or 1. Since the fitness function is weakly monotonic, we have, for any given $x \in \{0,1\}^{n-1}$,

$$f(\rho(x,1)) \geq f(\rho(x,0)).$$

Now let us suppose that $m_k = 1$ and

$$f(\rho(x,1) \oplus m) \geq f(\rho(x,1)).$$

That is, we suppose the offspring produce by mutating $\rho(x,1)$ is accepted by the algorithm, even though bit k is flipped from a 1 to a 0. Again, by monotonicity:

$$f(\rho(x,0) \oplus m) \geq f(\rho(x,1) \oplus m)$$

and therefore it holds that

$$f(\rho(x,0) \oplus m) \geq f(\rho(x,0)).$$

That is, for any x and m such that mutating bit k of $\rho(x,1)$ gets accepted, it also happens that mutating bit k of $\rho(x,0)$ gets accepted. Therefore

$$\Pr(1 \mapsto 0 \mid M(t) = m) \leq \Pr(0 \mapsto 1 \mid M(t) = m) \implies \Pr(1 \mapsto 0) \leq \Pr(0 \mapsto 1)$$

as required. □

We can now state our main result for our proposed scheme for hidden subset problems:

Theorem 4. *Consider a hidden subset problem on N bits, where the embedded function is weakly monotonic with a unique optimum. Suppose a $(1+1)EA$ with any reasonable mutation scheme has expected time at most τ to find an optimal solution. We run $\lambda = c \ln N$ copies of this algorithm in parallel (for some suitable constant c). We confirm a bit to have a particular value if we ever see that value in that bit position in at least $3/4$ of the runs at any particular time. All the relevant bits in the embedded problem will be confirmed correctly in $O(\tau \log \log N)$ iterations, with high probability, giving an overall cost of $O(\tau \log N \log \log N)$.*

Proof. Neutral bits, at each iteration, have a probability of $1/2$ of being either 0 or 1. Therefore the chance of finding $3/4$ of the population having the same value at a neutral bit position, at any time t, is, by Hoeffding's inequality, no more than $2/N^{c/8}$. The union bound implies that this will not happen, with high probability, for any of the neutral bits, at any time $t \geq 0$.

For the relevant bits, we first consider if it is possible to confirm an incorrect value in any bit position. Lemma 2 implies that the probability of finding an incorrect value in a relevant bit position at any time t is at most $1/2$. Therefore, again by Hoeffding's inequality, the probability of an incorrect value being confirmed at a particular position is at most $1/N^{c/8}$, and so we correctly identify all relevant bits with high probability.

The expected time for any one copy of the algorithm to find an optimal solution is at most τ, so by Lemma 1, the time for all λ copies to find an optimal solution is $O(\tau \log \log N)$ with high probability. Since the embedded function has a unique optimum, all copies of the algorithm agree on all relevant bits at this point. □

We still have a dependency on N for our overall runtime. However, it is difficult to see how to avoid this. Finding the relevant bits for the hidden ONEMAX problem allows us to find the unknown subset S. Since there are $\binom{N}{n}$ subsets of size n, we need a minimum of $\log_2 \binom{N}{n}$ bits of information to specify such a subset. Each iteration of each instance of the algorithm reveals one bit of information. Therefore it would seem we need a running time of $\Omega(n \log N)$ to confirm all n bits in a hidden subset problem.

5 Non-monotonic Functions

We have seen that the concept of confirming a bit value makes sense for the $(1+1)$EA running on weakly monotonic functions with a unique optimum (which includes strictly monotonic functions), and for some (but not all) weakly monotonic functions with a neutral plateau of global optima. We now consider whether it can also make sense for non-monotonic functions and, if so, to provide an efficient mechanism. We answer the question in the affirmative by considering the $(1+1)$EA, with mutation probability $1/n$ on the JUMP problem [10]:

$$\text{JUMP}_m(x) = \begin{cases} m + \sum_i x_i & \text{if } \sum_i x_i \leq n - m \text{ or } \sum_i x_i = n \\ n - \sum_i x_i & \text{otherwise} \end{cases}$$

where we take $m \leq n/8$. Note that this has a unique global optimum at 1^n, but local optima for all strings containing exactly m zeros, meaning it is non-monotonic. This also means that the focussed mutation method will not work on this problem, since given a string with m zeros, a mutation focussed on a bit position with value 0 will be rejected. Instead, we use the method of parallel runs from the previous section. It is worth noting that a similar mechanism enables EDAs and voting algorithms to solve JUMP efficiently [17,18].

Theorem 5. *Consider the $(1+1)EA$ with mutation probability $1/n$ running on JUMP with gap size $m < n/8$. We run $\lambda = c \ln n$ copies of this algorithm in parallel (for some suitable constant c), and confirm a bit value if we see at least 3/4 of the population agreeing on that value at any time. With high probability, we will never incorrectly confirm a bit value, and we will correctly confirm all bit values after $O(n \log^2 n \log \log n)$ function evaluations.*

Proof. We first consider a single run of the algorithm and bound the expected number of zeros in any string accepted by the algorithm. Notice that the number of zeros can never increase from one time step to the next. Let $X(t)$ be the current string at time t. Then

$$E[|X(t)|_0] = E[|X(t)|_0 \mid |X(0)|_0 > 3n/5] \Pr(|X(0)|_0 > 3n/5)$$
$$+ E[|X(t)|_0 \mid |X(0)|_0 \leq 3n/5] \Pr(|X(0)|_0 \leq 3n/5).$$

Hoeffding's inequality tells us that $\Pr(|X(0)|_0 > 3n/5) \leq \exp(-n/50)$. We also know that if $|X(0)|_0 \leq 3n/5$, then $|X(t)|_0 \leq 3n/5$ for all $t > 0$. Therefore

$$E[|X(t)|_0] \leq n \exp(-n/50) + 3n/5 < 0.65 \text{ n}$$

if $n \geq 150$. Now if i and j are any bit positions, then we have $\Pr(X(t)_i = 0) = \Pr(X(t)_j = 0)$ for all $t \geq 0$ due to the symmetry of the fitness function. That is, the probability that a bit is zero at a particular time is the same for all bits.

Suppose for some time t and some bit position i that $\Pr(X(t)_i = 0) > 0.65$, then since this must also be the case for all relevant bit positions, we have $E[|X(t)|_0] > 0.65$ n. But this contradicts what we have just stated. Therefore it must in fact be the case that for all t and any bit position, i, that $\Pr(X(t)_i = 0) \leq 0.65$. Given we are running $\lambda = c \ln n$ copies of the algorithm independently, the probability that we confirm the wrong value in any relevant bit position, at any time, by finding $3\lambda/4$ zeros at that position, is, by Hoeffding's inequality, at most $\exp(-0.02\lambda) = 1/n^{0.02c}$. Therefore, with high probability, we do not incorrectly confirm any of the relevant bits, given an appropriate choice of c.

Now any one run of the $(1+1)$EA on JUMP problem will find a string with exactly m zeros in expected time $O(n \log n)$. By Lemma 1, all λ copies will have found such a string in $O(n \log n \log \log n)$ iterations, with high probability. As the runs are independent, the probability of finding a 1 in a particular bit position after this time is $(n-m)/n \geq 7/8$ (since we consider $m \leq n/8$). So, again using Hoeffding's inequality and the union bound, we have that all bits have their values correctly confirmed within this time, with high probability, at a cost of

$O(n \log^2 n \log \log n)$ function evaluations. Note that we again have a situation in which all the correct bit values are known and confirmed much faster that the optimum string itself is discovered. □

6 Experiments

We now illustrate our theoretical results with some experiments. Figures 2a and 2b show how the focussed mutation method is able to confirm bit values for the $(1+1)$EA on ONEMAX and LEADINGONES respectively. The number of bits is fixed at $n = 100$ and we show the effect of varying the mutation rate, c (giving a bitwise probability of c/n). In the case of ONEMAX, while the average fitness starts of at 50, we do not know which of these bits are the correct ones, hence the number of confirmed bits starts at zero. Once the algorithm has found the correct solution, it takes a further en iterations to confirm the final bit. For LEADINGONES, the expected initial fitness is 2, and again the number of confirmed bits starts at zero. This time, however, the $(1+1)$EA typically finds new correct bits one or two at a time and they are confirmed at about the same rate once they are found. Hence the curve for the number of confirmed bits follows the fitness values closely.

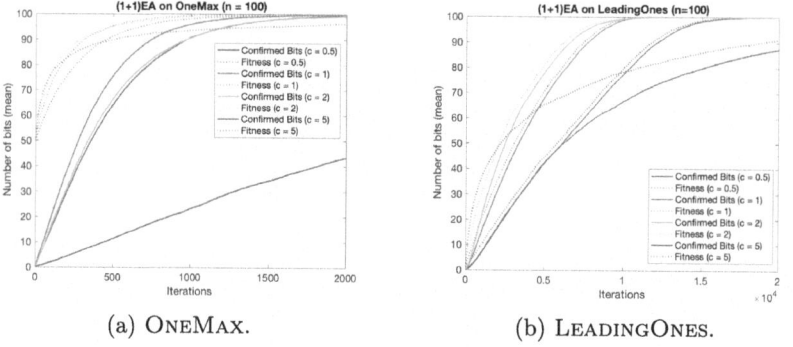

(a) ONEMAX. (b) LEADINGONES.

Fig. 2. The focussed mutation method for ONEMAX and LEADINGONES with $n = 100$. A range of mutation rates c/n are considered, with $c = 0.5, 1, 2$ and 5. Results are averaged over 100 runs.

Figures 2a and 2b illustrate the population-based method, running on hidden versions of ONEMAX and LEADINGONES respectively. Here, the string size is fixed at $N = 1000$, and the number of hidden bits, n, is varied. We use a population size of $10 \ln N$. Recall that the average run-time to find the optimum depends on n but not on N, when we used the specialist mutation method. We implement this mutation method by first choosing a uniform random number, u, between 0 and 1, and then using $p = 2^{-1/u}$ as the mutation probability. The value p is freshly generated in each iteration. This is a slight variant of the scheme proposed in [6], but it has the same required theoretical properties, since we have

$$\Pr\left(\frac{1}{2n} < p < \frac{1}{n}\right) = \frac{1}{\log_2 n} - \frac{1}{\log_2 2n} = \frac{1}{(\log_2 n)(\log_2 2n)} \geq \frac{1}{(\log_2 2n)^2}.$$

Using the same argument as in [6], we get that the time to optimise hidden ONEMAX is $O(n \log^3 n)$, and the time to optimise hidden LEADINGONES is $O(n^2 \log^2 n)$. In Fig. 3a, we see that for hidden ONEMAX, after an initial period, bit values are confirmed more quickly than the average fitness. This is because a bit value will be confirmed when 3/4 of the population agree on a value. For hidden LEADINGONES (Fig. 3b), the confirmed bits again trail slightly behind the correct bits as they are discovered, until right at the end when the remaining few are confirmed.

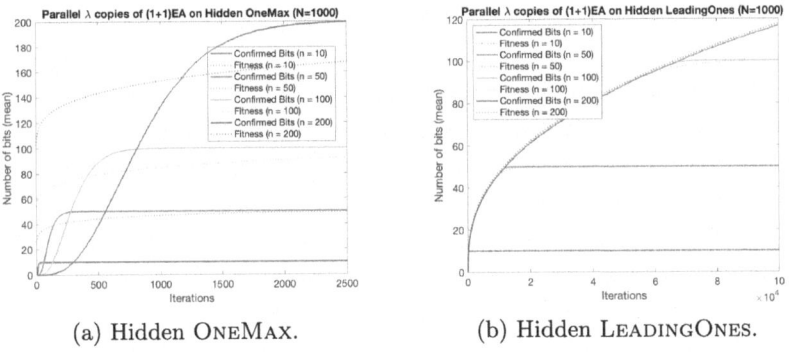

(a) Hidden ONEMAX. (b) Hidden LEADINGONES.

Fig. 3. The population-based method for hidden ONEMAX and LEADINGONES with $N = 1000$ and $n = 10, 50, 100$ and 200. Results are averaged over 100 runs.

7 Conclusions

We have given two methods for determining whether bit values can be considered as correct, even if an evolutionary algorithm is stopped before the optimum is found. The focussed mutation method works well for weakly monotonic problems with a unique optimum. The population-based approach has advantages, for example in the case of hidden subset problems, but comes at a cost in the size of the population. We suggest a useful line of further research would be to look into using such an approach with a population-based algorithm, such as the $(\mu + 1)$EA. Comparisons with EDAs, which seek to learn problem structure would also be of interest (see [12] and, in particular, [9]).

References

1. Bossek, J., Sudholt, D.: Do additional target points speed up evolutionary algorithms? Theor. Comput. Sci. (2023)
2. Böttcher, S., Doerr, B., Neumann, F.: Optimal fixed and adaptive mutation rates for the leadingones problem. In: Schaefer, R., Cotta, C., Kołodziej, J., Rudolph, G. (eds.) Parallel Problem Solving from Nature. PPSN XI, pp. 1–10. Springer, Berlin Heidelberg (2010)
3. Cathabard, S., Lehre, P.K., Yao, X.: Non-uniform mutation rates for problems with unknown solution lengths. In: FOGA 2011: Proceedings of the 11th Workshop Proceedings on Foundations of Genetic Algorithms. ACM (2011)
4. Ciesielski, V., Scerri, P.: An anytime algorithm for scheduling of aircraft landing times using genetic algorithms. Aust. J. Intell. Inf. Process. Syst. 206–213 (1997)
5. Dean, T., Boddy, M.: An analysis of time-dependent planning. In: AAAI 1988: Proceedings of the Seventh National Conference on Artificial Intelligence, pp. 49–54 (1988)
6. Doerr, B., Doerr, C., Kötzing, T.: Solving problems with unknown solution length at almost no extra cost. Algorithmica **81**, 703–748 (2019)
7. Doerr, B.: Probabilistic tools for the analysis of randomized optimization heuristics. In: Doerr, B., Neumann, F. (eds.) Theory of Evolutionary Computation. NCS, pp. 1–87. Springer, Cham (2020). https://doi.org/10.1007/978-3-030-29414-4_1
8. Doerr, B., Jansen, T., Sudholt, D., Winzen, C., Zarges, C.: Optimizing monotone functions can be difficult. In: Parallel Problem Solving from Nature, PPSN XI, pp. 42–51 (2010)
9. Doerr, B., Krejca, M.S.: Significance-based estimation-of-distribution algorithms. IEEE Trans. Evol. Comput. **24**(6), 1025–1034 (2020)
10. Droste, S., Jansen, T., Wegener, I.: On the analysis of the $(1+1)$ evolutionary algorithm. Theoret. Comput. Sci. **276**, 51–81 (2002)
11. Garnier, J., Kallel, L., Schoenauer, M.: Rigourous hitting times for binary mutations. Evol. Comput. 173–203 (1999)
12. Hauschild, M., Pelikan, M.: An introduction and survey of estimation of distribution algorithms. Swarm Evol. Comput. **1**(3), 111–128 (2011)
13. Jansen, T., Zarges, C.: Fixed budget computations: a different perspective on run time analysis. In: GECCO 2012: Proceedings of the 14th Annual Conference on Genetic and Evolutionary Computation, pp. 1325–1332 (2012)
14. Jansen, T., Zarges, C.: Performance analysis of randomised search heuristics operating with a fixed budget. Theoret. Comput. Sci. **545**, 39–58 (2014)
15. Lehre, P.K., Sudholt, D.: Parallel black-box complexity with tail bounds. IEEE Trans. Evol. Comput. (2019)
16. Lengler, J., Steger, A.: Drift analysis and evolutionary algorithms revisited. In: Combinatorics, Probability and Computing, pp. 643–666 (2018)
17. Rowe, J.E., Aishwaryaprajna. The benefits and limitations of voting mechanisms in evolutionary optimisation. In: Proceedings of the 15th ACM/SIGEVO Conference on Foundations of Genetic Algorithms, FOGA 2019, pp. 34–42. Association for Computing Machinery (2019)
18. Witt, C.: How majority-vote crossover and estimation-of-distribution algorithms cope with fitness valleys. Theoret. Comput. Sci. **940**, 18–42 (2023)

Studies on Survival Strategies to Protect Expert Knowledge in Evolutionary Algorithms for Interactive Role Mining

Simon Anderer[1(✉)], Nicolas Justen[1], Bernd Scheuermann[2], and Sanaz Mostaghim[3]

[1] Pointsharp GmbH, Grünhutstraße 6, 76187 Karlsruhe, Germany
{Simon.Anderer,Nicolas.Justen}@pointsharp.com
[2] Karlsruhe University of Applied Sciences, 76133 Karlsruhe, Germany
Bernd.Scheuermann@h-ka.de
[3] Otto-von-Guericke-Universität Magdeburg, Magdeburg, Germany
Sanaz.Mostaghim@ovgu.de

Abstract. To maintain the integrity of information technology infrastructures within enterprises and organizations, it is imperative to implement robust and dependable access control mechanisms. A prevalent method is Role-Based Access Control (RBAC), wherein permissions are groped into roles, which are subsequently assigned to the users of an IT system. Such assignments are referred to as role concepts. The objective of the Role Mining Problem (RMP), whose decision variant is NP-complete, is to find a role concept with minimal number of roles and evolutionary algorithms have been applied as effective meta-heuristic techniques to approximate optimal solutions. Recent studies have demonstrated that the integration of expert knowledge through user interaction with running evolutionary algorithms can substantially enhance the optimization process. In this paper, expert knowledge is integrated into a evolutionary role mining algorithm by injecting favorable roles into the individuals representing role concepts. The impact of such role injections on the optimization progress is investigated and several survival strategies are presented to ensure that individuals with injected roles survive long enough to exert their beneficial effect on the optimization process. The proposed survival strategies are evaluated in a series of experiments.

Keywords: Evolutionary algorithms · Role based access control · Interactive role mining · Expert knowledge · Survival strategies

1 Introduction

In the realm of cybersecurity for corporate and organizational IT infrastructures, it is imperative to safeguard against both external threats, such as malware and phishing, and internal risks, which include erroneous or fraudulent activities by insiders. According to the latest Global Economic Crime and Fraud Survey by

PricewaterhouseCoopers, nearly 50% of the surveyed 1,200 entities encountered fraud and more than half of the reported cases involved insiders [16]. A measure to mitigate such risks is the implementation of access control mechanisms that restrict user access to sensitive information. In this context, Role-Based Access Control (RBAC) is a prevalent method wherein permissions are grouped into roles, which are then assigned to users. The corresponding optimization problem, which aims at finding a minimal set of roles based on a given assignment of permissions to users, is called the Role Mining Problem (RMP). The decision version of the RMP was shown to be NP-complete [21] and evolutionary algorithms (EAs) have emerged as an effective and competitive approach for rapidly identifying high-quality solutions.

It has been shown that integration of expert knowledge, specifically through the injection of roles presumed favorable for a given RMP specification, can significantly accelerate the role mining process [1]. However, since the fitness of an individual corresponds to its number of roles, individuals arising from the injection of favorable roles typically exhibit poorer initial fitness compared to unmodified individuals, in which the expert knowledge is not included. Consequently, the modified individuals are often eliminated by the EA's replacement method before their positive contributions can manifest in the optimization process. A straightforward approach might therefore involve exclusively preserving modified individuals. Yet modifications may not always yield improvements and could even deteriorate the optimization process. It is not guaranteed that roles that have proven to be beneficial in past role mining scenarios will also be favorable, for example, in different industries. Hence, it is advisable to maintain both modified and unmodified individuals until they reveal their impact.

To address this, in [2], survival strategies were introduced to protect expert knowledge which was injected into in EAs for role mining. However, these strategies were described briefly and have not been evaluated. This paper describes survival strategies, characterized by alterations in the EA's fitness function or the addition of supplementary populations, The proposed strategies are then evaluated in several experiments and compared with each other.

2 The Role Mining Problem

In the following, the main elements of the RMP are introduced. Subsequently, a formal definition is provided : $U = \{u_1, u_2, ..., u_M\}$ is a set of $M = |U|$ users, $P = \{p_1, p_2, ..., p_N\}$ is a set of $N = |P|$ permissions and $R = \{r_1, r_2, ..., r_K\}$ is a set of $K = |R|$ roles. Moreover, $UPA \in \{0,1\}^{M \times N}$ denotes a permission-to-user assignment matrix, where $UPA_{i,j} = 1$ implies that permission p_j must be assigned to user u_i, $UA \in \{0,1\}^{M \times K}$ is a role-to-user assignment matrix and $PA \in \{0,1\}^{K \times N}$ is a permission-to-role assignment matrix.

The Basic Role Mining Problem. *Given a set of users U, a set of permissions P and a permission-to-user assignment matrix UPA, find a role concept $\pi^* := \langle R^*, UA^*, PA^* \rangle$ comprising a minimal set of Roles R^*, a corresponding* role-to-user assignment matrix UA^* and a permission-to-role assignment matrix PA^*, such that each user is assigned exactly the permissions granted by UPA.

$$\textbf{Basic RMP} = \begin{cases} \min & |R|, \\ \text{s.t.} & \|UPA - UA \otimes PA\| = 0, \end{cases} \quad (1)$$

At this, $\|A\|$ denotes the sum of absolute values of elements of a matrix $A \in \{0,1\}^{m \times n}$: $\|A\| := \sum_{i=1}^{m} \sum_{j=1}^{n} |A_{i,j}|$ and $\otimes$ is the Boolean Matrix Multiplication: $(UA \otimes PA)_{ij} = \bigvee_{l=1}^{k}(UA_{il} \wedge PA_{lj})$. The constraint in (1) ensures that each user is assigned exactly the permissions required to perform his or her tasks within the company according to UPA. In this case, a role concept is called *0-consistent*. This excludes, for example, role concepts in which every user is assigned each permission on no permission at all. The minimization of the number of roles further prevents trivial role concepts like the creation of a unique role for each user or the creation of one role from each permission. However, this may provide a good strategy for initializing the individuals of an evolutionary algorithm.

3 Evolutionary Algorithms for the Role Mining Problem

There are many different solution strategies for the RMP. A widely used method is permission grouping, where permissions are grouped into a set of candidate roles. Thereafter, the candidate roles are assigned to users, mostly using greedy approaches [6,15,23]. Another solution strategy consists of mapping the RMP to other problems in data science, e.g. the Set Cover Problem [11], the Minimum Tiling Problem [22] or the Minimum Biclique Cover Problem [9] and using established solution approaches. A detailed overview of different solution strategies is provided by Mitra [14]. In recent years, the use of evolutionary algorithms (EAs) for the RMP has gained popularity. A role concept is represented by an individual I comprising a role-to-user assignment matrix $UA^{(I)}$, a permission-to-role assignment matrix $PA^{(I)}$, and the corresponding set of roles $R^{(I)}$. However, when employing standard crossover and mutation techniques, such as one-point crossover and bitflip mutation, $UA^{(I)}$ and $PA^{(I)}$ may be changed in such a way that additional permissions are assigned to users, or permissions needed are withdrawn. Consequently, these methods may produce solutions that violate the 0-consistency constraint, e.g. [8,17–19]. To address this, the *addRole-EA* was introduced, see Algorithm 1. It comprises novel operators for crossover and mutation, ensuring the generation of 0-consistent individuals. Its role-centric approach facilitates the integration of expert knowledge by injecting favorable roles, which makes it in particular suitable for evaluating the developed survival strategies, as discussed in Sect. 6. For a comprehensive description, refer to [4].

Pre- and Post-processing. To decrease the size of the RMP, different pre-processing steps have been identified. One of these includes, for example, aggregating users who share the same set of permissions into user classes. The post-processing step adjusts the derived role concepts to align with the original problem size after the addRole-EA has concluded.

Algorithm 1: The addRole-EA

Input: users U, permissions P, permission-to-user assignment UPA
Output: role concept $\pi^* = \langle R^*, UA^*, PA^* \rangle$

1 $doPreProcessing(\ )$;
2 $Pop := \{\ \}$;
3 $doInitialization(Pop)$;
4 $doEvaluation(Pop)$;
5 **while** *stopping condition not met* **do**
6 $\quad doSelectionCrossoverAndMutation(Pop)$;
7 $\quad doEvaluationAndReplacement(Pop)$;
8 **end**
9 $doPostProcessing(Pop)$;

Initialization and Evaluation. Initially, a seed individual, denoted as I_0, is established by $UA^{(I_0)} := UPA$ and $PA^{(I_0)} := I_N$ (N-dimensional identity matrix). I_0 is 0-consistent, since $UA^{(I_0)} \otimes PA^{(I_0)} = UPA \otimes I_N = UPA$. Further individuals are then generated from I_0 through the application of a random series of mutation operators, aimed at populating and diversifying the initial population. As stipulated by the Basic RMP, the fitness of an individual I equals the number of roles of the represented role concept, $fitness(I) := |R^{(I)}|$.

Selection, Crossover and Mutation. Selection of individuals for crossover and mutation is conducted randomly. The crossover involves exchanging roles between selected individuals. During mutation, new roles are generated via set-theoretic operations and added to an individual. The addition of new roles employs the addRole-method, which not only adds a new role to $R^{(I)}$, $UA^{(I)}$, and $PA^{(I)}$ but also examines the existing roles to identify any redundancies. Redundant roles, if identified, are subsequently removed from $R^{(I)}$, $UA^{(I)}$, and $PA^{(I)}$, ideally resulting in a reduced number of roles.

Replacement and Stopping Condition. The addRole-EA operates as a steady-state evolutionary algorithm, utilizing an elitist selection scheme for the replacement process. The addRole-EA concludes when either a predefined maximum number of iterations has been reached or when no further improvement is observed over a specified number of consecutive iterations.

4 Interactive Role Mining

Since numerous role mining projects today are still carried out by human experts, a large amount of expert knowledge is available and should be integrated into EAs. In [1], interaction possibilities are described whereby experts can transfer their knowledge into a role mining algorithm e.g. by adding and deleting roles from individuals or by editing UA, PA, and UPA. Each action of an expert triggers an interaction event which must be integrated into the EA. Commonly, consultants and role specialists are engaged in implementing role concepts, such that, over time, they develop substantial expertise. These experts can identify roles that are particularly effective in specific contexts, potentially enhancing the optimization process. Such roles are referred to as *good* roles, and their injection into the addRole-EA represents the integration of expert knowledge throughout this paper. Due to their iterative nature, EAs offer a significant advantage to promptly respond to interaction events triggered by experts. At the start of each iteration, it can be checked whether an interaction event is currently pending. If this is the case, the corresponding event-handling method can be executed to adjust the individuals in the current population of the EA. Algorithm 2 outlines the required modifications to the evolutionary loop in the addRole-EA.

Algorithm 2: Evolutionary loop for event handling

```
1 while stopping condition not met do
2     if Interaction Event pending then
3         doEventHandling(Event, Pop);
4     end
5     doSelectionCrossoverAndMutation(Pop);
6     doEvaluationAndReplacement(Pop);
7 end
```

As existing role mining benchmarks lack a definition of *good* roles, the examination of the integration of expert knowledge is currently not viable. Therefore, a methodology is proposed in [1] to augment the instances of RMPlib, a publicly available RMP benchmark library [5], with the definition of *good* roles. This involves repeated application of the addRole-EA on each instance to select the best role concepts. Roles are classified to be *good* roles, if they appear in all selected role concepts, resulting in a set of *good* roles R^+ for each instance. To simulate the interaction of an expert with role mining software, these roles can then dynamically be added to the role mining process. The resulting interaction events are then handled by the addRole-method of the addRole-EA. To evaluate the impact of expert knowledge due to human interaction, the integration of *good* roles is examined on three benchmark instances of *RMPlib*: *PLAIN_small_02* (PS02), *PLAIN_small_05* (PS05), and *PLAIN_medium_01* (PM01). Different numbers of roles are injected at different points in time. Before role injection, the population is split into two identical populations Pop_0 and Pop_{I01}. In Pop_{I01}, roles are added to all individuals using the addRole-method. In Pop_0 the individuals remain unmodified, i.e. no roles are added. Thereafter, further iterations of the addRole-EA are executed on both populations. Figure 1 shows the numbers of roles of the best individual of Pop_{I01} resp. Pop_0 over iterations for PS02, (injection of 5 *good* roles at iteration 5,000).

It can be seen, that the inclusion of expert knowledge by injecting *good* roles significantly enhances the optimization process. In [1], this is further examined in by defining various fitness levels (numbers of roles) and measuring the required iterations and computation time for the best individual in each population to achieve them. Results reveal that injecting *good* roles consistently reduces both. However, a temporary decrease in fitness, denoted as φ, is observed for modified individuals in Pop_{I01}, due to an increased number of roles immediately after the addition of roles. As expert-suggested roles may not always be suitable in practice or may even hinder the optimization process, maintaining both modified

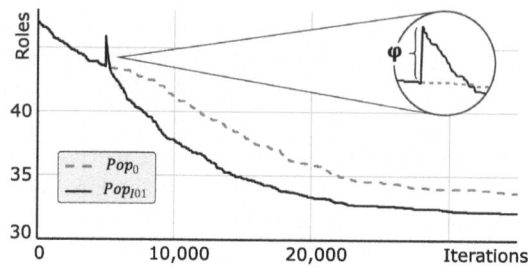

Fig. 1. Roles over iterations for the addition of *good* roles on PS02, based on [1].

and unmodified individuals for a certain time is advantageous. Yet, due to the elitist replacement method of the addRole-EA, modified individuals might not advance to the next generation, due to the initial increase of φ. Hence, developing effective survival strategies is crucial to ensure modified and unmodified individuals survive long enough to realize their optimization potential.

5 Survival Strategies

The previous section demonstrated that injecting *good* roles can enhance the role mining process. However, it was further observed that, initially, individuals with additional roles often exhibit poorer fitness compared to unmodified individuals. This section introduces effective survival strategies aimed at ensuring modified individuals persist long enough within a combined population Pop to evaluate whether: (a) The *good* roles added are actually good roles, such that they enhance the optimization process in terms accelerating the reduction of the number of roles in the individuals. Hence, the individuals, which comprise the additional roles, should be retained. (b) The *good* roles added are actually not good roles, i.e. they do not contribute to the enhancement of the optimization process. In this case, the injection of roles also worsens the individuals' fitness initially and the subsequent role optimization tends to decelerate. Thus, the modified individuals have permanently poorer fitness, which is why the unmodified individuals prevail automatically, such that this case is not further investigated at this point.

5.1 Survival Strategies in Literature

Even though the integration of expert knowledge and interaction events into EAs has already been examined, the survival of individuals modified by such events is not particularly well studied. Only Knecht [13] discusses this within multi-objective optimization, noting that if an individual's fitness worsens with respect to optimization objectives, it may no longer remain on the Pareto front and could be removed from future populations. This aligns with findings from role mining discussed in the previous section. However, the injection and handling of *good* roles in the context of role mining has not yet been investigated.

In the following, different concepts are presented that constitute the foundation for the survival strategies proposed in this paper. A common method in EAs involves implementing a penalty term to penalize constraint violations and guide the search towards feasible solutions or improve diversity among individuals, e.g. [7,10,24]. Karimi et al. [12] vary the population size of their swarm dynamically, based on the age of the individuals as well as the occurrence of dynamic events. Zhao et al. [27] propose the Multiobjective Cooperative Coevolutionary Algorithm, where several populations are developed separately. At regular intervals, the best individual of each population as well as some random individuals are combined using crossover operators. Additionally, the best individuals in terms of a crowding measure are copied into the different populations to enable further information exchange. Wang et al. [25] developed the Estimation of Evolvability Genetic Algorithm. In addition to the fitness of an individual, also its evolvability, which corresponds to the ability to generate offspring of high fitness, is considered. At the beginning of each iteration, the individuals are grouped into different sub-populations, such that crossover and mutation is executed separately. Eventually, the resulting individuals are reunited in a common population. Zhang et al. [26] also use several populations. An explorer population searches for promising regions in the solution space and stores corresponding solutions in an archive. An exploiter population uses the stored solutions and searches for better solutions in their local environment.

5.2 Fitness Protection

An intuitive idea to ensure the survival of the individuals modified due to the injection of *good* roles in a joint population *Pop* is to adjust the definition of the fitness within the addRole-EA. Similar to [7,10,24], a penalty term is introduced and added to the fitness of an individual that does not contain the injected roles. This aligns the fitness of modified and unmodified individuals to ensure they have similar selection probabilities for the next generation. Assuming $R^+_{inj} := \{r^+_1, ..., r^+_n\}$ represents the set of injected *good* roles, the presence of these roles in an individual I can be assessed as follows:

$$mod(I) := \frac{mod_1(I) + ... + mod_n(I)}{n}, \qquad (2)$$

where $mod_i(I) = 1$ in case the injected role r^+_i is contained in individual I: $r^+_i \in R^{(I)}$. In case $r^+_i \notin R^{(I)}$, $mod_i(I) = 0$. Hence, $mod(I)$ corresponds to the percentage of the injected roles contained in I. Based on that, the fitness function of the addRole-EA is replaced by a new fitness function $fitness^{FP}(I)$:

$$fitness^{FP}(I) := |R^{(I)}| + \alpha \cdot n \cdot (1 - mod(I)), \qquad (3)$$

where $\alpha \geq 0$ can be used to control whether either modified ($\alpha > 1$) or unmodified individuals ($\alpha < 1$) should be preferred during replacement. For $\alpha = 1$, the penalty term $\alpha \cdot n \cdot (1 - mod(I))$ corresponds exactly to the number of injected roles that are not contained in individual I.

5.3 Incubator Protection

The following two survival strategies draw on the concept of multiple populations, as seen in [25–27]. Unlike the *Fitness Protection* strategy, which modifies the fitness function of the addRole-EA to protect modified individuals, this approach introduces an additional population, Pop_{inc}, or incubator population, alongside the joint population Pop whenever *good* roles are injected. The best PS_{inc} modified individuals from Pop, based on fitness (number of roles), are copied to Pop_{inc}. The addRole-EA's crossover and mutation operators are then applied separately to individuals in Pop and Pop_{inc}. The selection for the next generation's populations is performed using the original elitist replacement method. The next generation's incubator population is chosen solely from the current Pop_{inc}, whereas the joint population considers all individuals. This setup allows individuals from the incubator population to integrate into the joint population if the modifications prove beneficial for optimization. Algorithm 3 outlines the adapted evolutionary loop for *Incubator Protection*. If *good* roles are injected at different times, the approach can be extended to support multiple simultaneous incubator populations.

Algorithm 3: *doGeneration_IP(populations Pop, Pop_{inc})*

1. $doSelectionCrossoverAndMutation(Pop_{inc})$;
2. $doSelectionCrossoverAndMutation(Pop)$;
3. $doEvaluationAndReplacement(Pop_{inc})$;
4. $Pop := Pop \cup Pop_{inc}$;
5. $doEvaluationAndReplacement(Pop)$;

5.4 Population Split Protection

While *Incubator Protection* focuses on protecting individuals modified due to the injection of *good* roles, *Population Split Protection* aims at protecting the modifications themselves. Here, all individuals are initially included in the joint population Pop. From this, two temporary populations are formed: (1) The regular population, Pop_{reg}, comprises the best individuals based on fitness. (2) An additional population, Pop_{add}, consists of the best PS_{add} individuals that include all injected roles in their chromosomes, ensuring $mod(I) = 1$ for all $I \in Pop_{add}$, see Algorithm 4.

Algorithm 4: *doGeneration_PSP(population Pop)*

1. $Pop_{reg} := \{\ \}$;
2. $Pop_{add} := \{\ \}$;
3. $doGeneration\_reg(Pop, Pop_{reg})$;
4. $doGeneration\_add(Pop, Pop_{add})$;
5. $Pop := Pop_{reg} \cup Pop_{add}$;

It is noticeable that this method allows individuals to be part of both Pop_{reg} and Pop_{add}. Thus, individuals in Pop_{add} can assert themselves in the joint population Pop, if the injected roles have contributed to the reduction of the total number of roles. The selection, crossover, mutation, and replacement operations of the addRole-EA are then applied separately to Pop_{reg} and Pop_{add}. The algorithmic process for the selection and update of the regular population is outlined in Algorithm 5:

Algorithm 5: *doGeneration_reg(populations Pop, Pop_{reg})*

1 **for** $i \in \{1, 2, ..., PS\}$ **do**
2 $\quad$ find best ind. $I^* \in Pop \setminus Pop_{reg}$;
3 $\quad$ $Pop_{reg} := Pop_{reg} \cup \{I^*\}$;
4 **end**
5 $doSelectionCrossoverAndMutation(Pop_{reg})$;
6 $doEvaluationAndReplacement(Pop_{reg})$;

Algorithm 6 details the update for the additional population Pop_{add}. Notably, individuals from crossover and mutation that no longer satisfy $mod(I) = 1$ are removed from Pop_{add}.

Algorithm 6: *doGeneration_add(populations Pop, Pop_{add})*

1 **for** $i \in \{1, 2, ..., PS_{add}\}$ **do**
2 $\quad$ find best ind. I^* in $Pop \setminus Pop_{add}$ for which $mod(I^*) = 1$;
3 $\quad$ $Pop_{add} := Pop_{add} \cup \{I^*\}$;
4 **end**
5 $doSelectionCrossoverAndMutation(Pop_{add})$;
6 **for** *individual* $I \in Pop_{add}: \quad mod(I) < 1$ **do**
7 $\quad$ $Pop_{add} := Pop_{add} \setminus \{I\}$;
8 **end**
9 $doEvaluationAndReplacement(Pop_{add})$;

Eventually, the original joint population Pop is replaced by individuals from both the regular and additional populations. Since the replacement method is executed separately for each population, it guarantees enough individuals in Pop satisfy $mod(I) = 1$, allowing for the creation of the additional population in the next iteration.

6 Evaluation and Experiments

In this section, the developed survival strategies are analyzed in more detail. First, the effects of the different parameters on the respective survival strategy are examined. Then, the survival strategies are compared with each other in terms of performance and computation times.

6.1 Experimental Setup

To analyze different survival strategies, three instances of the *PLAIN_x* benchmark of *RMPlib* were selected: PS02 (small, high density), PS05 (larger, lower density), and PM01 (medium-sized, low density) [5]. Following Sect. 5, the set of *good* roles R^+ was identified for each instance. The addRole-EA ran 200 times per instance, and the top 20 role concepts were selected for each instance. A role was classified a *good* role if present in all 20 role concepts. R^+ included 10 roles for PS02, 46 roles for PS05, and 144 roles for PM01. For each instance, 20% of the number of roles used for the creation of the respective benchmark instance (see [5]) were selected randomly from R^+ and injected at iteration $t = 5,000$. The joint population *Pop* contained both original (not including added roles) and modified individuals (including added roles). Individuals in *Pop* were then processed according to the selected survival strategy outlined in Sect. 5. Unlike [1], where modified and unmodified individuals were contained in two separate populations, here, both types of individuals are kept in the joint population. This allows the survival strategies to adequately allow the effects of injecting *good* roles to manifest. All tests performed for the different evaluation scenarios were run on a computer with the following specifications: processor Intel Core i5-4570S, 2.90 GHz, 16 GB RAM. The addRole-EA as well as the different survival strategies were implemented in Java. Its settings (population size = 20, mutation rate = 1.0, crossover rate = 0.1) were adopted from [4].

6.2 Evaluation of Fitness Protection

In order to evaluate survival strategy *Fitness Protection*, the influence of weight parameter α was investigated. For this purpose, 20 runs of the addRole-EA were performed on each benchmark instance for $\alpha \in \{0.25, 0.5, 1.0\}$ with different random seeds. Results were averaged. Figure 2(a) shows the progression of the number of roles of the best individual over iterations considering the different values of α for instance PS02 compared to the case in which no survival strategy was used. As described above, 5 *good* roles were injected at iteration $t = 5,000$. Since similar results were obtained for PS05 and PM01 in all test scenarios, the results obtained for PS02 are shown as a representative throughout this section.

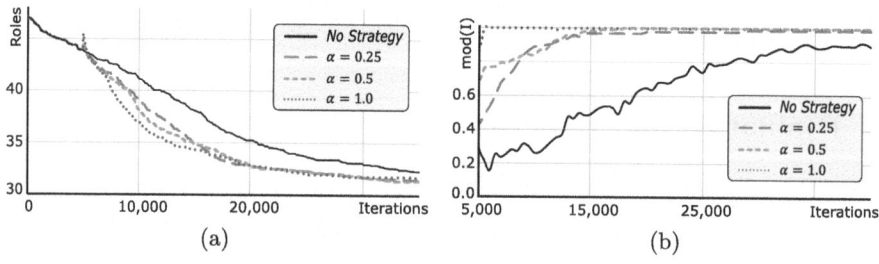

Fig. 2. Fitness Protection: (a) Roles over iterations (b) $mod(I)$ over iterations.

It is evident that *Fitness Protection* markedly accelerates the optimization process, particularly in short-term consideration. Larger α values enhance this early acceleration, as evidenced by Tables 1 (a) and (b), which detail the iterations and computation time needed by the addRole-EA to achieve a pre-defined role level k after the injection of *good* roles at iteration $t = 5,000$. Higher values of α tend to reduce iteration counts required to reach specified role levels. This short-term improvement is particularly relevant in scenarios like multi-objective role mining [13, 20, 27] or dynamic role mining, where the optimization problem is constantly changing [3]. To also consider the long-term effects on the fitness after the convergence of the addRole-EA, Table 1(c) lists the number of roles of the best individual after the execution of 100,000 iterations. It can be seen that the application of the survival strategy has no significant impact on the number of roles. In order to investigate the inclusion of the injected good roles into the chromosomes of the individuals, Fig. 2(b) shows the average value of $mod(I)$, as defined in Eq. 2, over all individuals of the current population, starting from the occurrence of the events at $t = 5,000$ on PS02. It can be seen that the value of α has a major impact on the progression of $mod(I)$. The higher the value for α is chosen, the faster good roles are included into the chromosomes of the individuals. For $\alpha = 1.0$, the added roles are immediately included into all individuals of the population, i.e. $mod(I) = 1$. Therefore, in case the *good* roles injected by the expert turn out to have negative effect on the optimization process, there are no more of the original, unmodified individuals to which the algorithm might return. The survival strategy, thus, corresponds to an extinction strategy for the unmodified individuals in that case.

6.3 Evaluation of Incubator Protection

The evaluation of survival strategy *Incubator Protection* was performed in the same way as for the previous strategy. Here, the influence of the population size of the incubator population $PS_{inc} \in \{2, 5, 10\}$ was investigated. Clearly, $PS_{inc} \geq 2$ needed to chosen to enable crossover also in the incubator population. Figure 3 shows the progression of the number of roles considering the different values of PS_{inc} compared to the case in which no survival strategy was used on benchmark instance PS02.

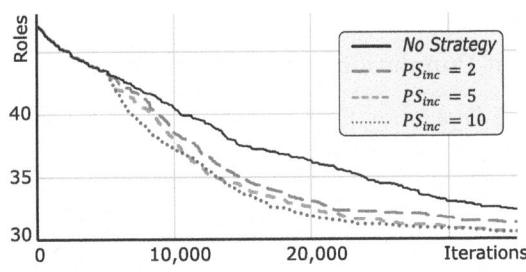

Fig. 3. Incubator Protection: Roles over iterations.

Figure 3 and Tables 1(a) and (b) show that the application of the survival strategy *Incubator Protection* accelerates the optimization process. It can be seen that increasing PS_{inc} leads to an acceleration of the optimization process in short-term consideration. Even if one would expect that the increase in population size would increase the computing time, it turns out that the reduction in computing time associated with the reduction in the number of roles outweighs this effect. Again, the application of *Incubator Population* shows no significant influence with regard to the number of roles after 100,000 iterations, see Table 1(c). Considering the progression of $mod(I)$ for the regular Population *Pop* in Fig. 4(a), it becomes evident that, the larger the size of the incubator population, the faster the added roles are included into the individuals of the regular population.

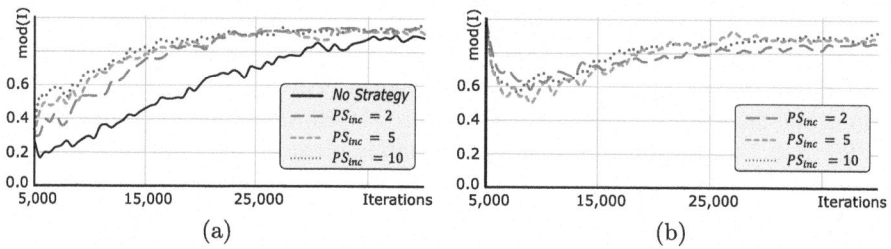

Fig. 4. Incubator Protection: $mod(I)$ over iterations (a) for *Pop* (b) Pop_{inc}.

This is also evident for the individuals of the incubator population, see Fig. 4(b). It is obvious from the design of the evaluation scenario and the functionality of *Incubator Protection* that after event occurrence all individuals of the incubator population comprise all of the injected roles, such that $mod(I) = 1$ at $t = 5,000$. In the further course, however, these are removed from the chromosomes of some individuals, such that $mod(I) < 1$. This is due to the operation principle of the addRole-method, in which obsolete roles are deleted. This also applies to the roles that were injected, in case they are not actually needed. In the later course, however, they can be re-included into the individuals through crossover.

6.4 Evaluation of Population Split Protection

For the evaluation of *Population Split Protection*, the influence of the population size $PS_{add} \in \{2, 5, 10\}$ of the additional population Pop_{add} was examined in the same test setup. Figure 5(a) shows the number of roles over iterations for the different populations while (b) shows $mod(I)$ within population *Pop* on PS02.

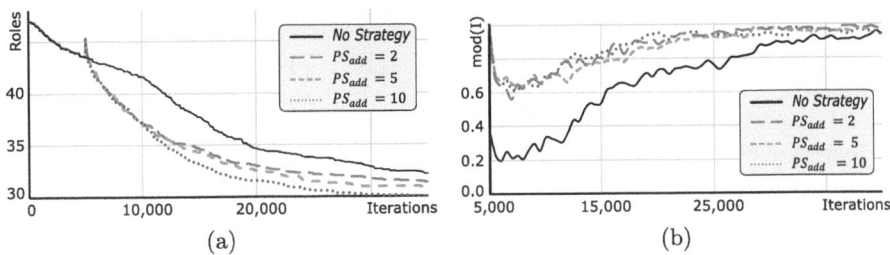

Fig. 5. Population Split: (a) Roles over iterations (b) $mod(I)$ over iterations.

The effects mirror the other survival strategies: increasing the size of the additional population accelerates short-term optimization but does not effect the number of roles after 100,000 iterations significantly. In contrast to the other survival strategies, larger populations significantly increase computation time, likely due to the time-intensive calculation of $mod(I)$. This is supported by the computation times for *Fitness Protection*, as indicated in Table 1 (b), which also involves the calculation of $mod(I)$. Figure 5(b) details the progression of $mod(I)$ within the joint population Pop on PS02, showing that modified individuals, which include the roles added by the expert, eventually prevail. However, unlike the other strategies, no obvious correlation exists with parameter PS_{add}. As $mod(I) = 1$ inherently applies to all individuals in Pop_{add}, due to the design of *Population Split Protection*, this is not shown at this point.

6.5 Comparison of Survival Strategies

All strategies cause for a significant improvement in short-term consideration. It is shown that increasing the respective parameter values results in an acceleration of the optimization process. Furthermore, for *Fitness Protection* and *Incubator Protection*, it can be seen that the value of the respective parameter controls the speed at which the modified individuals spread in the population and thus displace the unmodified individuals. This is particularly advantageous if it cannot be guaranteed with certainty that the roles injected by the expert are favorable in the role mining scenario under consideration. In terms of performance (Table 1(a)), computing time (Table 1(b)) and the number of roles attained after 100,000 iterations (Table 1(c)), *Incubator Population* outperforms the other strategies independent of the choice of its parameter in most of the scenarios under investigation.

Table 1. Evaluation results for the different survival strategies.

	Roles k	No Strategy	Fitness Protection $\alpha =$			Incubator Protection $PS_{inc} =$			Population Split $PS_{add} =$		
			0.25	0.5	1.0	2	5	10	2	5	10
(a) Number of iterations needed to obtain an individual with at most k roles.											
PS02	40	9,330	6,770	6,530	6,450	6,820	5,700	**5,640**	7,180	7,280	7,090
	38	12,337	9,190	9,070	7,590	8,930	8,400	**6,890**	9,130	9,000	9,050
	36	15,403	11,090	10,140	8,930	10,820	9,950	**8,850**	11,560	11,160	10,940
PS05	75	5,917	5,500	5,330	5,350	5,260	5,210	**5,140**	5,420	5,390	5,380
	70	7,640	6,500	5,820	5,890	5,960	5,670	**5,600**	6,050	5,950	5,950
	65	9,197	7,520	6,600	6,820	6,650	**6,170**	6,290	6,820	6,930	6,920
PM01	420	5,500	5,270	5,270	**5,260**	5,440	5,390	5,310	5,360	5,420	5,400
	400	6,210	5,730	**5,720**	**5,720**	6,030	5,890	5,830	5,860	5,880	5,890
	380	6,870	6,160	**6,170**	6,180	6,530	6,360	6,250	6,290	6,330	6,320
(b) Computation time (s) needed to obtain an individual with at most k roles.											
PS02	40	221,96	180,16	172,75	169,18	179,70	**155,25**	156,53	190,03	203,17	212,07
	38	277,92	234,32	230,18	**193,73**	224,42	222,28	194,09	242,11	255,60	284,05
	36	327,08	272,60	252,02	**220,33**	259,32	255,84	246,97	301,82	316,70	348,61
PS05	75	79,55	80,75	76,42	77,00	75,66	75,26	**74,73**	79,01	80,36	84,04
	70	98,18	103,93	87,75	89,54	84,12	**81,68**	82,34	97,21	99,69	109,19
	65	113,62	126,55	104,86	110,06	91,79	**88,03**	93,06	118,51	132,26	149,92
PM01	420	1.216,33	1.277,42	1.277,18	1.270,65	1.250,77	1.252,94	**1.245,99**	1.332,93	1.434,91	1.518,47
	400	1.335,82	1.539,37	1.533,93	1.533,74	**1.352,52**	1.353,36	1.366,73	1.630,36	1.777,20	2.002,31
	380	1.438,56	1.775,41	1.781,32	1.787,43	**1.433,62**	1.441,70	1.460,35	1.880,46	2.100,54	2.413,05
(c) Average number of roles of the best individual I^* of population Pop after 100,000 iterations.											
PS02		30.33	30.10	30.35	30.10	29.75	29.25	**29.05**	30.15	30.10	29.45
PS05		50.05	50.40	49.90	49.75	49.90	**49.65**	49.85	50.25	49.95	50.10
PM01		151.75	151.80	152.20	152.35	151.70	**151.00**	151.85	152,20	152,55	152.75

7 Conclusion and Future Works

In this paper, different survival strategies for the integration of expert knowledge in terms of injecting good roles into evolutionary algorithms for role mining were presented and evaluated. It could be shown that all strategies significantly enhance the optimization process. At this, the strategy *Incubator Protection* was superior to the other strategies in most cases, both in terms of the acceleration in the reduction of roles as well as the computation time. The size of the incubator population can be used to regulate the rate of modified and unmodified individuals in the regular population. This is valuable in cases where it is not guaranteed that the included expert knowledge is actually favorable for the further optimization. Hence, Incubator Protection is recommended as the preferable survival strategy in practical scenarios. However, it would be interesting to analyze whether combinations of the methods presented could lead to better results. Since the proposed survival strategies are independent of the specifications of the algorithm used, as long as it offers the possibility to inject roles, it would be interesting to evaluate the proposed strategies in the context of other role mining algorithms. An obvious next step in future research consists of applying the presented survival strategies also for other events stemming from user interaction as the deletion of roles or the editing of the different matrices of the RMP: UA, PA or UPA. It could be shown that the injected roles stemming from the inclusion of expert knowledge do not require a lot of time to unfold their effects onto the optimization process, such that it could be interesting to investigate scenarios, in which the presented survival strategies are applied only for only a limited number of iterations. By utilizing and adapting different benchmark instances of RMPlib, the effects of injecting good roles on synthetic data have been comprehensively analyzed. However, since the survival strategies were designed for the application in real-world use cases, the related effects, which arise, for example, from the injection of rather bad or obstructive roles, should be investigated in the future.

Acknowledgments. This study was funded by the German Ministry of Education and Research under grant number 16KIS1000.

Disclosure of Interests. The authors declare that they have no conflict of interest.

References

1. Anderer, S., Justen, N., Scheuermann, B., Mostaghim, S.: Interactive role mining including expert knowledge into evolutionary algorithms. In: Proceedings of the 15th International Joint Conference on Computational Intelligence - ECTA, pp. 151–162. INSTICC, SciTePress (2023). https://doi.org/10.5220/0012153000003595
2. Anderer, S., Justen, N., Scheuermann, B., Mostaghim, S.: Survival strategies for evolutionary role mining algorithms using expert knowledge. In: Proceedings of the Genetic and Evolutionary Computation Conference Companion, GECCO 2024, pp. 623–626. Companion, Association for Computing Machinery, New York (2024). https://doi.org/10.1145/3638530.3654183
3. Anderer, S., Kempter, T., Scheuermann, B., Mostaghim, S.: The dynamic role mining problem: role mining in dynamically changing business environments. In: Bäck, T., Wagner, C., Garibaldi, J.M., Lam, H.K., Cottrell, M., Merelo, J.J., Warwick, K. (eds.) Proceedings of the 13th International Joint Conference on Computational Intelligence, IJCCI 2021, Online Streaming, 25–27 October 2021, pp. 37–48. SCITEPRESS (2021)
4. Anderer, S., Kreppein, D., Scheuermann, B., Mostaghim, S.: The addrole-EA: a new evolutionary algorithm for the role mining problem. In: Guervós, J.J.M., Garibaldi, J.M., Wagner, C., Bäck, T., Madani, K., Warwick, K. (eds.) Proceedings of the 12th International Joint Conference on Computational Intelligence, IJCCI 2020, Budapest, Hungary, 2–4 November 2020, pp. 155–166. SCITEPRESS (2020)
5. Anderer, S., Scheuermann, B., Mostaghim, S., Bauerle, P., Beil, M.: Rmplib: a library of benchmarks for the role mining problem. In: Lobo, J., Pietro, R.D., Chowdhury, O., Hu, H. (eds.) SACMAT 2021: Proceedings of the 26th ACM Symposium on Access Control Models and Technologies, Virtual Event, Spain, 16–18 June 2021, pp. 3–13. ACM (2021)
6. Blundo, C., Cimato, S.: A simple role mining algorithm. In: Shin, S.Y., Ossowski, S., Schumacher, M., Palakal, M.J., Hung, C.C. (eds.) Proceedings of the 2010 ACM

Symposium on Applied Computing (SAC), Sierre, Switzerland, 22–26 March 2010, pp. 1958–1962. ACM, New York (2010)
7. Deb, K., Datta, R.: A bi-objective constrained optimization algorithm using a hybrid evolutionary and penalty function approach. Eng. Optim. **45**(5), 503–527 (2013)
8. Du, X., Chang, X.: Performance of AI algorithms for mining meaningful roles. In: Proceedings of the IEEE Congress on Evolutionary Computation, CEC 2014, Beijing, China, 6–11 July 2014, pp. 2070–2076. IEEE (2014)
9. Ene, A., Horne, W.G., Milosavljevic, N., Rao, P., Schrei-ber, R., Tarjan, R.E.: Fast exact and heuristic methods for role minimization problems. In: Ray, I., Li, N. (eds.) Proceedings of the 13th ACM Symposium on Access Control Models and Technologies, SACMAT 2008, Estes Park, CO, USA, 11–13 June 2008, pp. 1–10. ACM, New York (2008)
10. Hu, Y.B., Wang, Y.P., Guo, F.Y.: A new penalty based genetic algorithm for constrained optimization problems. In: 2005 International Conference on Machine Learning and Cybernetics, vol. 5, pp. 3025–3029. IEEE (2005)
11. Huang, H., Shang, F., Liu, J., Du, H.: Handling least privilege problem and role mining in RBAC. J. Comb. Optim. **30**, 63–86 (2015)
12. Karimi, J., Nobahari, H., Pourtakdoust, S.H.: A new hybrid approach for dynamic continuous optimization problems. Appl. Soft Comput. **12**, 1158–1167 (2012)
13. Knecht, K.: Grundriss-generierung mit k-dimensionalen baumstruktur-en. In: Donath, D., König, R. (eds.) Arbeitspapiere Nr. 9 Informatik in der Architektur. Bauhaus-Universität Weimar, Professur Informatik in der Architektur, Weimar (2011)
14. Mitra, B., Sural, S., Vaidya, J., Atluri, V.: A survey of role mining. ACM Comput. Surv. **48**, 1–37 (2016)
15. Molloy, I.M., Li, N., Li, T., Mao, Z., Wang, Q., Lobo, J.: Evaluating role mining algorithms. In: Carminati, B., Joshi, J. (eds.) 14th ACM Symposium on Access Control Models and Technologies, SACMAT 2009, Stresa, Italy, 3–5 June 2009, Proceedings, pp. 95–104. ACM, New York (2009)
16. PwC: PwC's Global Economic Crime and Fraud Survey 2022. Pricewaterhouse-Coopers (2022)
17. Saenko, I., Kotenko, I.: Genetic algorithms for role mining problem. In: Cotronis, Y., Danelutto, M., Papadopoulos, G.A. (eds.) Proceedings of the 19th International Euromicro Conference on Parallel, Distributed and Network-based Processing, PDP 2011, Ayia Napa, Cyprus, 9–11 February 2011, pp. 646–650. IEEE, New York (2011)
18. Saenko, I., Kotenko, I.: Design and performance evaluation of improved genetic algorithm for role mining problem. In: Stotzka, R., Schiffers, M., Cotronis, Y. (eds.) Proceedings of the 20th Euromicro International Conference on Parallel, Distributed and Network-Based Processing, PDP 2012, Munich, Germany, 15–17 February 2012, pp. 269–274. IEEE, New York (2012)
19. Saenko, I., Kotenko, I.: Using genetic algorithms for design and reconfiguration of RBAC schemes. In: Proceedings of the 1st International Workshop on AI for Privacy and Security, PrAISe@ECAI 2016, The Hague, Netherlands, 29–30 August 2016, pp. 1–9. ACM, New York (2016)
20. Schaffer, J.D.: Multiple objective optimization with vector evaluated genetic algorithms. In: Grefenstette, J.J. (ed.) Proceedings of the 1st International Conference on Genetic Algorithms, Pittsburgh, PA, USA, July 1985, pp. 93–100. L. Erlbaum Associates, Hillsdale, New Jersey (1985)

21. Vaidya, J., Atluri, V., Guo, Q.: The role mining problem. In: Lotz, V., Thuraisingham, B. (eds.) Proceedings of the 12th ACM Symposium on Access Control Models and Technologies, SACMAT 2007, Sophia Antipolis, France, 20–22 June 2007, pp. 175–184. ACM, New York (2007)
22. Vaidya, J., Atluri, V., Guo, Q., Lu, H.: Role mining in the presence of noise. In: Foresti, S., Jajodia, S. (eds.) Data and Applications Security and Privacy XXIV - Proceedings of the 24th Annual IFIP WG 11.3 Working Conference, Rome, Italy, 21–23 June 2010. Lecture Notes in Computer Science, vol. 6166, pp. 97–112. Springer, Heidelberg (2010)
23. Vaidya, J., Atluri, V., Warner, J., Guo, Q.: Role engineering via prioritized subset enumeration. IEEE Trans. Dependable Secure Comput. **7**, 300–314 (2010)
24. Volkovas, R., Fairbank, M., Perez-Liebana, D.: Diversity maintenance using a population of repelling random-mutation hill climbers. In: 2017 9th Computer Science and Electronic Engineering (CEEC), pp. 37–42. IEEE (2017)
25. Wang, Y., Wineberg, M.: Estimation of evolvability genetic algorithm and dynamic environments. In: Genetic Programming and Evolvable Machines, vol. 7, pp. 355–382. Springer, Heidelberg (2006)
26. Zhang, W., Zhang, M., Zhang, W., Meng, Y., Wu, H.G.: Innate-adaptive response and memory based artificial immune system for dynamic optimization. Int. J. Performability Eng. **14**, 2048–2055 (2018)
27. Zhao, W., Alam, S., Abbass, H.A.: Mocca-II: a multi-objective co-operative co-evolutionary algorithm. Appl. Soft Comput. **23**, 407–416 (2014)

Diversification Through Candidate Sampling for a Non-iterated Lin-Kernighan-Helsgaun Algorithm

Tarek Boufar[✉][iD], Omar Rifki[iD], and Matthieu Basseur[iD]

Univ. Littoral Côte d'Opale, LISIC, 62100 Calais, France
{tarek.boufar,omar.rifki,matthieu.basseur}@univ-littoral.fr

Abstract. The Traveling Salesman Problem (TSP) is a famous NP-hard problem that has been extensively studied. The heuristic first introduced by Lin and Kernighan, is considered one of the most effective method for solving the TSP and has undergone numerous modifications and extensions. This study examines the fundamental version of Helsgaun's adjustments (LKH-1) due to the significant improvements it has brought to the heuristic. A key adjustment implemented is the disposition of a candidate set to each city using the α-measure, recognized for its efficacy in approximating the lower bound of the tour cost derived from the minimum 1-tree. To enhance diversification and accelerate the process, Helsgaun applied two main techniques across different trials: a judicious initial tour and dynamic candidate set reordering. In this paper, we propose a candidate selection strategy designed to replace these techniques. Our approach employs a randomized sampling selection strategy integrated directly into the k-opt search in the sense of a Partial Neighborhood Local Strategy (PNLS). Compared with LKH-1.3 limited to 2-opt operator, our approach shows improved results for large instances and comparable results for others. The goal of this work is to maintain a high level of solution quality while simplifying the overall algorithm.

Keywords: Travelling salesman problem · Lin-Kernighan-Helsgaun algorithm · local search

1 Introduction

For a given number of cities, the Traveling Salesman Problem (TSP) aims to find the shortest Hamiltonian cycle for the complete weighted graph $G = (V, E)$ with $n = |V|$ nodes. Given the NP-hard nature of the problem, heuristics come into play, as they are known for their ability to provide convincing results within a reasonable time frame. Their use has been widely adopted in addressing combinatorial optimization problems such as the TSP.

Supplementary Information The online version contains supplementary material available at https://doi.org/10.1007/978-3-031-86849-8_4.

In 1973, the Lin-Kernighan (LK) [9] local search heuristic was introduced, leading to increased interest in heuristic methods for solving the TSP. One of the most notable advancements to LK is Helsgaun's [6] enhancement (LKH), which could be applied as well to other routing problem variants based on the TSP [6]. This is known as the latest version, LKH-3 [8]. His adjustments primarily focused on the edge selection strategy for inclusion in the new tour during the k-opt exchange move. Previously, Lin and Kernighan used the nearest neighbors distance metric[1] to compare edges for exchanges. Helsgaun introduced an α-measure [3] method to determine candidate sets, using an α-value defined based on the minimum spanning 1-tree as a metric for evaluating edge quality. In essence, the idea is to calculate the additional cost incurred if edge (i,j) is included in the minimum 1-tree considered as a lower bound. Focusing on the most promising areas of the landscape involves retaining only a limited number of edges adjacent to each node—particularly those with the lowest additional costs. A lower α-value increases the likelihood that an edge will be part of the minimum 1-tree of the graph, thereby raising its chances of inclusion in the optimal tour. This technic was inspired by Stewart [11], who demonstrated how minimum spanning trees could be used to accelerate 3-opt heuristics.

Helsgaun suggests that a more effective edge selection can be achieved by reorganizing the candidate set for each node. Whenever a shorter tour is discovered, all edges common to this new tour and the previously shortest tour are prioritized as the first two candidate edges for their respective end nodes. By adopting this strategy, LKH more thoroughly investigates the edges of the tours considered optimal at that point in the search. Moreover, to expedite the process, selecting an initial tour can be advantageous. Preference is given to initial tours that share edges with the best-known tour and have a lower α-value cost.

Although this improvement mechanism produces high-quality solutions, it is quite complex. It involves incorporating the k-opt move into a loop of numerous trials, each of which aims to adjust the order of candidates and reset the initial tour. In addition, iterating over the candidate sets to select an improving edge to add to the solution is not flexible, as it may overlook edges with high α-values that have the potential to be part of the optimal. Other mechanisms have been added, such as node activation/deactivation and allowing non-sequential moves, such as `Gain23` with positive gain. In our proposal, we tackle the challenges of achieving high-quality solutions in the algorithm from the perspective of Partial Neighborhood Local Search (PNLS) method [12], without all these previous improvement mechanisms.

Our goal is to replace the pre- and post-adjustment steps applied when entering k-opt moves. We have eliminated dynamic reordering, initial tour reconstruction, and the node deactivation notion. Instead, we opt for a single trial within a defined time frame. The goal of diversifying the search will be directly integrated into the k-opt moves through a randomized selection of candidates. By *'randomized'*, we mean a selection that does not strictly follow the order dictated by the original candidate rankings. The objective is to consider candidates that may

---

[1] Regarding the type of the metric used: Euclidean, geographical, Manhattan, ...

not be top-ranked, providing long-term benefits and preventing stagnation during the search. One way to implement randomness is through *'biased'* sampling, where selection, while random, still partially respects the order of the original candidate set.

Using this approach, we retain the efficiency provided by the α-measure while also giving appropriate consideration to candidates that are not top-ranked. Our biased neighborhood exploration produces slightly better results than LKH, especially on large instances—with identical parameter settings.

The remainder of this paper is structured as follows: Sect. 2 provides an overview of the TSP and outlines the LKH algorithm. Section 3 presents our contribution, which involves integrating PNLS into LKH. Section 4 reports on the comparative experimental results and analyses. Finally, Sect. 5 concludes the paper.

2 Overview of TSP and the LKH Algorithm

Given an undirected, complete, and weighted graph G, the TSP consists of finding the minimum valuation of a *Hamiltonian cycle* in G. By valuation, we mean the sum of the costs of the edges $(c_1, c_2, \ldots, c_n)$ that constitute it. An alternative definition that aligns with our subject consists in defining the TSP as a minimum 1-tree:

Given a selected vertex v, for instance vertex 1, the 1-tree structure consists a spanning tree covering the vertices $\{2, 3, \ldots, n\}$ to which we add two distinct edges connecting vertex 1. A 1-tree contains exactly one cycle that is formed by the inclusion of vertex 1. A minimum-weight 1-tree is a 1-tree with the minimum weight among all possible 1-trees. To find a minimum-weight 1-tree of the graph, we start by determining the minimum spanning tree of the vertices $\{2, 3, \ldots, n\}$. Then, we add the two edges with the lowest weight incident to vertex 1. This process is applied for every vertex v in the graph, and the result will be the one with the minimal cost.

Any TSP tour is a 1-tree (with an arbitrary vertex 1) where each vertex has a degree of 2. If a minimum-weight 1-tree is also a tour, then it is the optimal tour for the TSP. Figure 1 illustrates this structure.

(a) Minimum 1-Tree Forming a Tour.   (b) Minimum 1-Tree not Forming a Tour.

Fig. 1. Construction of a Tour from a 1-Tree.

Therefore, the minimum 1-tree provides a lower bound for the length of the optimal TSP tour. This concept is fundamental for using the α-measure in Helsgaun's candidate selection. We will delve into this further in Sect. 2.3.

2.1 The LKH Algorithm

Since Helsgaun started to add his modifications in the LK heuristic, different versions have been considered with the following main features:

- **LKH-1** solves a TSP using sequential 5-opt move as the basic move, which is the one considered in this study, especially LKH-1.3 [6].
- **LKH-2** [7] allows all movements that can be decomposed into a sequence of k-exchanges for any k where $2 \leq k \leq n$, both sequential and non-sequential moves integrated directly into the ordinary search.
- **LKH-3** [8] is an extension of LKH-2 to solve variants of the TSP and vehicle routing problem.

As can be seen, the evolution of the LKH algorithm since its initial versions demonstrates its adaptability and continuous improvement in addressing the challenges posed by the TSP. A deep understanding of the internal mechanisms of the base algorithm (LKH-1) enables us to present our contribution. In the following, we describe its fundamental components:

2.2 K-Opt Moves in LKH

LKH is based on an interchange strategy. It aims to exclude a certain number k of $X = \{x_1, \ldots, x_k\}$ edges belonging to the current tour and to include k of $Y = \{y_1, \ldots, y_k\}$ edges not belonging to the tour, such that the new tour has a lower cost. This interchange of edges is referred to as a k-opt move. The sets X and Y are built sequentially. Initially, both X and Y are empty. In each step $i \leq k$, a pair of edges, x_i and y_i, are added to X and Y, respectively. A resulting sequence will be of the form $\{x_1, y_1, x_2, y_2, ..., x_k, y_k\}$ which constitutes a chain (cycle) of adjoining edges. Adjacency is obtained by x_i and y_i sharing an endpoint, and so must y_i and x_{i+1}. If t_1 denotes one of the two endpoints of x_1, we have in the iteration $i \geq 1$ the following constraints: $x_i = (t_{2i-1}, t_{2i})$, $y_i = (t_{2i}, t_{2i+1})$ and in $i = k$, $t_{2i+1} = t_1$ for closing the tour. See the description in Appendix A.

Feasibility of the moves is crucial and should be checked at each iteration, ensuring that there is always a way to close the tour. We denote this constraint by **C-I** for $i \geq 2$. During the exchange process, $Gain_i$ represents the cumulative gain from exchanging all pairs of x_j and y_j where $1 \leq j \leq i$. This is expressed as $Gain_i = g_1 + g_2 + \cdots + g_i$, where $g_i = x_i - y_i$. $Gain_i$ must always remain positive, so negative partial sums (g) may be accepted as long as $Gain_i \geq 0$. We denote this constraint by **C-II**.

Consider a node t_1 randomly chosen from the current tour, with t_1^1 and t_1^2 being its adjacent nodes. Let $C(S)$ represent the cost of the solution S. S includes all edges as node pairs. We define $cand(S, t_1)$ as the set of candidates for node t_1. Algorithm 1 [17] outlines the k-opt movement. The entry point is: k-OptMove($S_{random}, t_1, \emptyset, 1, 5$). To illustrate the algorithm execution, refer to Appendix B.

Algorithm 1: K-OptMove

input : S_{in} input solution; t_1 starting city; t sequence of the involved cities; i current search depth, k_{max} maximum search depth
output: S_{out} output solution; t sequence of the involved cities

1 **for** $h \leftarrow 1 : 2$ **do**
2     $t_{2i} \leftarrow t_{2i-1}^h$;
3   **if** t_{2i} does not satisfy **C-I then continue**;
4   **if** $i \geq 2 \wedge \sum_{j=1}^{i} d(t_{2j-1}, t_{2j}) > \sum_{j=1}^{i-1} d(t_{2j}, t_{2j+1}) + d(t_{2i}, t_1)$ **then**
5        $S_{out} \leftarrow S_{in}$;
6   **for** $j \leftarrow 1 : i$ **do**
7   RemoveEdge(t_{2j-1}, t_{2j}) from S_{out};
8   **for** $j \leftarrow 1 : i - 1$ **do**
9   AddEdge(t_{2j}, t_{2j+1}) into S_{out};
10   AddEdge(t_{2i}, t_1) into S_{out};
11   **return** (S_{out}, t)
12   **if** $i = k_{max}$ **then return** ($S_{in}, \emptyset$);
13   **for** $j \leftarrow 1 :$ MaxCandidates **do**
14        $t_{2i+1} \leftarrow$ the j-th candidate in $cand(S_{in}, t_{2i})$;
15   **if** t_{2i+1} does not satisfy **C-II then continue**;
16        $(S_{temp}, t') \leftarrow$ K-OptMove($S_{in}, t_1, t \cup \{t_{2i}, t_{2i+1}\}, i+1, k_{max}$) ;
17   **if** $C(S_{temp}) < C(S_{in})$ **then**
18   **return** (S_{temp}, t');

19 **return** ($S_{in}, \emptyset$);

2.3 Candidates Set

As previously stated, the LKH algorithm was able to outperform the basic version of Lin and Kernighan by changing the metric for selecting candidate nodes t_{2i+1}. The measure of approximation from one node to another is based on its proximity in the graph's minimum 1-tree. The α-value of an edge (i, j) is defined as the quantity

$$\alpha(i,j) = L(T^+(i,j)) - L(T), \tag{1}$$

where T is a minimum 1-tree (the lower bound) of length $L(T)$ and $T^+(i,j)$ is a minimum 1-tree required to contain edge (i, j). This means that the calculation of all the α-values for all possible edges (i, j) is relative to the quality of the graph's minimum 1-tree. This is limited for each node i to a value MaxCandidates.

Enhancing the quality of the α-values can be achieved by maximizing the lower bound of the optimal TSP by adding penalties (π-values) [4]. $\pi = [\pi_1, \ldots, \pi_n]$ is a vector of quantities to be added to both adjacent edges for each node. If $L(T_\pi)$ is the length of the minimum 1-tree after adding the π-values, then the lower bound $w(\pi)$ of the optimal solution can be calculated by:

$$w(\pi) = L(T_\pi) - 2\sum_{i=1}^{n} \pi_i, \tag{2}$$

using a sub-gradient optimization method to maximize it (inspired by [5] and [2]). Note that this modification does not change the optimal solution of the TSP, but it changes the minimum 1-tree. After adding the π-values, the α-value is further improved for determining the candidate set. We denote the α-value after adding the π-values as α_π-value. Finally, the candidate set of each city records MaxCandidates cities with the smallest α_π-value in ascending order.

2.4 LKH Improvement Toolbox

In order to better guide its search, improve the diversification of solutions and avoid redundancy, LKH employs several mechanisms, outlined as follows:

- **Initial Tour:** As noted by Helsgaun, substantial runtime reductions can be achieved by selecting near-optimal initial tours that favor promising edges. This construction method is efficient, and the diversity of initial solutions is sufficient for k-OptMove to yield high-quality final solutions (the construction heuristic is described in Appendix C).
- **Activation/Deactivation Mechanism:** In the initial state, all nodes are active. Upon entering k-OptMove, node t_1 is deactivated. Each exchange reactivates nodes that are part of the potential improvement sequence[2]. At a certain point in the search, if no nodes remain active, LKH ceases execution.
- **Multiple Trials:** LKH involves multiple trials to achieve good results. Each TRIAL reconsiders the initial tour and reorders the candidate set of each node. This is done in the same manner as Iterated Local Search (ILS).
- **Dynamic Reordering:** This step occurs after each TRIAL. For each node t a reordering of the candidate set $cand(S,t)$ is performed. This is done by prioritizing in $cand(S,t)$ the nodes belonging to the best and second-best tours (see Appendix D for an example.).
- **Gain23:** When all nodes become inactive (no further improvements are found), a positive non-sequential move is applied, representing a perturbation (or kick) that yields a strictly positive gain. This causes new nodes to be reactivated and passed through the k-OptMove function again (see Gain23 in LKH [6]).

3 LKH with a Partial Sampled Candidates Selection

Candidate selection follows the order obtained from the modified minimum 1-tree T_π (see Algorithm 1 line 13) as described in Sect. 2.3. Although the size of this set is a parameter (MAX_CANDIDATES), the optimal solutions for instances known in the literature (TSPLIB [15] library) are always almost achieved with a set size of 5. Increasing MAX_CANDIDATES only prolongs the trial time without significant improvements. The previous mechanisms play an important role in guiding the search without the need to expand MAX_CANDIDATES.

---

[2] This is referred to as a Flip, which performs a 2-opt move and activates four nodes.

We propose to control the number and selection of candidates through the prism of Partial Neighborhood Local Search (PNLS), which can be seen as an intuitive paradigm for handling diversification. In the following, we provide a brief description of PNLS, followed by our proposed heuristic.

3.1 Partial Neighborhood Local Search

A local search algorithm explores the search space by moving from one solution x_i to another $x_{i+1} \in N(x_i)$ for a given neighborhood N, guided by an evaluation function f. In order to strike a balance between intensification and diversification, PNLS was introduced [12].

This search operates by selecting a solution at each step from a sample of neighbors of size λ which is configurable and directly impacts the balance between intensification and diversification. A fair value of λ should allow for adequate exploitation while also exploring significant portions of the search space. Studies in [14] describe the impact of two different selection strategies for PLNSs, which are as follows:

Sample Walk (SW): Proposed by [12], this method selects the best neighbors among a sample of λ_{SW} neighbors, irrespective of whether it enhances or degrades the current solution. It can be viewed as an $(1, \lambda)$ evolution strategy[3] (with $\lambda = \lambda_{SW}$).

Intensification/Diversification Walk (ID): Introduced in [10], this approach selects the first improving solution encountered among the λ neighbors. If no such improvement is found, two variants are proposed: either choose the *best* deteriorating neighbor or choose *any* deteriorating one.

One motivation of using the PNLS principle resides in its extreme simplicity, that allows the possibility to integrate the partial neighborhood exploration principle into various complex and dedicated search heuristics. Despite the effectiveness of the LKH algorithm, its complexity arises from various components, particularly its improvement mechanisms presented in Sect. 2.4. For instance, to obtain good solutions, MAX_TRIALS must be set to a high value to further guide the search towards promising areas in the search landscape. Rather than treating the problem as a pre- or post-adjustment step that modifies its state, our approach aims to replace the LKH improvement mechanisms by integrating a candidate selection strategy directly into the main procedure k-OptMove. The next section provides a detailed description of our approach.

3.2 LKH-RSS (Randomized Sampled Selection)

The exploration of LKH can be seen as a partial depth-first search tree of movements, where each node (a FLIP) has a maximum of MaxCandidates children, and

---

[3] In Evolution Strategy (ES), the number of offspring (λ) controls the exploration-exploitation tradeoff. A smaller λ promotes more exploitation (by focusing on improving existing solutions), while a larger λ favors exploration (by trying new possibilities).

the depth varies and is limited to the value of MAX_SWAPS[4]. Unlike other heuristics (2-opt, 3-opt, ...), it does not generate a neighborhood or aims to select the best (BEST) or the first best (FIRST) among them. However, it applies the movement as soon as it finds an improving one, which causes the depth to vary.

We choose the following approach to implement PLNS in LKH, which is to apply sampling to the candidate set of nodes t_{2i} (involved in k-OptMove), as this is where moves are differentiated. Random sampling is performed by uniformly selecting λ candidates from the set. Each candidate is then processed in k-OptMove. The sample should be at most equal to MAX_CANDIDATES. If λ = MAX_CANDIDATES, the imposed order by the set will be intentionally ignored.

However, random sampling contradicts the reason that led Helsgaun to use the α-metric. A more 'strategic' sampling that carefully and diversely selects candidates will potentially lead to greater success in finding better solutions. One possible approach involves implementing a biased sampling strategy, where the likelihood of selecting a candidate is determined by its rank within the candidate set, as assessed by the α-metric. Therefore, we assign a geometric probability that decreases by half for each subsequent candidate order: $2^{(-x)}$, where x represents the candidate's order. This way, candidates with higher rankings are more likely to be chosen, while lower-ranked candidates still have some possibility of being picked, thereby sustaining diversity in the search process. It is important to emphasize that the last two candidates share the same probability to ensure that the total probability of all candidates sums to 1. Figure 2 illustrates an example with the following parameters: MaxCandidates = 5, λ = 3 (the small boxes above the candidates represent the selection probability).

The figure also describes how node selection takes into account previously selected ones. When a node is selected, it will not be considered in subsequent selections. The probability remains relative to the order and not the candidate itself. Thus, distant candidates have an increasingly larger chance of being selected as more candidates are chosen in the sample.

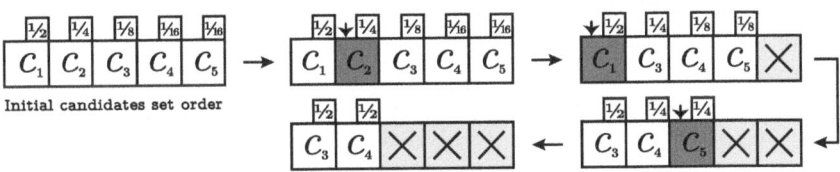

Fig. 2. Selection Example with our Biased Sampling.

Let us recall that k-OptMove does not allow deteriorating moves[5], as stated in constraint **C-II** of Sect. 2.2. This means that there may come a point when

---

[4] It is important to distinguish between MAX_SWAPS and k depth. MAX_SWAPS is the limit to execute best (negative) swap when candidates do not improve the current solution.

[5] When all candidates yield negative gains (g_i), it selects the best worst candidate, subject to 2 constraints: firstly, the best negative cost of an edge emanating from node t_{2i} must: $Gain_i + c(t_{2i-1}, t_{2i}) - c(t_{2i}, t_{2i+1}) \geq 0$; secondly, current SWAPS $\leq$ MAX_SWAPS.

all nodes become inactive and `Gain23` will be performed. In our approach, since candidate selection is only partially performed, it is not justified to deactivate a node that does not bring a positive gain ($Gain$). Instead, we aim to provide another opportunity to select previously missed candidates due to sampling. With a sufficient time window, we ensure that all nodes are examined, including those not involved in an exchange. By doing so, we can omit the `Gain23` and remain within a standard descent framework.

Another benefit of using RSS is that it prioritizes selecting improving candidates that are not necessarily ranked first according to the α-metric. For example, in an LKH-like ordered selection, the 1st candidate is automatically chosen (if it brings a positive gain) by LKH, while the 3rd candidate, with a lower α-value, may lead to better long-term advantages. Our algorithm, through biased and random probabilities, gives the 3rd-ranked candidate the chance to be selected first. In doing so, we prevent search stagnation when navigating the search landscape (See Fig. 3 for an illustration).

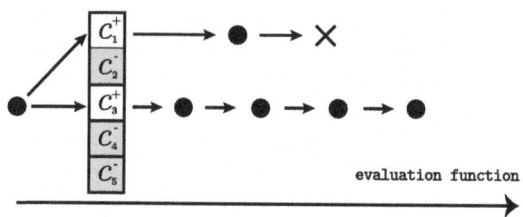

Fig. 3. Benefit of Sampled Selection

4 Experimental Results and Analyses for LKH-RSS

In this section, we first describe the experimental setup conducted with LKH-RSS, including the benchmark instances and parameter settings used. Next, we compare the computational results obtained for each strategy. Following this, we analyze the best-performing strategy of LKH-RSS (along with its optimal parameter combination) in comparison to the original LKH. Finally, we present the runtime performance relative to LKH-RSS and LKH.

4.1 Experiments Setup

The proposed method was incorporated into the LKH-1.3 codebase[6], exclusively employing the basic move `MOVE_TYPE = 2`. `Gain23` is disabled as previously stated. The three strategies considered are: *Ordered*, *Biased*, and *Random*. It is important to note that our methods are applied within a non-iterative framework due to the mechanism of always reactivating nodes with non-positive gain, thus setting `MAX_TRIALS = 1`. The time window `TRIAL_TIME` is set to 200 s for *small*

---

[6] http://akira.ruc.dk/~keld/research/LKH/LKH-1.3.tgz.

instances (> 1,000 nodes), 500 s for *medium* instances (1,000 − 10,000 nodes) and 1000 s for *large* ones (> 1,000 nodes). These durations were selected based on instance size and provide a sufficient time frame to reach peak performance, as observed in our preliminary experiments[7]. All instances from the TSPLIB, VLSI [16], and National TSP [1] benchmarks were tested. For optimal visualization in this paper and based on observed trends, we selected a representative set of 20 *small* and *medium* instances, and 30 *large* instances. Each instance involves 40 runs of different random initial tours. For each run all tested strategies have the same initial tour. Various combinations of (λ, MAX_CANDIDATES) are evaluated. All the experiments were run on a server using an AMD EPYC™ 7643 2.3 GHz 48-core CPU and 512 GB RAM, running Debian 5.10 Linux operation system.

4.2 Comparison of 3 Strategies: Ordered, Biased and Random

In this section, our primary objective is not to achieve optimality, but rather to draw comparisons between various sampled selection strategies under identical parameter configurations. Six combinations were tested. λ^8 is always < MAX_CANDIDATES. We chose values of $\lambda \leq 7$: specifically $\{2, 3, 4, 5, 6, 7\}$. For MAX_CANDIDATES, we used $\{15, 30\}$, which are large enough to diversify the selection. We omit displaying the combinations $(\lambda, 30)$ for *Ordered* and *Biased*, and $(\lambda, 15)$ for *Random*, because:

- For *Ordered*, λ is necessarily $\leq$ MAX_CANDIDATES so it would not matter.
- For the *Biased* strategy, MAX_CANDIDATES is not expected to have a substantial impact, as biased selection is less likely to choose very distant candidates. Therefore, setting MAX_CANDIDATES to a larger value does not negatively impact the algorithm's efficacy; thus, we opt for 30.
- For *Random*, depending on the value of λ and the size of the instance, MAX_CANDIDATES can influence the results by allowing the selection of distant candidates. We choose to present those limited to 30 (this aspect will be discussed later.).

Table 2 shows the gap between the mean of the total runs and the best known (BK) tour, defined as $Gap_{Avg}\% = (\text{CostSum/nbrRuns} − \text{BK})/\text{BK} \times 100$, where CostSum represents the sum of all nbrRuns run costs. We include statistical dominance among strategies, according to the legend described in the Table 1(a). Note that the comparison across multiple strategies has not been corrected. From the analysis of Table 2, we can exhibit some observations:

- At first reading, the gap increases accordingly to the size of the instances and decreases when λ increases, which is rather intuitive.
- For the *Ordered* strategy, a larger value of λ results in better performance, which is logical since more candidates are tested (see complete data in supplementary material).

---

[7] These time windows are sufficient; the point at which the L-shaped curve flattens into a horizontal tail can be achieved at around 0.5 s for small instances, 2–7 s for medium ones, and 10–50 s for large instances.
[8] The λ candidates are always considered, regardless of their feasibility, as for LKH.

Table 1. Caption of the results tables reporting statistical dominance (with p-value of 0.05) according to a pair-wise binomial tests. The shades of grey indicate the number of times a method statistically dominates the other strategies.

(a) For 13 strategies.

	Dominate t methods												
t	12	11	10	9	8	7	6	5	4	3	2	1	0
color													

(b) For 3 strategies.

	Dominate t methods		
t	2	1	0
color			

- The *Random* strategy shows a marked dependence on MAX_CANDIDATES. When comparing two combinations with the same λ, it appears that a high MAX_CANDIDATES leads to less cautious selection, often resulting in an irrational choice of candidate ranks. The opposite trend can be observed for larger instances.
- The *Random* strategy is consistently outperformed by the *Biased* strategy, although the average gaps are often similar. Additionally, it can be observed that a smaller value of λ can perform significantly better compared to the *Random* strategy with a larger λ. This trend is particularly observable in *small* instances.
- For the *Biased* strategy, we observe that increasing the value of λ to 5 (or 6) enhances efficiency, but further increases beyond that point result in diminished efficiency. This indicates that a value of 5 offers a suitable balance between intensification and diversification. Selecting a higher value of λ, such as 7, would include more nodes ranked first, which will lead to partially lose the diversification aspect (same behavior as [13] - Sect. 3.1).
- With respect to instance size, an *Ordered* selection of candidates performs poorly compared to our strategies, highlighting the impact of not using Helsgaun's improvement mechanisms. In comparison with other strategies, it can be concluded that the primary cause is search stagnation resulting from selecting the top candidates in a fixed order.

Our approach, however, is more generic and applicable to various problems (especially TSP variants), based on the idea that continuous and random selection allows for easy discovery and movement within interesting areas of the search landscape. This is strongly related to λ and n.

4.3 Comparison Between LKH-RSS-Biased and LKH-1.3

In a second step, we compare the results obtained with the *Biased* strategy (LKH-RSS(2opt, Biased)) using the best combination $\lambda = 5$ to those of an LKH using: (2opt, NoGain23), (2opt, Gain23) and (5opt, Gain23).

To have a fair comparison with RSS, we must adhere to RSS's 1 trial execution time. Therefore, we will limit the time of each trial for LKH so that: TRIAL_TIME$_{\text{LKH}}$ × MAX_TRIALS = TRIAL_TIME$_{\text{RSS}}$. For example, for *small* instances: $1.66s \times 60 = 100s$. A value of MAX_TRIALS = 60 is sufficient to

Table 2. Comparison results of three strategies (*Ordered*, *Random*, and *Biased*) based on $Gap_{Avg}\%$ across 50 instances over 40 runs. A smaller $Gap_{Avg}\%$ indicates better performance. Darker cells represent higher number of statistical dominance over other strategies.

		Ordered	Biased						Random					
(λ, MAX_CANDIDATES)		(7, 15)	(2, 30)	(3, 30)	(4, 30)	(5, 30)	(6, 30)	(7, 30)	(2, 30)	(3, 30)	(4, 30)	(5, 30)	(6, 30)	(7, 30)
Instance	Dimension													
gr24	24	0.42	0.00	0.00	0.00	0.40	0.60	0.49	0.00	0.00	0.00	0.00	0.00	0.00
st70	70	1.26	0.27	0.37	0.24	0.37	0.27	0.41	0.56	0.56	0.47	0.47	0.32	0.47
kroB100	100	1.04	0.36	0.41	0.32	0.35	0.36	0.36	0.42	0.38	0.30	0.28	0.26	0.27
bier127	127	1.10	0.59	0.36	0.41	0.45	0.44	0.65	1.14	0.97	0.85	0.83	0.74	0.76
xqf131	131	2.10	0.30	0.36	0.32	0.30	0.72	0.91	0.71	0.57	0.51	0.56	0.45	0.42
kroA150	150	1.27	0.36	0.28	0.36	0.29	0.31	0.34	0.35	0.34	0.31	0.27	0.14	0.17
u159	159	1.75	0.12	0.14	0.21	0.30	0.52	0.79	1.03	1.04	0.91	0.70	0.74	0.77
brg180	180	1.90	0.61	0.53	0.71	0.49	0.46	0.42	0.38	0.20	0.31	0.29	0.26	0.14
a280	280	2.02	1.15	0.69	0.81	0.74	0.82	1.04	2.18	1.37	0.97	1.02	0.93	0.98
pbn423	423	3.56	1.68	1.05	1.20	1.29	1.29	1.13	1.97	1.88	1.43	1.19	1.21	1.06
xql662	662	3.91	1.29	1.05	0.99	1.03	1.02	1.16	2.37	1.81	1.38	1.29	1.19	1.25
rbu737	737	2.94	1.06	0.89	0.76	0.77	0.93	0.86	2.09	1.71	1.47	1.20	1.12	1.14
rat783	783	1.33	0.85	0.68	0.64	0.62	0.56	0.70	1.70	1.51	1.32	1.17	1.06	0.94
fra1488	1488	7.25	1.80	1.41	1.24	1.16	1.10	1.28	2.42	2.01	1.75	1.65	1.47	1.55
rl1889	1889	8.23	1.96	1.45	1.33	1.49	1.48	1.65	2.43	2.20	1.82	1.62	1.45	1.54
dcb2086	2086	7.41	1.80	1.55	1.43	1.38	1.38	1.33	2.63	2.25	2.14	1.78	1.68	1.60
dea2382	2382	13.87	1.40	1.10	0.94	0.96	1.00	1.06	2.03	1.66	1.44	1.34	1.24	1.14
dbj2924	2924	6.18	1.83	1.40	1.24	1.27	1.33	1.30	2.53	2.16	1.90	1.75	1.67	1.61
bgf4475	4475	11.92	1.50	1.17	1.19	1.07	1.16	1.26	2.67	2.17	1.94	1.71	1.50	1.44
dga9698	9698	11.96	1.97	1.50	1.30	1.22	1.28	1.27	3.15	2.60	2.32	2.03	1.81	1.76
brd14051	14051	2.59	1.23	0.86	0.74	0.72	0.71	0.78	2.05	1.65	1.41	1.27	1.15	1.06
xrb14233	14233	14.22	1.84	1.44	1.28	1.16	1.19	1.23	2.93	2.48	2.09	1.91	1.70	1.66
ho14473	14473	154.04	3.85	2.65	1.91	1.79	1.45	1.42	2.21	1.68	1.43	1.32	1.17	1.07
it16862	16862	10.17	2.19	1.73	1.38	1.31	1.45	1.51	2.86	2.33	1.94	1.81	1.63	1.56
xia16928	16928	21.03	2.02	1.58	1.37	1.36	1.39	1.45	3.01	2.57	2.25	2.02	1.83	1.77
pjh17845	17845	15.38	2.06	1.61	1.43	1.37	1.40	1.40	3.03	2.58	2.22	2.03	1.84	1.74
d18512	18512	1.96	1.09	0.82	0.73	0.70	0.70	0.74	2.02	1.65	1.43	1.25	1.15	1.07
frh19289	19289	16.66	1.88	1.47	1.32	1.29	1.26	1.35	2.96	2.51	2.18	1.97	1.76	1.68
fnc19402	19402	17.54	2.00	1.53	1.37	1.34	1.33	1.43	3.00	2.56	2.22	2.03	1.85	1.77
ido21215	21215	15.23	1.93	1.50	1.37	1.30	1.31	1.36	3.00	2.54	2.18	1.97	1.79	1.68
fma21553	21553	23.63	2.27	1.75	1.59	1.55	1.62	1.61	3.38	2.85	2.48	2.22	2.09	2.00
vm22775	22775	16.05	3.12	2.10	1.43	1.52	1.49	1.70	3.66	2.98	2.23	2.07	1.81	1.90
lsb22777	22777	17.16	2.02	1.58	1.41	1.32	1.34	1.36	3.16	2.61	2.31	2.05	1.85	1.75
xrh24104	24104	22.41	2.18	1.61	1.42	1.42	1.41	1.47	3.16	2.63	2.31	2.08	1.91	1.80
sw24978	24978	9.88	1.82	1.17	0.92	0.91	1.02	1.02	2.57	2.06	1.73	1.51	1.37	1.27
bbz25234	25234	20.15	2.11	1.64	1.42	1.37	1.40	1.42	3.14	2.64	2.30	2.04	1.88	1.75
fyg28534	28534	18.97	2.15	1.61	1.37	1.32	1.34	1.38	3.40	2.85	2.42	2.17	1.97	1.82
icx28698	28698	22.17	2.14	1.67	1.47	1.46	1.42	1.45	3.27	2.77	2.40	2.17	2.00	1.84
boa28924	28924	25.15	1.90	1.47	1.31	1.24	1.27	1.28	2.98	2.45	2.15	1.93	1.78	1.65
ird29514	29514	20.19	2.08	1.63	1.45	1.40	1.41	1.43	3.32	2.79	2.38	2.15	1.98	1.85
xib32892	32892	26.55	2.13	1.58	1.43	1.39	1.41	1.42	3.12	2.61	2.29	2.04	1.89	1.74
bm33708	33708	9.55	1.82	1.17	0.99	0.93	0.93	0.97	2.54	1.95	1.70	1.46	1.31	1.25
pla33810	33810	11.54	2.61	1.62	1.39	1.27	1.22	1.27	2.96	2.44	2.09	1.87	1.72	1.66
bby34656	34656	19.97	2.11	1.58	1.44	1.34	1.34	1.42	3.22	2.75	2.34	2.08	1.94	1.81
pba38478	38478	22.43	2.21	1.71	1.47	1.42	1.40	1.43	3.41	2.88	2.51	2.21	2.00	1.88
ics39603	39603	23.01	2.29	1.68	1.47	1.41	1.39	1.45	3.40	2.84	2.46	2.19	1.97	1.83
rbz43748	43748	23.30	2.13	1.59	1.38	1.30	1.33	1.27	3.27	2.75	2.38	2.08	1.92	1.79
fna52057	52057	29.75	2.56	1.81	1.53	1.42	1.44	1.51	3.45	2.86	2.52	2.23	2.03	1.90
dan59296	59296	28.70	2.36	1.71	1.48	1.43	1.41	1.42	3.35	2.85	2.43	2.19	1.98	1.89
ch71009	71009	7.98	1.82	1.13	0.94	0.95	1.13	1.41	2.59	2.21	1.87	1.56	1.44	1.33

guarantee overall performance across the tested instances, while also providing enough time for each trial to execute k-OptMove[9]. Results for (5opt, Gain23) are

---

[9] As MAX_TRIALS increases, we are forced to spend less time on k-OptMove. This is particularly noticeable in larger instances. See supplementary material for details about this observation.

provided as a reference to illustrate the best way for achieving optimal results with Helsgaun's algorithm. From the observed results in Table 3, the following conclusions can be made:

Table 3. Comparison results of three algorithms: LKH(2opt, NoGain23), LKH(2opt, Gain23), LKH-RSS(2opt, Biased, $\lambda = 5$) for on 50 instances for 40 runs. LKH(5opt, Gain23) is given as a reference. Statistical dominance among strategies are included according to the legend described in the Table 1(b).

Instance	Dimension	LKH (5opt, Gain23)	LKH (2opt, NoGain23)	LKH (2opt, Gain23)	LKH-RSS (2opt, Biased, $\lambda = 5$)
gr24	24	0.00	**0.00**	**0.00**	0.40
st70	70	0.00	**0.00**	**0.00**	0.37
kroB100	100	0.00	**0.00**	**0.00**	0.35
bier127	127	0.00	**0.00**	**0.00**	0.45
xqf131	131	0.00	0.13	**0.00**	0.30
kroA150	150	0.00	0.15	**0.00**	0.29
u159	159	0.00	0.05	**0.00**	0.30
brg180	180	0.00	**0.00**	**0.00**	0.49
a280	280	0.00	0.01	**0.00**	0.74
pbn423	423	0.00	0.37	**0.05**	1.29
xql662	662	0.05	1.03	**0.17**	1.03
rbu737	737	0.04	0.53	**0.16**	0.77
rat783	783	0.00	0.28	**0.11**	0.62
fra1488	1488	0.00	1.68	**0.26**	1.16
rl1889	1889	0.08	2.14	**0.27**	1.49
dcb2086	2086	0.05	2.72	**0.53**	1.38
dea2382	2382	0.04	3.83	**0.47**	0.96
dbj2924	2924	0.11	3.22	**0.83**	1.27
bgf4475	4475	0.10	4.10	**0.72**	1.07
dga9698	9698	0.10	6.45	**1.20**	1.22
brd14051	14051	0.15	1.68	0.79	**0.72**
xrb14233	14233	0.11	7.43	1.23	**1.16**
ho14473	14473	0.56	21.32	**1.47**	1.79
it16862	16862	0.16	4.53	**1.02**	1.31
xia16928	16928	0.14	11.26	**1.34**	1.36
pjh17845	17845	0.14	10.14	**1.35**	1.37
d18512	18512	0.15	1.71	0.83	**0.70**
frh19289	19289	0.14	9.97	1.36	**1.29**
fnc19402	19402	0.14	10.30	1.37	**1.34**
ido21215	21215	0.15	9.56	1.37	**1.30**
fma21553	21553	0.17	12.84	**1.41**	1.55
vm22775	22775	0.08	7.19	**0.88**	1.52
lsb22777	22777	0.14	11.72	1.40	**1.32**
xrh24104	24104	0.16	12.84	1.44	**1.42**
sw24978	24978	0.08	5.22	0.99	**0.91**
bbz25234	25234	0.17	13.08	1.48	**1.37**
fyg28534	28534	0.14	12.05	1.53	**1.32**
icx28698	28698	0.17	14.72	1.57	**1.46**
boa28924	28924	0.14	15.37	1.42	**1.24**
ird29514	29514	0.19	12.93	1.52	**1.40**
xib32892	32892	0.21	16.60	1.51	**1.39**
bm33708	33708	0.08	4.95	1.00	**0.93**
pla33810	33810	0.37	6.81	1.47	**1.27**
bby34656	34656	0.14	14.15	1.47	**1.34**
pba38478	38478	0.15	15.43	1.52	**1.42**
ics39603	39603	0.16	16.44	1.63	**1.41**
rbz43748	43748	0.15	16.66	1.52	**1.30**
fna52057	52057	0.17	22.20	1.61	**1.42**
dan59296	59296	0.20	21.29	1.64	**1.43**
ch71009	71009	0.12	4.76	**0.95**	**0.95**

- For instances of large size ($> 14,000$), our approach yields slightly better performance to that of a complete LKH(2opt, Gain23).
- If Gain23 is not taken into account, our diversification method significantly outperforms LKH improvement mechanisms (re-initialization of the initial tour and reordering of candidates).

It is important to recall that our method is non-iterative, requiring only the initialization of the time frame and λ. To improve proximity to the optimum, sampling can be reapplied to MOVE_TYPE = 5 at each level k. Although MAX_CANDIDATES = 5 is deemed optimal for LKH, selecting $\lambda < 5$ can further decrease CPU time by specifying λ_k for each depth. This will achieve the performance guaranteed by a larger operator while also saving computation time.

4.4 Processing Time

In this step, we present execution traces for the previously discussed algorithms, focusing on large instances where LKH-RSS(2opt, Biased, $\lambda = 5$) achieves superior results. For example, comparison results for the instance boa28924 are shown in Fig. 4. For clarity, we have selected a single representative run out of 40, as its performance is indicative of the general trend.

A key point to consider is that, since LKH operates in an iterative mode, we present here all the trials that contributed, in one way or another, to obtaining the best result (represented by the dashed line). It is noteworthy that the best result (in LKH(2opt, Gain23)) is achieved much later compared to LKH-RSS.

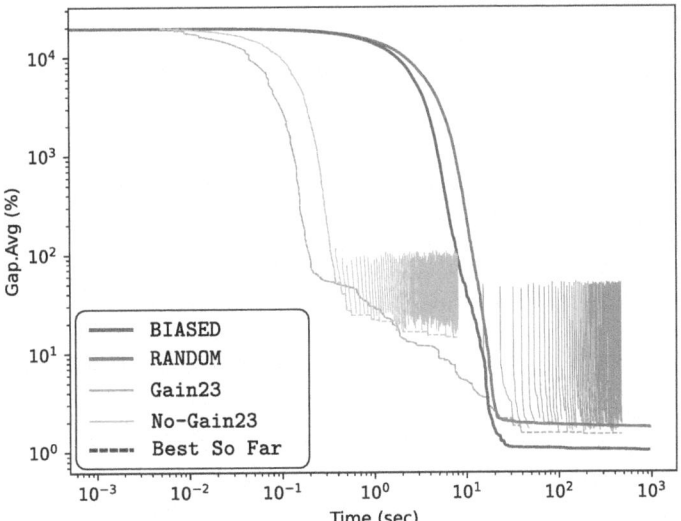

Fig. 4. Comparison of 1 Runtime of 4 Algorithms: LKH-RSS(2opt, Biased, $\lambda = 5$), LKH-RSS(2opt, Random, $\lambda = 5$), LKH(2opt, Gain23) and LKH(2opt, NoGain23) in a logarithmic scale for the boa28924 instance.

5 Conclusion

This paper proposes a candidate selection strategy aimed at simplifying improvement mechanisms implemented in the LKH algorithm. The latter involves allowing non-sequential moves, continuous node activation/deactivation during the search, in addition to iterating over several trials for resetting the initial solution, and applying numerous adjustments to the order of candidate nodes. This diversification mechanism is rather complex. A clear fine-tuning between intensification and diversification is not straightforward in this context. The approach we propose employs a biased sampling strategy integrated directly into the k-opt search while deactivating the improvement mechanisms of LKH. This allows to directly control the level of diversification in the algorithm, in the manner of partial neighborhood local search methods.

Compared to LKH-1.3 limited to 2-opt operators, the proposed method demonstrates improved results when `Gain23` is disabled, and slightly better results otherwise, especially for the large instances. The goal of this work is to maintain high solution quality while simplifying the overall algorithm. Future research could focus on generalizing the sampling to more efficient operators, such as 5-opt, and on gaining precise control over the values of λ at each level of the search depth.

Acknowledgments. Experiments presented in this paper were carried out using the CALCULCO computing platform, supported by SCoSI/ULCO (Service COmmun du Système d'Information de l'Université du Littoral Côte d'Opale).

References

1. National traveling salesman problem (2022). https://www.math.uwaterloo.ca/tsp/world/countries.html
2. Hansen, K.H., Krarup, J.: Improvements of the held - karp algorithm for the symmetric traveling-salesman problem. Math. Program. **7**(1), 87–96 (1974). https://doi.org/10.1007/BF01585505
3. Held, M., Karp, R.M.: The traveling-salesman problem and minimum spanning trees. Oper. Res. **18**(6), 1138–1162 (1970). https://doi.org/10.1287/opre.18.6.1138
4. Held, M., Karp, R.M.: The traveling-salesman problem and minimum spanning trees: Part II. Math. Program. **1**(1), 6–25 (1971). https://doi.org/10.1007/BF01584070
5. Held, M., Wolfe, P., Crowder, H.P.: Validation of subgradient optimization. Math. Program. **6**(1), 62–88 (1974). https://doi.org/10.1007/BF01580223
6. Helsgaun, K.: An effective implementation of the lin-kernighan traveling salesman heuristic. Eur. J. Oper. Res. **126**(1), 106–130 (2000). https://doi.org/10.1016/S0377-2217(99)00284-2
7. Helsgaun, K.: General k-opt submoves for the lin-kernighan TSP heuristic. Math. Program. Comput. **1**(2–3), 119–163 (2009). https://doi.org/10.1007/s12532-009-0004-6
8. Helsgaun, K.: An extension of the Lin-Kernighan-Helsgaun TSP solver for constrained traveling salesman and vehicle routing problems. Department of Computer Science, Roskilde University (2017)

9. Lin, S., Kernighan, B.W.: An effective heuristic algorithm for the traveling-salesman problem. Oper. Res. **21**(2), 498–516 (1973). https://doi.org/10.1287/opre.21.2.498
10. Neveu, B., Trombettoni, G., Glover, F.: ID walk: a candidate list strategy with a simple diversification device. In: Wallace, M. (ed.) CP 2004. LNCS, vol. 3258, pp. 423–437. Springer, Heidelberg (2004). https://doi.org/10.1007/978-3-540-30201-8_32
11. Stewart, W.R., Jr.: Accelerated branch exchange heuristics for symmetric traveling salesman problems. Networks **17**(4), 423–437 (1987). https://doi.org/10.1002/net.3230170405
12. Tari, S., Basseur, M., Goëffon, A.: Sampled walk and binary fitness landscapes exploration. In: Lutton, E., Legrand, P., Parrend, P., Monmarché, N., Schoenauer, M. (eds.) EA 2017. LNCS, vol. 10764, pp. 47–57. Springer, Cham (2018). https://doi.org/10.1007/978-3-319-78133-4_4
13. Tari, S., Basseur, M., Goëffon, A.: On the use of $(1, \lambda)$-evolution strategy as efficient local search mechanism for discrete optimization: a behavioral analysis. Nat. Comput. **20**(3), 345–361 (2021). https://doi.org/10.1007/s11047-020-09822-2
14. Tari, S., Basseur, M., Goëffon, A.: Partial neighborhood local searches. Int. Trans. Oper. Res. **29**(5), 2761–2788 (2022). https://doi.org/10.1111/itor.12983
15. TSPLIB: A library of sample instances for the traveling salesman problem (TSP) (1995). http://comopt.ifi.uni-heidelberg.de/software/TSPLIB95
16. VLSI: A library of sample instances for the traveling salesman problem (2013). https://www.math.uwaterloo.ca/tsp/vlsi
17. Zheng, J., He, K., Zhou, J., Jin, Y., Li, C.: Reinforced lin-kernighan-helsgaun algorithms for the traveling salesman problems. Knowl. Based Syst. **260**, 110144 (2023). https://doi.org/10.1016/j.knosys.2022.110144

Instance Space Analysis and Algorithm Selection for a Parallel Batch Scheduling Problem

Francesca Da Ros[1](✉)[iD], Luca Di Gaspero[1][iD], Marie-Louise Lackner[2][iD], and Nysret Musliu[2][iD]

[1] Università degli Studi di Udine, Udine, Italy
{francesca.daros,luca.digaspero}@uniud.it
[2] Christian Doppler Laboratory for Artificial Intelligence and Optimization for Planning and Scheduling, TU Wien, Vienna, Austria
{marie-louise.lackner,nysret.musliu}@tuwien.ac.at

Abstract. This paper addresses the Oven Scheduling Problem (OSP), a parallel batch scheduling problem in semiconductor manufacturing, and identifies strengths and weaknesses of solution methods using the Instance Space Analysis (ISA) methodology. We propose a comprehensive feature set to effectively characterize OSP instances and generate more diverse instances compared to the literature. The performance of two state-of-the-art algorithms for the OSP – Simulated Annealing and Large Neighborhood Search – is analyzed using ISA, revealing distinct regions of superior or inferior performance for each, as well as areas of equal performance. Finally, we propose an automated algorithm selection approach that outperforms any single algorithm.

Keywords: Simulated annealing · Large neighborhood search · Oven scheduling problem · Algorithm visualization · Empirical analysis

1 Introduction

Several parallel batch scheduling problems have been proposed in the literature [12] with the aim of optimizing resource usage by grouping compatible tasks and processing them in batches. Among these formulations, the Oven Scheduling Problem (OSP) has recently been introduced within the electronic component manufacturing industry [24]. The OSP stands out from other similar problems due to its complex combination of objective components and constraints [7, Table 3]. Traditional analyses of stochastic algorithms, such as gap analysis and critical difference plots, have identified two algorithms as the most effective for solving the OSP: Simulated Annealing (SA) [7] and Large Neighborhood Search (LNS) [5]. Both have shown to deliver high-quality solutions across a range of instances, yet their performance can vary depending on the specific instance characteristics. Given the significant impact that scheduling quality can have on overall system efficiency, selecting the most suitable algorithm for a given problem instance becomes essential. One promising methodology for achieving this is the Instance Space Analysis (ISA) [42], which allows for a comprehensive evaluation of algorithms across a diverse range of problem instances. ISA connects algorithms' performance to instance characteristics, identifying areas where each algorithm excels or struggles.

In this work, we aim to analyze the instance space of the OSP and offer an overview of the performance patterns of SA and LNS considering the ISA framework. Moreover, we investigate the applicability of automated algorithm selection. The main contributions are: (i) A new set of 1396 instances that span a broader range of variations compared to those present in the literature. (ii) A set of 165 features that define the characteristics of the instances of parallel batch scheduling problems. Out of the complete set, 7 are selected by ISA. (iii) The identification of areas of the instance space in which one algorithm outperforms the other. (iv) The analysis of the performance of the algorithms w.r.t. instance features highlighting their strengths and weaknesses. (v) The development of an algorithm portfolio through automated algorithm selection using four Machine Learning (ML) models, each outperforming any single algorithm.

The remainder of this work is organized as follows. Section 2 describes the parallel batch scheduling problem at hand, namely the OSP. Section 3 recalls the ISA framework, while Sect. 4 presents the ISA scheme for the OSP. Section 5 discusses our experimental results. Finally, Sect. 6 concludes the work.

2 The Oven Scheduling Problem

2.1 Problem Description

The OSP is an NP-hard parallel batch scheduling problem [24]. We report an informal description and refer the reader to the Supplementary Material for the Integer Linear Programming (ILP) formulation.

An instance of the OSP consists of a set of *machines* (also referred to as *ovens*), a set of *jobs*, a set of *attributes*, representing the temperatures at which jobs need to be processed, and the length of the *scheduling horizon*, representing the total time available for processing.

Each machine is characterized by its *initial attribute* (temperature), a set of *availability* intervals indicating when it can operate, and a *maximum processing capacity*. Each job is characterized by its *release time*, *due date*, *minimum* and *maximum processing time*, specific *attribute* (temperature), *size*, and a subset of *eligible machines* for where it can be processed. Changing the oven temperature to match a job attribute incurs economic and time costs, represented by matrices describing *setup costs* and *setup times* for switching settings.

The aim of the OSP is to group compatible jobs into *batches* and devise a processing *schedule* on the set of machines. The optimization goal of the OSP consists of three parts: minimizing the cumulative processing time of the batches across machines (p), minimizing the cumulative setup costs associated with these batches (sc), and minimizing the count of tardy jobs (t). The objective function obj is a linear combination of these components:

$$\text{obj} = (w_p \cdot \tilde{p} + w_{sc} \cdot \tilde{sc} + w_t \cdot \tilde{t})/(w_p + w_{sc} + w_t) \quad \in [0,1] \subset \mathbb{R} \quad (1)$$

where $\tilde{p}$, $\tilde{sc}$, and $\tilde{t}$ are the normalized versions of p, sc, and t, and w_p, w_{sc}, and w_t are the associated weights. We consider the objectives in a lexicographic manner: minimizing processing time takes precedence over minimizing tardiness, which, in turn, is prioritized over minimizing setup costs. This is achieved by setting $w_p = w_t \cdot (n+1)$, $w_t = n \cdot \max_{SC} + 1$, and $w_{sc} = 1$, where n is the number of jobs and $\max_{SC}$ is the maximal setup cost (see [24, Use Case 2]).

A solution to the OSP must satisfy the following constraints: (i) Each job is assigned to exactly one batch on one of its eligible machines. (ii) Each batch must be processed within the scheduling horizon. (iii) All jobs in one batch have the same start and end time. (iv) All jobs assigned to one batch must share the same attribute. (v) The batch size must not exceed the capacity of the machine it is assigned to. (vi) A batch may not start before the release time of any job in that batch. (vii) The processing time of each batch must lie between the minimum and maximum processing times of all assigned jobs. Preemption is not allowed. (viii) Batches on the same machine may not overlap, and setup times must be accounted for between consecutive batches. (ix) Both the batch processing time and the preceding setup time must lie entirely within one of the machine's availability intervals.

2.2 Related Work and State of the Art

The OSP was initially proposed by Lackner et al. [24] in collaboration with an industrial partner from Central Europe. In the original work, the problem was addressed using a construction heuristic, ILP, and Constraint Programming (CP). While these approaches produced good and optimal results for small instances, they struggled to scale to medium and large instances. Consequently, a set of Local Search (LS) [6,7,25] and LNS approaches [5] were proposed. Overall, the current state-of-the-art results are provided by SA and LNS as demonstrated by Da Ros et al. [7] and Da Ros et al. [5]. Da Ros et al. [8] introduced theoretically calculated lower bounds for all three objective components of the OSP. These lower bounds are computed in a limited amount of time (less than 3s on average) and their quality is comparable to those obtained by commercial tools.

The literature on (parallel) batch scheduling spans across different industries [4,12,29,38,44]. Many solution methods have been proposed to tackle such problems, spanning from exact methods [1,20,44] to metaheuristic approaches [2,17,21,28,47]. These methods cannot be directly applied to the OSP, since it presents a more complex and diverse set of objectives and constraints compared to previous problems [12]. For example, several studies focus solely on minimizing tardiness [3] or exclusively on minimizing makespan [28,45].

3 Instance Space Analysis

ISA is a methodology developed by Smith-Miles and Muñoz [42] that builds on the Algorithm Selection Problem introduced by Rice [39]. It represents the

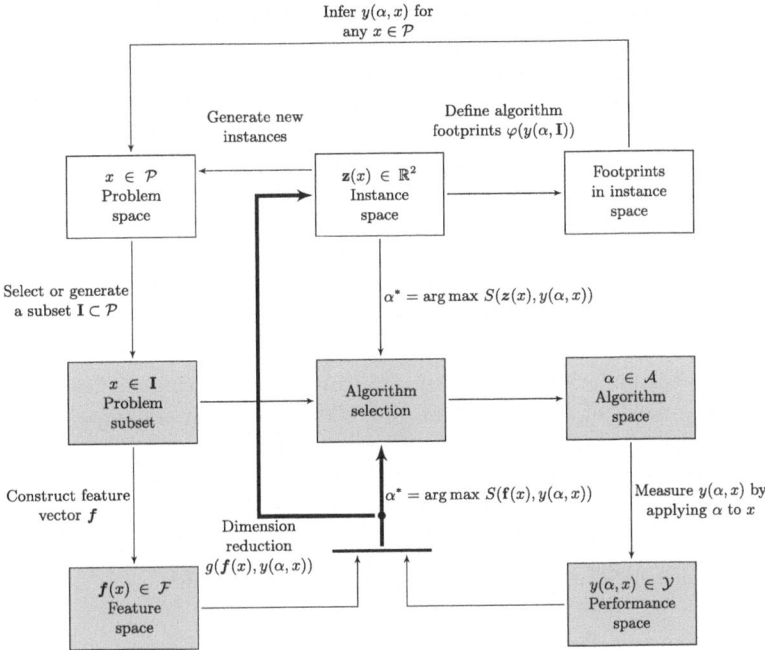

Fig. 1. ISA framework [42] extending the Algorithm Selection Problem framework by Rice [39]. The blocks in common with Rice [39] are reported in gray.

relationship between problem instance properties and algorithm performance. Specifically, instances are projected in a 2D space, so that one can: (i) Visualize the distribution and diversity of instances. (ii) Assess instance features. (iii) Identify and measure algorithm strengths and weaknesses (algorithm footprints). (iv) Locate areas where additional instances could offer more insights.

Figure 1 presents the general framework for developing an ISA as detailed by Smith-Miles and Muñoz [42]. The building blocks related to the work of Rice [39] are highlighted in gray. The problem space, denoted by $\mathcal{P}$, represents all possible problem instances. However, computational results are typically only available for a subset, denoted as $\mathbf{I}$. The feature space, $\mathcal{F}$, consists of vectors f that contain measures describing and characterizing the instances in $\mathbf{I}$. These instances are solved using a set of algorithms constituting the algorithm space $\mathcal{A}$. The algorithms' performance is captured in the performance space, denoted by $\mathcal{Y}$, which is the set of metrics $y(\alpha, x)$ that measure the performance of algorithm $\alpha \in \mathcal{A}$ when solving instance $x \in \mathbf{I}$. The algorithms' performance, together with the instance features, forms what is known as metadata, which is used to learn a mapping $g(f(x), y(\alpha, x))$. This mapping projects an instance x from the high-dimensional feature space to a 2D space, referred to as the instance space.

The ISA framework is sufficiently general and can be tailored to various domains by customizing elements such as instance features, performance metrics, and other relevant parameters. In this regard, ISA has been applied across various fields, including combinatorial optimization [10,14,19,43], ML [31,32], Black-Box Optimization (BBO) [30,34], and software engineering [35,36]. To the best of our knowledge, ISA has not been applied to batch scheduling problems. Details on the ISA in the context of the OSP are reported in Sect. 4.

4 Instance Space Analysis for the OSP

4.1 Problem Subset

Our ISA includes a total of 2190 instances: 794 coming from the literature and real-world cases, and 1396 newly generated specifically for this work.

Existing Instances. We include in our ISA the 80 original instances for the OSP from the work of Lackner et al. [24] (OSP-1), which encompass a diverse set of parameters: a varying number of jobs (10, 25, 50, or 100), machines (2 or 5), and attributes (2 or 5) [23]. We include 40 larger instances (OSP-2) and 25 tuning instances (indicated as TU) provided by y Da Ros et al. [6], which scale up to 500 jobs. We also consider 49 real-world instances from an industrial partner (RL) and 600 instances related to a similar problem [9] (DA).

Newly Generated Instances. We generate 1396 additional instances (indicated as NEW) using the random instance generator described by Lackner et al. [24] and publicly available through a Command Line Interface (CLI).[1] These new instances serve two objectives: (i) Better cover the instance space. (ii) Train the models used for algorithm selection. To ensure the diversity of this new instance set, the parameter values chosen for the generation cover a broader range than those of the existing instances. These parameter values include the number of jobs, machines, and attributes, as well as maximum processing times, machine capacity limits, etc. The generation process is repeated multiple times until no (large) areas of the instance space remain empty.

4.2 Algorithm Space

We analyze the two state-of-the-art algorithms from the literature (SA and LNS).

Simulated Annealing. The first algorithm is the multi-neighborhood SA proposed by Da Ros et al. [7]. This approach uses the Metropolis acceptance criterion and a geometric cooling scheme, along with a cutoff mechanism to reduce computational costs during the early stages of the search [13,18]. It employs the construction heuristic by Lackner et al. [24] as the initial solution and utilizes a union of four neighborhoods: swapping two batches, creating unary or larger batches, and adding jobs to an existing batch. We use the implementation from the original work, which is available in public repositories.[2] The algorithm was implemented in C++ and the code is compiled using Clang++18.

---
[1] See https://github.com/marielouiselackner/OvenSchedulingCLI.
[2] See https://github.com/iolab-uniud/osp-ls.

Large Neighborhood Search. The second algorithm is the LNS proposed by Da Ros et al. [5]. This approach iteratively destroys and repairs parts of an existing solution [40]. It employs the construction heuristic by Lackner et al. [24] as the initial solution and uses three destroy methods: destroying random batches, destroying random batches with the same attribute, and destroying random batches on the same machine. The repair operator is based on a CP model utilizing interval variables and two levels for the optimality tolerance. The size of the destruction is increased after several idle iterations and reset to the initial size whenever an improvement in the objective function is found. We use the implementation from the original work, which is available in public repositories.[3] The control structure and the destroy operators were implemented in Python (v.3.10.12); the repair operator is run with `CP Optimizer` (v.22.1.0) [22].

Experimental Setup. The algorithms' executions are scheduled on a machine featuring 2x Intel Xeon Platinum 8368 2.4GHz 38C, 8x64GB RDIMM. We run each algorithm 5 times with a different seed (i.e., from 42 to 46). Both methods are stopped with a 1-hour time limit or after 10000 idle iterations. The tuning is the same as in the original papers, as the employed machines are comparable.

4.3 Performance Space

As the primary performance metric for our ISA, we employ the value of the objective function obj, where lower values indicate better performance. In addition, following the approach suggested by y Smith-Miles et al. [41], we apply two supplementary methods to analyze the results. First, we consider an algorithm to be *good* for a given instance if its performance is within $(1+\epsilon)$ times the best performance on that instance. Formally, this is expressed by $\widehat{\text{obj}}_\alpha \leq 1+\epsilon$, where $\widehat{\text{obj}}_\alpha(x) = \frac{\text{obj}_\alpha(x)}{\min_{\alpha' \in \mathcal{A}} \text{obj}_{\alpha'}(x)}$ is the scaled cost of algorithm α on instance x, relative to the best known cost for x. We set $\epsilon = 0.01$ (i.e., a 1% gap). Moreover, an algorithm is considered a *winner* for an instance if it is among the best-performing algorithms, meaning it achieves the best performance with $\epsilon = 0$. If an algorithm is the sole winner for an instance, it is classified as a *unique winner*.

4.4 Feature Space

We propose 165 features to describe the instances, including 115 direct features, 9 features connected to lower bounds, and 41 graph-based features. While the majority of direct features are also discussed by Strassl and Musliu [43], to the best of our knowledge, we are the first to extensively use theoretically calculated lower bounds in the context of ISA and to build a compatibility graph (graph representations are used in the ISA literature though [10,43]). The complete description of the features is reported in the Supplementary Material.

---

[3] See https://github.com/iolab-uniud/osp-lns.

Direct Features. Direct features can be extracted directly from the instances through simple counts or examination of instance entities, possibly involving only minor calculations. For example, we include the number of jobs, the number of machines, the number of operations, calculated as the product between the number of jobs and the number of machines, the number of jobs per machine considering the concept of eligible machines, the job sizes, and the normalization weights for each cost component. These features are extracted employing the numpy (v.1.26.4) Python package [16] and the above-mentioned CLI.

Lower Bounds as Features. The feature set includes theoretical lower bounds for the OSP by Da Ros et al. [8] and related metrics: the bound on the number of batches, lower bounds on cost components, the upper bound on processing time reduction from batching, and the time to compute these bounds. These features are extracted using the OSP CLI.

Graph Features. Graph features are based on the undirected job compatibility graph. In this graph, each node corresponds to a job, and arcs connect pairs of nodes (jobs) that can be processed together under the following conditions: (i) They share the same attribute. (ii) They can be processed on at least one common machine. (iii) Their processing times are compatible. (iv) Their release and due dates are compatible. Given that job tardiness is already considered in the objective function obj, we also introduce a relaxed version of the graph where condition (iv) is omitted. For each graph (complete and relaxed), we extract features such as the number of edges, density, and isolated nodes using the NetworkX (v.3.1) Python package [15].

Key Statistical Figures. Some features are represented by a single value, while others require aggregation to capture their statistical distribution. In such cases, we compute the minimum (min), maximum (max), range ($range = \max - \min$), mean, quartiles ($q1, q2, q3$), and standard deviation (std). For instance, the total number of jobs is a single value, whereas the jobs' sizes are aggregated into average, standard deviation, etc.

4.5 Instance Space Construction

Feature Processing. First, the feature values are preprocessed by (i) Bounding outliers to the median plus/minus five times the inter-quartile range. (ii) Normalizing the values so they are normally distributed by applying a Box-Cox transformation and a z-score standardization.

For each algorithm, the most relevant features are identified by calculating their Pearson correlation with the algorithm's performance, retaining only those with a correlation above a minimum threshold of 0.45. The full set of features is then grouped using k-means clustering (with $k = 7$ clusters, a maximum of 1000 iterations, and 100 repeats), based on similarity in Pearson correlation values.

This grouping allows us to select a representative feature from each cluster, simplifying and enhancing the effectiveness of subsequent visualizations. The representative feature is selected by initially projecting the features onto a 2D space using a Principle Component Analysis (PCA). Next, a Random Forest (RF) model (with 100 trees) is applied to classify each instance as either *easy* or *hard* for each algorithm. Finally, a Genetic Algorithm (GA) is employed to guide the feature selection process, identifying the combination of features that produces the most accurate RF predictions (see Smith-Miles and Muñoz [42, Section 3.2.2]). The selected features are projected into a 2D space according to:

$$\begin{pmatrix} z_1 \\ z_2 \end{pmatrix} = \begin{pmatrix} c_1^1 & c_2^1 \\ \ldots & \\ c_1^k & c_2^k \end{pmatrix}^T \cdot \begin{pmatrix} \text{feature}_1 \\ \ldots \\ \text{feature}_k \end{pmatrix} \qquad (2)$$

The coefficients $c_1^1, \ldots, c_2^k$ are determined numerically by solving the problem defined in Smith-Miles and Muñoz [42, Section 3.2.3]. The 2D projection serves as the primary tool for exploring the instance space.

Instance Space Boundaries. The coverage of the instance space is assessed by identifying two boundaries: an experimental and a theoretical one. The experimental boundary is defined by the convex hull of the point cloud, representing the area with empirical evidence of existing instances. The theoretical boundary, in contrast, is derived from the hypercube formed using the minimum and maximum values of each selected feature. This hypercube is refined by eliminating improbable feature combinations: for any two features f and g with a correlation $\rho_{f,g} > 0.75$, vertices containing $\min(f)$ and $\max(g)$, or $\min(g)$ and $\max(f)$ are removed. The same pruning method is applied to features with strong negative correlations. The remaining vertices are projected into the 2D space using the previously determined coefficients, and their convex hull defines the theoretical boundary. Note that this procedure yields an approximate boundary, which may significantly over- or underestimate the true extent of the space.

Strengths and Weaknesses of Algorithms. The instance space is divided into regions of strengths and weaknesses for each algorithm, enabling automated recommendations. The ISA procedure trains a separate Support Vector Machine (SVM) for each algorithm using normalized coordinates as inputs.

We use MATLAB native SVM implementation with a polynomial kernel and automatic tuning via 10-fold stratified cross-validation. This approach predicts the best algorithm for new instances, prioritizing the model with the highest precision, and if needed, the one with the highest average performance. Instances without a clear winner are marked as having no strong preference.

While SVM provides useful insights, the analysis should be complemented by considering the concept of an *algorithm footprint*, which represents the generalized area of the instance space where the algorithm performs well or best. ISA constructs the algorithm footprint by employing Density-Based Spatial Clustering of Applications with Noise (DBSCAN) to identify high-density clusters of instances where each algorithm performs well. For each identified cluster, it generalizes the convex hull concept to form a polygon that tightly encloses all points within the cluster. We set a purity threshold of 0.75, meaning that at least 75% of all instances where the algorithm shows good or best performance must be enclosed within the footprint. These footprints offer a more nuanced assessment of algorithm performance, providing metrics such as the footprint area (a), density (d, calculated as the number of instances within the footprint per unit area), and purity (p).

Experimental Setup. The instance space is constructed using the ISA implementation available from the public repository[4] on a MacBook Air M3 with 16GB of memory and macOS Sonoma 14.3. All parameters not explicitly mentioned are set to default values. To enhance result visualization, the ISA images are reproduced using the Python package seaborn (v. 0.13.2) [46].

4.6 Algorithm Selection

Algorithm selection aims to predict the best algorithm (the winner) for a given instance, treating it as a classification problem with classes representing the algorithms. This contrasts with the ISA approach, which builds separate models for each algorithm and combines predictions, resolving ties using precision and performance probability. We apply our approach to validate the ISA outcomes (Sect. 5.1) and to explore the benefits of an algorithm portfolio (Sect. 5.2).

Dataset. We focus on instances with a unique winner ($\epsilon = 0.0$), excluding ties. Thus, instances where SA and LNS perform equally are omitted, as misclassifying a tie still yields an optimal result. For experiments validating ISA results (Sect. 5.1), we use only the features identified by ISA. In contrast, for algorithm selection (Sect. 5.2), we use the full feature set, following related research [10,43].

Classification Algorithms. The following standard classification algorithms were used for algorithm selection: (i) Most Frequent (MF), always predicting the most frequent class as a baseline (i.e., SA). (ii) Random Forest (RF), with 170 trees for Sect. 5.1 and 200 trees for Sect. 5.2, reflecting the use of two datasets. (iii) Decision Tree (DT), using the Gini criterion, a maximum depth of 8, minimum samples to split set to 14, and minimum samples at a leaf set to 10. (iv) K-Nearest Neighbors (KNN), with 5 neighbors. (v) Support Vector Machine (SVM), with a linear kernel and regularization parameter $C = 1$. All models are implemented using scikit-learn (v.1.5.1) [37] and evaluated via 10-times iterated stratified 10-fold cross-validation, using accuracy as the primary metric. Parameter tuning was performed for the specified parameters (see Supplementary Material), with default settings otherwise. Experiments are run on the same Mac Book described in Sect. 4.5.

---
[4] See https://github.com/andremun/InstanceSpace and Mu noz et al. [33].

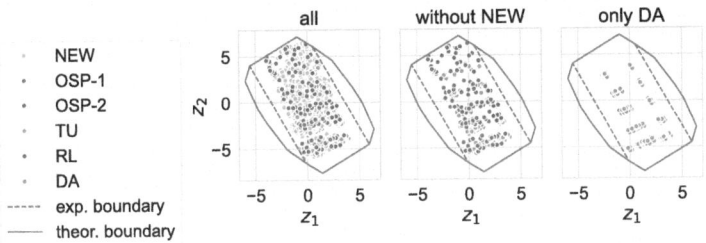

Fig. 2. Instance space organized by the sources of the instances. (Color figure online)

Performance Measures. We evaluate the models using: (i) *Accuracy*, the proportion of correct classifications; (ii) *Recall*, the proportion of actual positives correctly identified; (iii) *Precision*, the proportion of true positives among all positive predictions. These metrics are also applied to assess ISA SVM models.

SHapley Additive ExPlanations. To validate the results of the ISA (Sect. 5.1), we train a RF model using the 7 features selected by the ISA procedure and analyze their influence on the model's predictions using SHapley Additive exPlanations (SHAP) [27]. SHAP is a game-theoretic approach for interpreting ML model outputs, where SHAP values quantify each feature's impact by assessing its contribution to the overall prediction [11]. These values are obtained using the shap (v.0.46.0) Python library [26].

5 Experimental Analysis

5.1 Instance Space

Projection Matrix. The projection matrix computed by the ISA procedure (Sect. 4.5) and used to map the processed features into the 2D space is as follows:

$$\begin{pmatrix} z_1 \\ z_2 \end{pmatrix} = \begin{pmatrix} 0.1342 & 0.4788 \\ -0.2584 & 0.6512 \\ -0.44 & 0.6504 \\ -0.48 & 0.3313 \\ -0.3202 & 0.5603 \\ -0.4475 & 0.3603 \\ 0.9907 & 0.6723 \end{pmatrix}^T \cdot \begin{pmatrix} \text{number of jobs} \\ \text{number of jobs per machine (q3)} \\ \text{number of operations} \\ \text{normalization weight processing time} \\ \text{time to compute lower bounds} \\ \text{lower bound number batches} \\ \text{upper bound reduction processing time} \end{pmatrix} \quad (3)$$

Four out of seven features selected are connected to the size of the instance: the *number of jobs*, the third quartile of the number of jobs per machine (*number of jobs per machine (q3)*), the *number of operations*, and the lexicographic weight of the processing time (w_p, indicated by *normalization weight processing time*). The remaining three features pertain to lower bounds: *time to compute lower bounds*, *lower bound number batches*, and *upper bound reduction processing time*. Notably, no features related to compatibility graphs were selected.

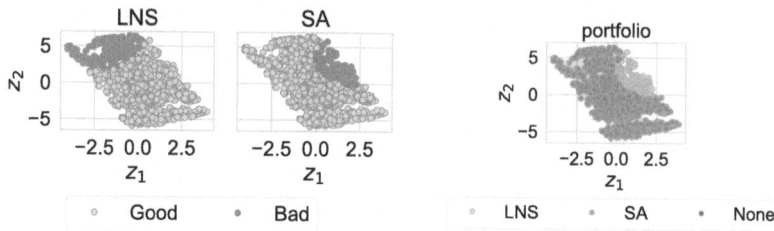

Fig. 3. Algorithm footprint of predicted SVM performance of LNS and SA.

Fig. 4. Recommended algorithms from SVM prediction models.

Distribution of Instances. Figure 2 shows the projection of the instances within the instance space, organized by source. The left plot displays all instances (2190), the middle plot shows only those previously used in the literature (794), and the right one only DA instances (600). Theoretical boundaries ("theor. boundary") are marked with a continuous red line, while experimental boundaries ("exp. boundary") are indicated by a dashed red line. The middle plot reveals that the existing instances only cover certain regions of the instance space. In particular, the upper-left part of the instance space is only sparsely covered. The NEW instances, generated for this study, cover the entire instance space, showing the ability of the generator [24] to produce instances with diverse characteristics.

Among the existing instances, the OSP-1 set is distributed across four lines between $-2.4 < z_1 < 2.4$ and $z_2 < 2.3$; these lines roughly correspond to the number of jobs (10, 25, 50, and 100) used in instance creation. In contrast, the OSP-2 instances occupy the upper portion of the space ($z_2 \geq 2.5$). The TU set spans regions that overlap with both OSP-1 and OSP-2, which can be expected since TU instances were specifically generated to exhibit similar characteristics [6]. The RL set shares the same z_2 range as OSP-1 but is more concentrated toward the right side of the instance space ($z_1 > 1.3$). Lastly, the DA instances form distinct clusters due to their generation process, where processing times, job sizes, and ready times were sampled from discrete uniform distributions [9].

Algorithm Footprints. For each algorithm, a SVM model is trained to predict regions where the algorithm is expected to perform well ($\epsilon = 0.01$ w.r.t. the winner). Figure 3 illustrates predictions for LNS (left) and SA (right). Instances where the algorithm performs well are cyan, while poor performance is shown in magenta. Both algorithms perform well in the lower and mid-left regions. SA has strengths in the top(-left) region, which is weak for LNS, while LNS is strong in the mid-right, where SA is outperformed. The clear distinction, especially in the topmost areas, suggests the benefit of an algorithm portfolio approach.

Figure 4 integrates the SVM models into a unified algorithm selection, with LNS in cyan and SA in magenta. Ties are resolved by choosing the SVM model with higher precision, assigning the lower region to SA. In a few instances (3,

Table 1. SVM performance metrics.

	Pr	accuracy	precision	recall
LNS	0.709	0.899	0.879	0.994
SA	0.815	0.869	0.894	0.952
portfolio	0.898	–	–	–

Table 2. Footprint metrics on the actual outcomes (not simulated from SVM).

	a_G	d_G	p_G	a_B	d_B	p_B
LNS	0.87	0.94	0.88	0.17	1.05	0.81
SA	0.74	1.03	0.93	0.18	1.10	0.98

labeled 'None' in black), neither algorithm is recommended due to limited statistical support, possibly requiring longer computation times or a hybrid approach.

Table 1 presents the performance metrics of the SVM models. The high precision values confirm that when an SVM model predicts good performance for an algorithm on a given instance, the prediction is reliable. Additionally, the high recall levels indicate that the model is unlikely to overlook instances where the algorithm performs well. The probability of correctly selecting a good algorithm (i.e., $Pr = Pr(\widehat{\mathrm{obj}}_\alpha(x) \leq 1 + \epsilon)$, $\epsilon = 0.01$) for a given instance is 0.709 when always using LNS and 0.815 when always using SA. The combination of the models in the algorithm portfolio (Fig. 4) increases this probability to 0.898.

Table 2 presents the footprint of the algorithms based on their actual performance, rather than the simulated data from the SVM models. It lists the normalized values for the areas (a), densities (d), and purities (p) of the footprints, categorized for both good (subscript G) and best (subscript B) performance. In the best performance category, ties are resolved randomly, as no definitive criterion (such as precision, which is used for breaking ties in the combined SVM prediction) is available to determine the superior choice.

Focusing on the best performance, both algorithms exhibit densities greater than 1, indicating that their regions of strong performance are denser than average. While there is considerable overlap in the areas where both algorithms perform well, each also has unique regions where it outperforms the other. Notably, the extent of these exclusive areas is nearly equal for both algorithms.

Distribution of the Features. Given that each algorithm has distinct strengths and weaknesses, our goal is now to explore how instance properties influence algorithm performance. Figure 5 shows the distribution of selected features across the instance space, with colors representing normalized values: cyan shades indicate smaller values, and magenta shades indicate larger ones. Most features shift from high to low values top-down, except for the *normalization weight of processing time* and the *lower bound to the number of batches* (top-left

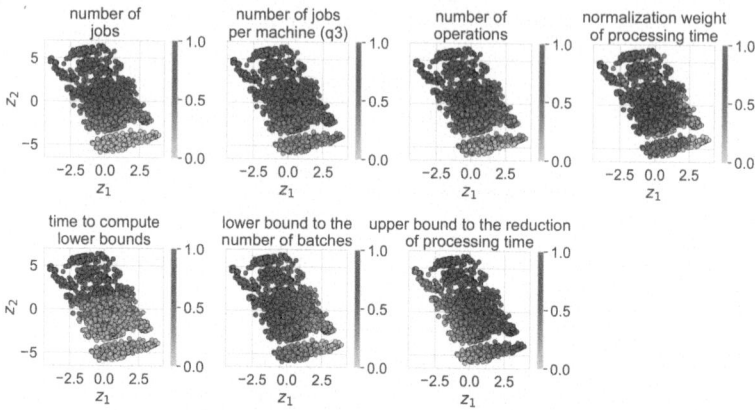

Fig. 5. Normalized feature distribution across instance space.

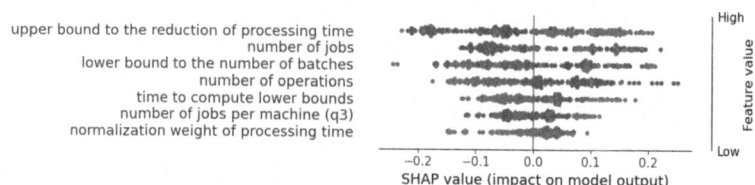

Fig. 6. SHAP values for RF predicting the best algorithm (positive class: SA).

to bottom-right) and the *upper bound to the reduction of processing time* (right to left). Instances easily solved by both algorithms generally have fewer jobs and operations, while SA performs better on larger instances, making instance size a key factor. Conversely, LNS is superior when there is more potential for processing time reduction and a smaller lower bound on the number of batches.

To validate these results, we trained a RF model with the 7 selected features and analyzed their influence using SHAP values. The model achieved an average accuracy of 0.831 ± 0.003, recall of 0.866 ± 0.004, and precision of 0.859 ± 0.005. Figure 6 shows the SHAP values for SA predictions (with symmetric results for LNS). Feature values are color-coded, pink for higher values and blue for lower. The x-axis represents the SHAP values. These results align with ISA findings. For example, lower *upper bound reduction of processing time* values favor SA, while higher values favor LNS. Likewise, a higher number of jobs is linked to SA, and fewer to LNS. This relatively clear distinction between feature values associated with SA and LNS is promising for feature-based algorithm selection. Indeed, the well-defined regions in the instance space suggest that predictive models can reliably guide algorithm selection decisions.

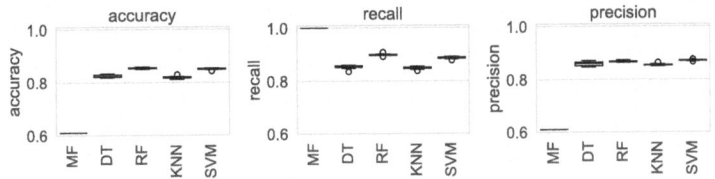

Fig. 7. Performance of the algorithm portfolio.

5.2 Algorithm Selection

Figure 7 shows the results of the algorithm selection, displaying accuracy, recall, and precision scores. Referring to accuracy and precision, all models outperform the MF model, with RF and SVM emerging as the top two options. The recall value for the MF model is 1, by definition, and is included here just for completeness. More specifically, RF achieves the highest accuracy at 0.855 ± 0.003, closely followed by SVM at 0.851 ± 0.003. For recall, RF led with 0.899 ± 0.004, while SVM followed at 0.886 ± 0.004. Finally, in terms of precision, SVM performed best at 0.871 ± 0.003, with RF slightly behind at 0.868 ± 0.003. Examining the algorithm selection process through the value of the objective function, MF shows an average gap of 1.5%. In contrast, all ML models achieve an average gap of less than 1%: RF at 0.3%, DT at 0.4%, SVM at 0.4%, and KNN at 0.6%.

These results indicate that using a combination of algorithms, selected through a suitable classification model (i.e., an algorithm portfolio), is more effective than consistently using MF (i.e., the most frequent winner, that is SA). The differences in accuracy and precision are minimal, with SVM and RF emerging as equally strong options, both significantly improving performance over the baseline.

6 Conclusions

We presented a comprehensive study of the OSP, evaluating the performance of two state-of-the-art algorithms, LNS and SA, using the ISA framework. The analysis spanned across 2190 instances described by 165 features. The results highlighted the strengths and weaknesses of each algorithm and emphasized the importance of identifying a collection of relevant features that define the instance space effectively. The introduction of novel instances (1396 instances) allowed us to achieve better coverage of the instance space, facilitating a more accurate understanding of the algorithms' behavior. By discovering the most relevant features, we could map regions in the projected space where algorithms performed differently. This distinction justifies an algorithm portfolio approach that, for any given instance, selects either LNS or SA, based on the instance's feature values. The results obtained from four well-known ML models confirmed the clear advantage of such a portfolio approach since it significantly enhances the performance compared to the use of any single algorithm.

Future research will explore the instance space for other variants of the OSP, such as those involving different objective weight settings. Additionally, the feature set could be expanded with more informative graph-based features, which have the potential to enrich the landscape of the instance space further. These enhancements could provide deeper insights into algorithm performance and reveal new opportunities for improving algorithm selection strategies.

Supplementary Material. https://github.com/francesdaros/isa-osp.

Acknowledgments. The paper is based upon work from a scholarship supported by SPECIES, the Society for the Promotion of Evolutionary Computation in Europe and its Surroundings. The financial support from the Austrian Federal Ministry of Labour and Economy, the National Foundation for Research, Technology and Development and the Christian Doppler Research Association is gratefully acknowledged.

References

1. Azizoglu, M., Webster, S.: Scheduling a batch processing machine with incompatible job families. Comput. Ind. Eng. **39**(3), 325–335 (2001). ISSN 0360-8352. https://doi.org/10.1016/S0360-8352(01)00009-2
2. Cheng, B., Wang, Q., Yang, S., Hu, X.: An improved ant colony optimization for scheduling identical parallel batching machines with arbitrary job sizes. Appl. Soft Comput. **13**(2), 765–772 (2013). ISSN 1568-4946. https://doi.org/10.1016/j.asoc.2012.10.021
3. Chou, F.D.: Minimising the total weighted tardiness for non-identical parallel batch processing machines with job release times and non-identical job sizes. Eur. J. Ind. Eng. **7**(5), 529–557 (2013). https://doi.org/10.1504/EJIE.2013.057380. pMID: 57380
4. Van De Rzee, D.J., Van Harten, A., Schuur, P.: Dynamic job assignment heuristics for multi-server batch operations – a cost based approach. Int. J. Prod. Res. **35**(11), 3063–3094 (1997). https://doi.org/10.1080/002075497194291
5. Da Ros, F., Di Gaspero, L., Lackner, M.L., Musliu, N.: Reducing energy consumption in electronic component manufacturing through large neighborhood search. In: Proceedings of the Genetic and Evolutionary Computation Conference Companion, GECCO 2024, pp. 1706–1714. Association for Computing Machinery, New York (2024). ISBN 9798400704956. https://doi.org/10.1145/3638530.3664132
6. Da Ros, F., Di Gaspero, L., Lackner, M.L., Musliu, N., Winter, F.: Local search algorithms for the oven scheduling problem. In: Proceedings of the Genetic and Evolutionary Computation Conference Companion, GECCO 2024, pp. 191–194. Association for Computing Machinery, New York (2024). ISBN 9798400704956. https://doi.org/10.1145/3638530.3654158
7. Da Ros, F., Di Gaspero, L., Lackner, M.L., Musliu, N., Winter, F.: Multi-Neighborhood Simulated Annealing for the Oven Scheduling Problem (2024). https://doi.org/10.2139/ssrn.4998899
8. Da Ros, F., Lackner, M.L., Musliu, N.: Theoretical lower bounds for the oven scheduling problem. In: Proceedings of the 14th International Conference on the Practice and Theory of Automated Timetabling (2024). https://doi.org/10.48550/arXiv.2410.01368

9. Damodaran, P., Vélez-Gallego, M.C.: A simulated annealing algorithm to minimize makespan of parallel batch processing machines with unequal job ready times. Expert Syst. Appl. **39**(1), 1451–1458 (2012). ISSN 0957-4174. https://doi.org/10.1016/j.eswa.2011.08.029
10. De Coster, A., Musliu, N., Schaerf, A., Schoisswohl, J., Smith-Miles, K.: Algorithm selection and instance space analysis for curriculum-based course timetabling. J. Sched. **25**(1), 35–58 (2022). ISSN 1099-1425. https://doi.org/10.1007/s10951-021-00701-x
11. Dwivedi, R., et al.: Explainable AI (XAI): core ideas, techniques, and solutions. ACM Comput. Surv. **55**(9) (2023). ISSN 0360-0300. https://doi.org/10.1145/3561048
12. Fowler, J.W., Mönch, L.: A survey of scheduling with parallel batch (p-batch) processing. Eur. J. Oper. Res. **298**(1), 1–24 (2022). ISSN 0377-2217. https://doi.org/10.1016/j.ejor.2021.06.012
13. Franzin, A., Stützle, T.: Revisiting simulated annealing: a component-based analysis. Comput. Oper. Res. **104**, 191–206 (2019). ISSN 0305-0548. https://doi.org/10.1016/j.cor.2018.12.015
14. Geibinger, T., Kletzander, L., Musliu, N.: Instance space analysis for the generalized assignment problem. In: Di Gaspero, L., Festa, P., Nakib, A., Pavone, M. (eds.) Metaheuristics, pp. 421–435. Springer, Cham (2023). ISBN 978-3-031-26504-4. https://doi.org/10.1007/978-3-031-26504-4_30
15. Hagberg, A.A., Schult, D.A., Swart, P.J.: Exploring network structure, dynamics, and function using NetworkX. In: Proceedings of the Python in Science Conferences, pp. 11–15 (2008). https://doi.org/10.25080/tcwv9851
16. Harris, C.R., et al.: Array programming with NumPy. Nature **585**(7825), 357–362 (2020). https://doi.org/10.1038/s41586-020-2649-2
17. Fowler, J.W., Phojanamongkolkij, N., Cochran, J.K., Montgomery, D.C.: Optimal batching in a wafer fabrication facility using a multiproduct G/G/C model with batch processing. Int. J. Prod. Res. **40**(2), 275–292 (2002). https://doi.org/10.1080/00207540110081489
18. Kirkpatrick, S., Gelatt, D., Vecchi, M.P.: Optimization by simulated annealing. Science **220**(4598), 671–680 (1983). https://doi.org/10.1126/science.220.4598.671
19. Kletzander, L., Musliu, N., Smith-Miles, K.: Instance space analysis for a personnel scheduling problem. Ann. Math. Artif. Intell. **89**(7), 617–637 (2021). ISSN 1573-7470. https://doi.org/10.1007/s10472-020-09695-2
20. Kosch, S., Beck, J.C.: A new MIP model for parallel-batch scheduling with non-identical job sizes. In: Simonis, H. (ed.) CPAIOR 2014. LNCS, vol. 8451, pp. 55–70. Springer, Cham (2014). https://doi.org/10.1007/978-3-319-07046-9_5
21. Krim, H., Zufferey, N., Potvin, J.Y., Benmansour, R., Duvivier, D.: Tabu search for a parallel-machine scheduling problem with periodic maintenance, job rejection and weighted sum of completion times. J. Sched. **25**(1), 89–105 (2022). ISSN 1099-1425. https://doi.org/10.1007/s10951-021-00711-9
22. Laborie, P., Rogerie, J., Shaw, P., Vilím, P.: IBM ILOG CP optimizer for scheduling. Constraints **23**(2), 210–250 (2018). ISSN 1572-9354. https://doi.org/10.1007/s10601-018-9281-x
23. Lackner, M.L., Mrkvicka, C., Musliu, N., Walkiewicz, D., Winter, F.: Benchmark instances and models for the Oven Scheduling Problem [Dataset] (2022). https://doi.org/10.5281/zenodo.7456937
24. Lackner, M.L., Mrkvicka, C., Musliu, N., Walkiewicz, D., Winter, F.: Exact methods for the oven scheduling problem. Constraints **28**(2), 320–361 (2023). ISSN 1572-9354. https://doi.org/10.1007/s10601-023-09347-2

25. Lackner, M.L., Musliu, N., Winter, F.: Solving an industrial oven scheduling problem with a simulated annealing approach. In: Proceedings of the 13th International Conference on the Practice and Theory of Automated Timetabling, pp. 115–120 (2022)
26. Lundberg, S.M., et al.: From local explanations to global understanding with explainable AI for trees. Nat. Mach. Intell. **2**(1), 56–67 (2020). ISSN 2522-5839. https://doi.org/10.1038/s42256-019-0138-9
27. Lundberg, S.M., Lee, S.I.: A unified approach to interpreting model predictions. In: Guyon, I., Luxburg, U.V., Bengio, S., Wallach, H., Fergus, R., Vishwanathan, S., Garnett, R. (eds.) Advances in Neural Information Processing Systems, vol. 30, Curran Associates, Inc. (2017). https://proceedings.neurips.cc/paper_files/paper/2017/file/8a20a8621978632d76c43dfd28b67767-Paper.pdf
28. Melouk, S., Damodaran, P., Chang, P.Y.: Minimizing makespan for single machine batch processing with non-identical job sizes using simulated annealing. Int. J. Prod. Econ. **87**(2), 141–147 (2004). ISSN 0925-5273. https://doi.org/10.1016/S0925-5273(03)00092-6
29. Mönch, L., Fowler, J.W., Mason, S.J.: Production planning and control for semiconductor wafer fabrication facilities: modeling, analysis, and systems, vol. 52. Springer (2012)
30. Muñoz, M.A., Smith-Miles, K.A.: Performance analysis of continuous black-box optimization algorithms via footprints in instance space. Evol. Comput. **25**(4), 529-554 (2017). ISSN 1063-6560. https://doi.org/10.1162/evco_a_00194
31. Muñoz, M.A., et al.: An instance space analysis of regression problems. ACM Trans. Knowl. Discov. Data **15**(2) (2021). ISSN 1556-4681. https://doi.org/10.1145/3436893
32. Muñoz, M.A., Villanova, L., Baatar, D., Smith-Miles, K.: Instance spaces for machine learning classification. Mach. Learn. **107**(1), 109–147 (2018). ISSN 1573-0565. https://doi.org/10.1007/s10994-017-5629-5
33. Muñoz, M.A., neelofarhassan, Christiansen, J.: andremun/instancespace: February 2023 (2023). https://doi.org/10.5281/zenodo.7700434
34. Muñoz, M.A., Smith-Miles, K.: Generating new space-filling test instances for continuous black-box optimization. Evol. Comput. **28**(3), 379–404 (2020). ISSN 1063-6560. https://doi.org/10.1162/evco_a_00262
35. Neelofar, N., Smith-Miles, K., Muñoz, M.A., Aleti, A.: Instance space analysis of search-based software testing. IEEE Trans. Software Eng. **49**(4), 2642–2660 (2023). https://doi.org/10.1109/TSE.2022.3228334
36. Oliveira, C., Aleti, A., Grunske, L., Smith-Miles, K.: Mapping the effectiveness of automated test suite generation techniques. IEEE Trans. Reliab. **67**(3), 771–785 (2018). https://doi.org/10.1109/TR.2018.2832072
37. Pedregosa, F., et al.: Scikit-learn: machine learning in Python. J. Mach. Learn. Res. **12**, 2825–2830 (2011). http://jmlr.org/papers/v12/pedregosa11a.html
38. Potts, C.N., Kovalyov, M.Y.: Scheduling with batching: a review. Eur. J. Oper. Res. **120**(2), 228–249 (2000). ISSN 0377-2217. https://doi.org/10.1016/S0377-2217(99)00153-8
39. Rice, J.R.: The algorithm selection problem. Adv. Comput. **15**, 65–118 (1976). ISSN 0065-2458. https://doi.org/10.1016/S0065-2458(08)60520-3
40. Shaw, P.: Using constraint programming and local search methods to solve vehicle routing problems. In: Proceedings of the 4th International Conference on Principles and Practice of Constraint Programming, CP 1998, pp. 417–431. Springer, Heidelberg (1998). ISBN 3540652248

41. Smith-Miles, K., Baatar, D., Wreford, B., Lewis, R.: Towards objective measures of algorithm performance across instance space. Comput. Oper. Res. **45**, 12–24 (2014). ISSN 0305-0548. https://doi.org/10.1016/j.cor.2013.11.015
42. Smith-Miles, K., Muñoz, M.A.: Instance space analysis for algorithm testing: methodology and software tools. ACM Comput. Surv. **55**(12) (2023). ISSN 0360-0300. https://doi.org/10.1145/3572895
43. Strassl, S., Musliu, N.: Instance space analysis and algorithm selection for the job shop scheduling problem. Comput. Oper. Res. **141**, 105661 (2022). ISSN 0305-0548. https://doi.org/10.1016/j.cor.2021.105661
44. Trindade, R.S., de Araújo, O., Fampa, M.: Arc-flow approach for parallel batch processing machine scheduling with non-identical job sizes. In: Baïou, M., Gendron, B., Günlük, O., Mahjoub, A.R. (eds.) ISCO 2020. LNCS, vol. 12176, pp. 179–190. Springer, Cham (2020). https://doi.org/10.1007/978-3-030-53262-8_15
45. Wang, H.M., Chou, F.D.: Solving the parallel batch-processing machines with different release times, job sizes, and capacity limits by metaheuristics. Expert Syst. Appl. **37**(2), 1510–1521 (2010). ISSN 0957-4174. https://doi.org/10.1016/j.eswa.2009.06.070
46. Waskom, M.L.: seaborn: statistical data visualization. J. Open Source Softw. **6**(60), 3021 (2021). https://doi.org/10.21105/joss.03021
47. Çelik, C., Saricicek, I.: Tabu search for parallel machine scheduling with job splitting. In: 2009 Sixth International Conference on Information Technology: New Generations, pp. 183–188 (2009). https://doi.org/10.1109/ITNG.2009.271

Meta-learning of Univariate Estimation-of-Distribution Algorithms for Pseudo-Boolean Problems

Olivier Goudet[1]([✉]), Adrien Goëffon[1], Frédéric Saubion[1], and Sébastien Verel[2]

[1] LERIA, Université d'Angers, Angers, France
{olivier.goudet,adrien.goeffon,frederic.saubion}@univ-angers.fr
[2] LISIC, Université du Littoral Côte d'Opale, Dunkirk, France
verel@univ-littoral.fr

Abstract. We propose a new framework for univariate Estimation-of-Distribution Algorithms (EDAs) in which neural networks are used at two levels: (i) for generating new solutions from a memory that encodes some information acquired since the beginning of the search, (ii) for updating this memory with the new solutions sampled at each iteration. These neural networks are then learned by neuro-evolution, so as to automatically discover new efficient strategies for solving pseudo-Boolean problems, without having any specific a priori on how to design them. The algorithms discovered demonstrate their competitiveness compared with existing univariate EDAs of the literature. On average, they achieve better scores for the same number of calls to the objective function, when tested on a variety of different distributions.

1 Introduction

Given a combinatorial optimization problem defined by a search space $\mathcal{X}$ and an objective function to optimize $f : \mathcal{X} \to \mathbb{R}$, an Estimation-of-Distribution Algorithm (EDA) [20,21,24] is a stochastic search algorithm whose solution sampling is based on a probabilistic model $\mathcal{D}_{\mathcal{X}} : \mathcal{X} \to [0, 1]$. It operates in three stages at each iteration: i) generation of a set of solutions $S \subseteq \{x \in \mathcal{X} \mid x \sim \mathcal{D}_{\mathcal{X}}\}$, ii) evaluation of these solutions with the objective function f, iii) updating of the probabilistic model $\mathcal{D}_{\mathcal{X}}$ according to the new information obtained.

It is worth emphasizing that stochastic search algorithms, such as EDAs, evolutionary and local search algorithms, are essentially heuristic approaches that aim at sampling solutions from $\mathcal{X}$ to identify high-quality outcomes.

Consequently, we may expect these algorithms to exhibit the capability to sample solutions with maximized fitness levels. In this paper, as in the related literature we consider on EDAs, we restrict our scope to binary problems, i.e., $\mathcal{X} = \{0, 1\}^n$ for a given size n. A simple and yet efficient approach for EDAs consists in considering a univariate model for estimating $\mathcal{D}_{\mathcal{X}}$, i.e., a Bernoulli distribution of parameter p_i is associated to each variable $x_i, 1 \leq i \leq n$ of a solution $x \in \mathcal{X}$. p_i is computed using an update strategy to gradually approach the best possible estimated distribution. Starting from the initial simple scheme proposed in [26], different variants of EDAs have been studied and generic algorithms have been proposed to provide a comprehensive view of their main principles and properties [3,4].

Given a problem instance $(\mathcal{X}, f)$ and a number T of iterations, an EDA can be formulated as a function π_θ that computes iteratively a set of sampled solutions $\pi_\theta(\mathcal{X}, f, T) = S \subseteq \mathcal{X}$ according to a strategy parameterized by θ. The expected output is hence $x^* = \text{argmax}_{x \in S} f(x)$. The parameter vector θ has different possible roles according to different versions of EDAs and it includes the update strategy of the probabilistic model.

Let us note that EDAs have demonstrated good performance compared to alternative approaches in numerous cases [23,27]. However, the univariate approach exhibits some limitations (see Sect. 2.2), and bivariate approaches have been developed [5,28] to improve the probabilistic model and attempt to capture more complex interactions between the variables that make up the solutions. A significant challenge is understanding how the features of the search space are reflected in the probabilistic models (e.g., Bayesian networks can be considered to learn more accurate models [6]). Following another research line, in this paper, we propose a general learning scheme based on simple neural networks to estimate better $\mathcal{D}_\mathcal{X}$ in the context of problems that include quadratic interactions between variables. Recent works have investigated the possible use of neural networks based techniques to enhance evolutionary search processes for continuous black-box optimization problems [22].

Our main idea is to replace the sampling process by a neural network that takes as input at generation t, the solutions that have been sampled at generation $t-1$ as well as a memory that represents an abstraction of past sampled sets. This information extracted from past generations is also managed by a neural network. These basic neural networks are trained before running the EDA but instead of involving a classic gradient-based learning process we consider here the setting of their weights as a continuous black-box optimization problems. Hence, we align our approach with neuro-evolution, a methodology proven successful in various learning contexts [17], particularly for reinforcement learning problems with non-Markovian rewards [8], which share similarities to our context.

Employing the CMA-ES algorithm [14], renowned for its efficiency in black-box continuous optimization problems, we leverage its capabilities for parameter tuning [19] within our framework. To thoroughly assess the performance of our algorithms through practical experience, we use PUBO$_i$ [31]–a versatile generator. PUBO$_i$ facilitates the exploration of diverse pseudoboolean functions, encompassing classical optimization problems expressible through Walsh functions. This ensures a comprehensive evaluation across a broad spectrum of benchmarks, contributing to a robust understanding of our algorithmic capabilities. Even though we adopt an univariate treatment of the variables within the neural networks, the experiments highlight that our approach offers a very interesting

Algorithm 1. The **UMDA** with parameters $\lambda \in \mathbb{N}, \theta = (\mu) \in [\![1, \lambda]\!]$

1: **Input**: an instance $(\mathcal{X}, f)$, with $\mathcal{X} = \{0,1\}^n$, $f : \mathcal{X} \to \mathbb{R}$, a number of iterations T
2: $p(0) = (\frac{1}{2}, \ldots, \frac{1}{2}) \in [0,1]^n$
3: **for** $t = 1, 2, \ldots, T$ **do**
4:     **for** $j = 1, 2, \ldots, \lambda$ **do**
5:         $x[j] \sim Sample(p(t-1))$
6:     **end for**
7:     Sort the individuals into $\tilde{x}[1], \ldots, \tilde{x}[\lambda]$ (in decreasing fitness order).
8:     %% Update the frequency
9:     $p(t) = \max(\frac{1}{n}, \min(1 - \frac{1}{n}, \frac{1}{\mu} \sum_{j=1}^{\mu} \tilde{x}[j]))$
10: **end for**
11: **Output**: the best solution found x^*

behaviour that allows to better explore the search space, in particular in presence of multiple and distant good solutions.

2 Related Work

As highlighted in surveys [21,27], most previous investigations into EDAs have predominantly focused on univariate methodologies. Nevertheless, EDAs can also be extended to multivariate algorithms [6,29] and non-binary problems [18]. Let us note that Based on the overview proposed in [3], we aim to discuss four EDA variants, one of which is the general EDA, encompassing all linear algorithms developed earlier, and the last one, recently proposed by [11] is an extension of this general EDA, with the use of a non-linear update function. As argued in [3, 11], the primary aim of these general linear and non-linear EDAs is to enable the simulation of diverse estimation-of-distribution management approaches through suitable parameter configurations.

2.1 Classic Univariate EDA Algorithms in the Literature

Introduced in [26], the Univariate Marginal Distribution Algorithm (**UMDA**) initially generates λ solutions $x[1], \ldots, x[\lambda]$ by a sampling process with respect to a frequency vector p. These solutions are sorted according to their fitness values to get $\tilde{x}[1], \ldots, \tilde{x}[\lambda]$ (with ties broken randomly). **UMDA** selects then the μ best solutions and computes the univariate marginal frequencies using the values $\tilde{x}[1]_i, \ldots, \tilde{x}[\mu]_i$ to get p_i and hence update the frequency vector p.

This process is sketched in Algorithm 1. We exclude λ from the parameter vector since it will not be considered in the calibration process, detailed in Sect. 5.1. Note that in this version we use a margin (line 8) that ensures that the values p_i are kept in the range $[\frac{1}{n}, 1 - \frac{1}{n}]$. As quoted in [4], using this restriction avoids frequencies from getting stuck in the limits 0 or 1. Note that the margins may have an impact on the properties of an EDA as studied in [7], where different EDAs are reformulated in the same n-Bernouilli-λ-EDA framework.

The initial Population-Based Incremental Learning (**PBIL**) algorithm, introduced by Baluja in [1], involved updating the probability vector based on the best-generated solutions, reflecting a shared concern with reinforcement learning techniques. In addition, it incorporated a mutation of this vector. In this context, our attention is directed towards reformulating the **PBIL** as presented in [3]. Compared to Algorithm 1, **PBIL** involves an extended parameter vector $\theta = (\mu, \rho)$ with $\rho \in [0,1]$. Then, the only difference concerns the update policy of the frequency vector (line 9 of the algorithm), replaced by: $p(t) = \max(\frac{1}{n}, \min(1 - \frac{1}{n}, (1-\rho) \cdot p(t-1) + \frac{\rho}{\mu} \sum_{j=1}^{\mu} \tilde{x}[j]))$.

We do not consider here all possible variations (e.g., Compact Genetic Algorithm (CGA) [16]) of the main principles of EDAs but we rather follow the lines proposed in [3] and its attempt to unify the different approaches, as described in the next subsection.

2.2 General Linear EDA Algorithm

In [3], the authors review different EDAs, and propose to unify these algorithms into a general one that can be parameterized to simulate different previously estimation of distribution approaches.

At the end of each generation t, this EDA updates the frequency $p(t)_i$ of each variable i by a linear recombination of the previous probability value $p(t-1)_i$ and the values of the λ individuals, $\tilde{x}[1]_i, \ldots, \tilde{x}[\lambda]_i$, sampled at that time step. Given a parameter vector $\theta = (\theta_\epsilon, \theta_0, \theta_1, \cdots, \theta_\lambda)$, it consists in replacing line 9 in Algorithm 1 by $p(t) = \max(\theta_\epsilon, \min(1 - \theta_\epsilon, \theta_0 \cdot p(t-1) + \sum_{j=1}^{\lambda} \theta_j \tilde{x}[j]))$, with $\theta_j \in \mathbb{R}$ such that $\sum_{j=0}^{\lambda} \theta_j = 1$ and $\theta_\epsilon \in [0, \frac{1}{2}[$.[1] This algorithm will be called **Linear-EDA** in the remainder of this paper.

Note that this algorithm corresponds to the **UMDA** when $\theta_\epsilon = \frac{1}{n}$, $\theta_0 = 0$, $\theta_j = \frac{1}{\mu}$ for $j \in [\![1, \mu]\!]$, and $\theta_j = 0$ otherwise. It corresponds to the **PBIL** when $\theta_\epsilon = \frac{1}{n}$, $\theta_0 = 1 - \rho$, $\theta_j = \frac{\rho}{\mu}$ for $j \in [\![1, \mu]\!]$, and $\theta_j = 0$ otherwise.

2.3 Univariate EDA with Non-linear Update Function

Based on **Linear-EDA**, [11] proposed Algorithm 2, called **Neuro-EDA**, which is again similar to Algorithm 1 except that line 9 is replaced by an update of the frequency vector using a non-linear function modeled by a feed-forward neural network $g_\theta : \mathbb{R}^{\lambda+1} \to \mathbb{R}$, parametrized by a vector of parameters θ (i.e., weights of the neural network). The classical frequency bounds $\frac{1}{n}$ and $1 - \frac{1}{n}$ are replaced by an activation function $\Phi : \mathbb{R} \to]0, 1[$, corresponding to the distribution function of the centered reduced normal distribution used in probit models.

---

[1] In the original version presented in [3], θ_ϵ is set to the value of 0, which corresponds to frequency bounds 0 and 1. However, applying these frequency bounds is often ineffective in solving pseudo-Boolean problems characterized by rugged landscapes, as we will demonstrate through our experimental analysis in Sect. 5. Therefore, in our study θ_ϵ will be calibrated for each type of problem instance in the range $[0, \frac{1}{2}[$.

Algorithm 2. Neuro-EDA with parameter $\lambda \in \mathbb{N}$ and neural network g_θ parameterized by a vector of parameters θ.

1: **Input**: an instance $(\mathcal{X}, f)$, with $\mathcal{X} = \{0,1\}^n$, $f : \mathcal{X} \to \mathbb{R}$, a number of iterations T
2: $p(0) = (\frac{1}{2}, \ldots, \frac{1}{2}) \in [0,1]^n$
3: **for** $t = 1, 2, \ldots, T$ **do**
4:     **for** $j = 1, 2, \ldots, \lambda$ **do**
5:        $x[j] \sim Sample(p(t-1))$
6:     **end for**
7:     Sort the individuals into $\tilde{x}[1], \ldots, \tilde{x}[\lambda]$ (in decreasing fitness order).
8:     %% Update the probability vector
9:     **for** $i = 1, 2, \ldots, n$ **do**
10:        $p(t)_i = \Phi(g_\theta([p(t-1)_i, \tilde{x}[1]_i, \tilde{x}[2]_i, \ldots, \tilde{x}[\lambda]_i]))$
11:     **end for**
12: **end for**
13: **Output**: the best solution found x^*

Note that in the existing univariate EDA proposed in the literature only one scalar probability value is learned per variable during the search, which may limit the algorithm's potential to learn information about past events encountered during the search. To overcome this limitation, we propose to replace each scalar probability value p_i for each variable by a vector m_i of size $l > 0$ for each variable. Each memory vector m_i can be seen as an encoding or a projection into a latent space of the information collected during the search for each variable i, that will allow to sample new promising solutions for the next iterations.

3 The Memory-EDA Algorithm

First, one can notice, that an equivalent formulation of **Neuro-EDA** of [11] and described in Algorithm 2 can be obtain, where for each variable $i = 1, \ldots, n$:

- the probability $p(t)_i \in [0,1]$ is replaced by a scalar value $m(t)_i \in \mathbb{R}$ such that $m(t)_i = \Phi^{-1}(p(t)_i)$;
- the sampling of the variable $x[j]_i$ for each individual j with the Bernoulli law (line 5) is replaced by the affectation $x[j]_i \leftarrow H(m(t)_i + \epsilon_{i,j})$, with $\epsilon_{i,j} \sim \mathcal{N}(0,1)$ and H the Heaviside function defined by $H(x) = 0$ if $x < 0$ and $H(x) = 1$ if $x \geq 0$;
- the update step (line 10) becomes $m(t)_i = g_\theta([\Phi(m(t-1)_i), \tilde{x}[1]_i, \ldots, \tilde{x}[\lambda]_i])$.

This new formulation enables a transition from a scalar memory unit to a vector memory unit per variable. The main motivation for introducing this memory vector is to allow more information extraction from past visited states of the distribution, and hence to possibly discover better update strategies, while keeping a univariate approach.

Let $\{d_\phi : \mathbb{R}^{l+1} \to \mathbb{R}, \phi \in \Phi\}$ and $\{e_\gamma : \mathbb{R}^{\lambda+l} \to \mathbb{R}^l, \gamma \in \Gamma\}$ denote two classes of functions, with $l \in \mathbb{N}^*$. In practice, d_ϕ and e_γ denote two neural networks, d_ϕ and e_γ playing respectively the role of the decoder and the encoder. The two sets of parameters of the neural networks ϕ and γ can be concatenated in a single vector of parameters $\theta = (\phi, \gamma)$ with $\theta \in \Theta$.

Memory-EDA extends **Neuro-EDA** with the use of a memory vector $m(t)_i \in \mathbb{R}^l$, with $l > 0$ for each variable $i \in [\![1, n]\!]$, which is dynamically updated during the search. It repeats a loop of T generations. Each generation involves the execution of two components:

(1) *decoding*: λ individuals are generated in parallel (as a batch of new solutions) with the decoder d_ϕ. For each individual $x[j]$ with $j \in [\![1, \lambda]\!]$, the value assigned to each variable $i \in [\![1, n]\!]$ is computed as $x[j]_i = H(d_\phi(m(t-1)_i, \epsilon_{i,j}))$. This neural network takes as input the memory vector $m(t-1)_i$ of size l concatenated with the value $\epsilon_{i,j} \sim \mathcal{N}(0, 1)$, and outputs a scalar real value which is transformed to a binary value with the Heaviside function H.

(2) *encoding*: for each variable i, the encoder e_γ takes as input the λ binary values $\tilde{x}[1]_i, \ldots, \tilde{x}[\lambda]_i$ and the previous memory vector $m(t-1)_i$, and outputs the new memory vector $m(t)_i$ of size l that will be used to sample new individuals at the next generation.

The pseudo-code of this algorithm is shown in Algorithm 3. At the beginning, all the memory values $m(0)_i$ are initialized with the value 0.

Algorithm 3. Algorithm **Memory-EDA** with with parameter $\lambda \in \mathbb{N}$, decoder and encoder neural networks d_ϕ, e_γ parameterized by the global parameter vector $\theta = (\phi, \gamma)$.

1: **Input**: an instance $(\mathcal{X}, f)$, with $\mathcal{X} = \{0, 1\}^n$, $f : \mathcal{X} \to \mathbb{R}$, a number of iterations T
2: **for** $i = 1, 2, \ldots, n$ **do**
3:     $m(0)_i = O_l$ %% Initialization with null vectors of size l.
4: **end for**
5: **for** $t = 1, 2, \ldots, T$ **do**
6:     **for** $j = 1, 2, \ldots, \lambda$ **do**
7:         **for** $i = 1, 2, \ldots, n$ **do**
8:             $\epsilon_{i,j} \sim \mathcal{N}(0, 1)$
9:             $x[j]_i = H(d_\phi(m(t-1)_i, \epsilon_{i,j}))$
10:        **end for**
11:    **end for**
12:    Sort the individuals into $\tilde{x}[1], \ldots, \tilde{x}[\lambda]$ (in decreasing fitness order).
13:    %% Update the memory
14:    **for** $i = 1, 2, \ldots, n$ **do**
15:        $m(t)_i = e_\gamma(m(t-1)_i, \tilde{x}[1]_i, \tilde{x}[2]_i, \ldots, \tilde{x}[\lambda]_i)$
16:    **end for**
17: **end for**

---

4 Discovering New Estimation-of-Distribution Algorithms with Neuro-Evolution

The three algorithms **Linear-EDA**, **Neuro-EDA** and **Memory-EDA** are parameterized by a vector θ whose values lie in $\mathbb{R}$. The choice of the values of θ is paramount in defining an efficient EDA algorithm. In this section, we first describe the learning process we have chosen for the emergence of new EDA strategies, followed by the continuous optimization algorithm used to learn the vector of parameters θ, which is itself an estimation-of-distribution algorithm (meta-learning).

4.1 EDA Strategy Learning for a Set of Instances

In this paper, the algorithms are tested on instances of the Quadratic Unconstrained Binary Optimization (QUBO) problem, which is a single objective pseudo-boolean optimization problem with quadratic interactions between binary variables. NP-hard and NP-complete combinatorial optimization problems can be easily transformed into QUBO [9,25]. The objective function $f : \{0,1\}^n \to \mathbb{R}$ to maximize is defined by: $f(x) = x^t Q x$ where Q is a real matrix of dimension $n \times n$ and x^t is the transposed vector of x. We consider a black-box optimization scenario with an unknown matrix Q.

The $PUBO_i$ generator of QUBO instances [31] can tune the density of matrix Q, as well as the importance of binary variables, and consequently the non-uniformity of the matrix. More formally, the fitness function is defined by $f(x) = \sum_{i=1}^{m} f_i(x_{i_1}, x_{i_2}, x_{i_3}, x_{i_4})$, where each sub-function f_i is a quadratic function randomly selected from a set $\{\varphi_1, \ldots, \varphi_4\}$ where φ_k has $2k$ symmetric local optima. In $PUBO_i$, the binary variables are divided into two classes of importance: important, and non-important variables. For each sub-function f_i, the four variables x_{i_j} are selected according to importance degree parameter d: the probability of selecting an important variable is proportional to the degree of importance. An additional parameter α, called importance co-appearance, tuned the co-variance of selecting two important variables for the same sub-function f_i. We refer the reader to [31] for more details.

EDAs parameterized by θ, such as those described in Sects. 2 and 3, are presented as stochastic algorithms π_θ. For each instance $(\mathcal{X}, f)$ drawn from $PUBO_i(n, m, d, \alpha)$, π_θ will be launched during T iterations. Remind that, at the end of this search, it returns the solution $x^* = \pi_\theta(\mathcal{X}, f, T)$ with the maximum fitness value $x^* \in \mathcal{X}$ encountered during the sampling process. An algorithm π_θ can indeed be viewed as a dynamic sampling policy.

To assess the performance of the stochastic policy π_θ on random instances generated as $(\mathcal{X}, f) = PUBO_i(n, m, d, \alpha)$ and solved with a budget of T iterations (corresponding to $T \times \lambda$ calls to the objective function f), we propose to evaluate the quantity $F(\pi_\theta, \mathcal{X}, f, T) = \mathbb{E}_{(\mathcal{X}, f) \sim PUBO_i(n,m,d,\alpha)}[f(\pi_\theta(\mathcal{X}, f, T))]$. However, since this expected value cannot be practically computed, we rely on an empirical estimator computed as an average of the score obtained by the policy π_θ for a finite number q of instances $(\mathcal{X}, f)_1, \ldots, (\mathcal{X}, f)_q$ randomly sampled from the generator $PUBO_i(n, m, d, \alpha)$, and a finite number r of restarts:

$$\bar{F}(\pi_\theta, \mathcal{X}, f, T) = \frac{1}{qr} \sum_{i=1}^{q} \sum_{j=1}^{r} f(\pi_\theta((\mathcal{X}, f)_i, T)). \tag{1}$$

4.2 Optimizing the Parameters with an Evolutionary Algorithm

Given an EDA π_θ, the generator of instances $PUBO_i(n, m, d, \alpha)$ and a budget of T generations, the optimization goal is find the vector of parameters θ in real-valued $\mathbb{R}^{|\theta|}$ search space that maximize the estimated score $\bar{F}(\pi_\theta, \mathcal{X}, f, T)$. Learning this set of parameters with gradient descent techniques is impossible in this case due to the non-derivable sampling of the binary variables in the solutions $x^* \in \mathcal{X}$ produced by the EDA π_θ. Therefore, to solve this problem we propose to use black-box optimization evolutionary algorithm, such as the covariance matrix adaptation evolution strategy (CMA-ES) [13,15], which was already successfully applied to learn neural networks for episodic reinforcement learning [17].

5 Experiments

The primary goal of this section is to empirically address two pivotal questions. Firstly, we seek to evaluate the efficiency of the new neural network strategy **Memory-EDA** (see Sect. 3). The second objective is to study the emerging strategy discovered by **Memory-EDA** at the end of the evolutionary process.

5.1 Experimental Settings

QUBO Datasets Generation. We consider instances of size $n \in \{32, 64, 128\}$ for the training, validation, and test phases. Larger instances of size $n = 256$ are considered for a final out-of-distribution test to check the robustness of the different learned EDAs. The number of sub-functions m tunes the density of the matrix (16% and 43% for uniform instances respectively for the two values of m). Three types of interaction mechanisms are used between variables. The instances I_{uni} have no specific important variables, i.e. I_{uni} instances are similar to QUBO problems with a random matrix. The instances I_{imp} have important variables: the marginal probability of having important variables is equal to $d = 10$ times the probability of non-important ones. Additionally, for the I_{ic} instances, the value of the co-appearance parameter is high, the selection of important variables is not independent, and the selection of important variables is concentrated. For each tuple of parameter values, we generate a training set $\mathcal{D}_{train}$ including 1,000 independent instances, a validation set $\mathcal{D}_{valid}$ containing 100 instances, and an additional test set $\mathcal{D}_{test}$ with 100 instances.

EDA Configurations. All EDAs presented in Sect. 2 and 3 generate $\lambda = 20$ solutions at each iteration. A sensitivity analysis regarding to this critical parameter (which is not presented here for space limitations) has been conducted. It reveals that increasing the value of λ beyond 20 does not significantly improve results for all EDAs and does not change the ranking of the different strategies among themselves for the instances of size up to $n = 256$ considered in this paper.

Note that, in our experiments, λ does not depend on the size n of the instance, as this is usually the case with EDAs [21,30]. We follow this choice to be able to test any EDA, learned from a distribution of instances of given size n, on other instances generated with a different size n' (see Sect. 5.3 on out-of-distribution testing).

Concerning **Neuro-EDA** and **Memory-EDA**, we always consider neural networks with one hidden layer of $h = 20$ neurons and a sigmoid activation function. We have deliberately chosen fairly small neural network sizes to limit their training time.

EDA Training Settings. To optimize the vector of parameters θ for the different EDAs, we used the CMA-ES algorithm from the `pycma` library [12]. The multivariate normal distribution of CMA-ES is initialized with mean parameter μ (randomly sampled according to a centered normal distribution with $\sigma_\mu = 0.1$ standard deviation) and initial standard deviation $\sigma_0 = 0.2$. The size of the population of CMA-ES is always set to $P = 30$ for all EDA training.

Settings of EDA Code. All the EDAs presented in this paper are implemented in `Python 3` with `Pytorch 2.1` library. We developed a massive parallel EDA training platform that will be made publicly available after publication, together with the benchmark instances and result files. During the training, validation, and test phases, each EDA is launched in parallel on a block of 100 different instances. At each iteration, the sampling of the λ individuals as well as the fitness computations are performed in parallel with tensor calculus on GPU hardware.

5.2 EDA Training and Validation Phase

For each of the 18 configurations of the $PUBO_i$ generator described in Sect. 5.1, we run two different training processes:

- For the baseline EDAs, namely **UMDA** and **PBIL**, we performed a grid search to seek the best set of parameter values on each type of training set, with parameters $\mu \in [\![1, \lambda]\!]$ and $\rho \in \{0, 0.1, 0.2, \ldots, 1\}$. In the remainder of this document, these versions of the EDAs are referred to as "Fine Tuned UMDA" and "Fine Tuned PBIL".
- For the **Linear-EDA**, **Neuro-EDA** and **Memory-EDA** variants, we run 10 independent training processes with CMA-ES, as described in Sect. 4.2. During this training phase, at each generation of CMA-ES, a batch of 100 training instances is randomly sampled without replacement in $\mathcal{D}_{train}$ and each configuration of the vector of parameters θ (which are indeed the individuals of the CMA-ES population) are evaluated on these same 100 instances. At each generation, the EDA configuration obtaining the best score on average on these 100 training instances is evaluated on the 100 instances of the validation set $\mathcal{D}_{valid}$.

An example of the evolution of this average validation score obtained by **Linear-EDA**, **Neuro-EDA**, and **Memory-EDA** for the instances generated by $PUBO_i$ with parameters $(n = 128, m = 0.20, I = I_{imp})$ is shown in Fig. 1. The red line corresponds to the average baseline average score obtained by the "Fine Tuned PBIL" algorithm on the same validation instances. Color ranges correspond to minimum and maximum average values obtained for the 10 independent runs. As highlighted in this figure, **Linear-EDA** learns faster toward average good score on the validation set. This phenomenon may be explained by the fact that this variant has fewer parameters to calibrate, compared to **Neuro-EDA** and **Memory-EDA**, which, on the other hand, achieve better results on average at the end of the training process (see next section for confirmation of this score difference using statistical tests).

We do not observe any over-fitting phenomenon for all strategies, which may be explained by the fact that we use a sufficiently diversified batch of training instances at each generation of the learning process. Let us also note that the range of colors, indicating minimum and maximum scores for the 10 independent training, are relatively tight at the end of the calibration process, which highlights that the training process is quite consistent for the different strategies. At the end of this training process, for each $PUBO_i$ configuration and each EDA variant, the configuration of parameters θ allowing to reach the best average score on the validation set is chosen for the test phase presented in the next subsection.

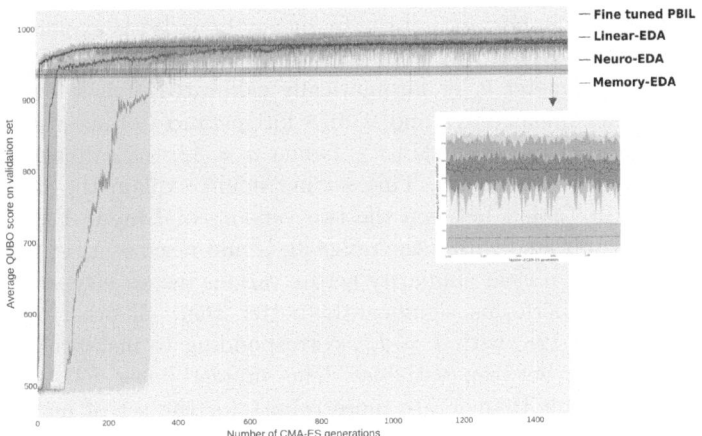

Fig. 1. Evolution of the average QUBO score obtained by **Linear EDA**, **Neuro-EDA** and **Memory-EDA** on the validation set over the generations of CMA-ES.

5.3 Test Phase

First, we perform evaluations to assess if the best strategies selected for each configuration of the validation set $\mathcal{D}_{valid}$ generated by $PUBO_i$, are still performing well on the corresponding test set $\mathcal{D}_{test}$, which has been independently sampled with the same $PUBO_i$ configuration. Secondly, we evaluate whether the best strategy learned for each specific $PUBO_i$ configuration still performs well for new test sets with different (n, m, I) configurations.

In-Distribution Tests. Table 1 summarizes the average scores obtained by the different EDAs' strategies, described in Sect. 2 and 3, on the test sets $\mathcal{D}_{test}$ with 100 independent restarts for each instance of the same type (10,000 runs), whose scores are of the same order of magnitude. The significantly better values of each variant in comparison with all others are underlined. The significance test is a Student t-test with p-value 0.001 and Bonferroni correction for multiple comparisons. The normality condition required for this test was first confirmed using a Shapiro-Wilk statistical test on the empirical distribution of the 10,000 scores obtained by each strategy.

For **Linear-EDA** (see Sect. 2.2), we report the results obtained for three different configurations of the frequency bounds, corresponding to $\theta_\epsilon = 0$, $\theta_\epsilon = \frac{1}{n}$, and a calibrated value of $\theta_\epsilon \in [0, \frac{1}{2}[$ obtained during the training phase. We also report the result obtained by a classic Tabu search algorithm (**TS**) with aspiration criterion [10]. In **TS**, after each flip, a move is set tabu during $\Delta = \beta + R$ iterations where R is a random integer in $[\![1, 10]\!]$ and β is an hyperparameter calibrated in the range $[\![1, n]\!]$ for each specific distribution of instances. We give **TS** the same budget of $T \times \lambda$ calls to the objective fitness function, as for all the EDAs.

Table 1 shows the importance of the frequency bounds (as explained in Subsect. 2.1) for **Linear-EDA**, which confirms the results observed in [2]. Interestingly, when the parameter θ_ϵ is automatically calibrated, it is set, on average, respectively to the values 0.017 and 0.0078 for instances of sizes $n = 64$ and $n = 128$. This almost corresponds to $\frac{1}{n}$ (when $n = 32$, the calibrated value is two times above this threshold). This setting might explain the proximity of results observed in Table 1 between the two versions of **Linear-EDA**. Specifically, when θ_ϵ is calibrated within the range $[0, \frac{1}{2}[$ and when θ_ϵ is set to the value $\frac{1}{n}$, the results exhibit a clear similarity across various instance types.

Memory-EDA performs significantly better than all other versions for datasets of size $n = 128$, with $I = I_{ic}$, corresponding to instances with more complex interactions between variables. The simpler linear EDAs, with frequency borders greater than 0, are more robust for the set of instances with simpler fitness landscapes (i.e., corresponding to $I = I_{uni}$). We also observe that **Memory-EDA** is better than **TS** for all types of instances (except for the datasets with I_{uni}) that exhibits less rough fitness landscapes). This highlights that this EDA strategies can be competitive with classic local search algorithms such as **TS**, solving QUBO as a black-box problem. As detailed in Sect. 5.4, this good performance could be explained by the ability of **Memory-EDA** to explore very distant areas of the search space during a run, which distinguish it from typical local search algorithms, which cannot explore such distant areas comprehensively.

Table 1. Average score (fitness values) obtained by different EDAs on test sets. The best scores are in bold. Underlined values correspond to significant better results (t-test with p-value 0.001 and Bonferroni correction for multiple comparisons).

Instances			Methods			Linear-EDA			Neuro-EDA	Memory-EDA
n	m	I	TS	UMDA	PBIL	$\theta_\epsilon = 0$	$\theta_\epsilon = \frac{1}{n}$	cal. θ_ϵ		
32	0.05	I_{uni}	65.5	64.6	64.6	64.8	**66.5**	**66.5**	**66.5**	**66.5**
32	0.05	I_{imp}	46.7	46.4	46.4	46.7	**46.8**	**46.8**	**46.8**	**46.8**
32	0.05	I_{ic}	42.0	41.9	41.8	42.0	**42.1**	42.0	**42.1**	**42.1**
32	0.20	I_{uni}	145.8	143.1	143.1	143.0	**147.5**	**147.5**	**147.5**	**147.5**
32	0.20	I_{imp}	107.5	106.1	106.4	107.3	**108.2**	**108.2**	**108.2**	**108.2**
32	0.20	I_{ic}	99.6	98.4	98.5	99.6	100.8	100.8	100.8	**100.9**
64	0.05	I_{uni}	208.6	202.4	202.4	200.8	209.4	209.6	209.6	**209.8**
64	0.05	I_{imp}	148.4	146.4	146.4	148.9	152.0	152.0	152.1	**152.2**
64	0.05	I_{ic}	140.2	140.3	140.3	142.0	144.9	144.9	145.2	**145.3**
64	0.20	I_{uni}	**446.7**	428.7	429.7	425.2	444.3	444.7	443.4	446.2
64	0.20	I_{imp}	334.0	324.5	326.5	332.8	338.0	339.7	340.8	**340.9**
64	0.20	I_{ic}	318.9	315.7	315.9	318.5	325.2	325.4	**328.2**	**328.2**
128	0.05	I_{uni}	**624.9**	605.5	605.4	593.3	624.5	**624.9**	621.6	623.4
128	0.05	I_{imp}	458.2	447.9	448.0	457.7	467.2	467.1	470.2	<u>**473.3**</u>
128	0.05	I_{ic}	433.1	427.5	428.4	430.4	439.9	440.5	445.7	**447.0**
128	0.20	I_{uni}	**1282.0**	1239.4	1239.3	1210.6	1278.1	1278.4	1271.8	1274.0
128	0.20	I_{imp}	964.8	938.4	939.3	963.2	970.0	979.5	991.7	**991.9**
128	0.20	I_{ic}	950.4	937.3	936.7	934.0	962.8	962.6	978.0	<u>**982.6**</u>

Out-of-Distribution Tests. Here, our aim is to check whether the best EDAs can also perform well on larger instances from distributions on which they have not been trained. We select the best EDAs trained on the distribution of different types with size $n = 128$, and we test them on 20 new instances of size $n = 256$ of different types. The results, reported in Table 2, are those obtained by a baseline finely-tuned tabu search, as well as by **Memory-EDA** and the best **Linear-EDA** version.

Each value reported in this Table corresponds to an average score computed for 20 different instances, each of them solved with 100 independent restarts. Average best results obtained by each strategy for each type of new instances are in bold. The underlined values appearing in this Table highlights that **Memory-EDA** trained on $PUBO_i$ instances generated with parameters ($n = 128, m = 0.20, I_{ic}$) obtained significant better results than the best **Linear-EDA** strategies trained on the different distributions. Interestingly, we observe that **Memory-EDA**, trained on the distribution of instances

$(n = 128, m = 0.20, I_{ic})$, always obtains the best results for all type of new distributions. Learning on the most complex set of instances generated by PUBO$_i$ leads indeed to the discovery of a more robust strategy that can be transferred to a wide range of instances of different types.

Table 2. Out-of-distribution test results obtained by **Linear-EDA** (with $\theta_\epsilon = \frac{1}{n}$) and **Memory-EDA**.

		Test instances					
		$(n,m) = (256, 0.05)$			$(n,m) = (256, 0.20)$		
		I_{uni}	I_{imp}	I_{ic}	I_{uni}	I_{imp}	I_{ic}
Fined tuned TS		1780	1331	1318	3628	2733	2750
Linear-EDA	$128, 0.05, I_{uni}$	1794	1340	1343	3643	2748	2792
	$128, 0.05, I_{imp}$	1664	1349	1290	3382	2746	2654
	$128, 0.05, I_{ic}$	1740	1350	1326	3533	2758	2734
	$128, 0.20, I_{uni}$	**1797**	1334	1342	**3650**	2744	**2799**
	$128, 0.20, I_{imp}$	1719	**1354**	1314	3491	**2762**	2706
	$128, 0.20, I_{ic}$	1794	1336	**1344**	3638	2747	2791
Memory-EDA	$128, 0.05, I_{uni}$	1477	1296	1224	2995	2626	2480
	$128, 0.05, I_{imp}$	1635	1353	1292	3326	2751	2648
	$128, 0.05, I_{ic}$	1516	1315	1245	3075	2672	2533
	$128, 0.20, I_{uni}$	1500	1305	1233	3041	2645	2504
	$128, 0.20, I_{imp}$	1617	1347	1281	3287	2739	2623
	$128, 0.20, I_{ic}$	**1814**	**1389**	**1368**	**3685**	**2840**	**2835**

5.4 Memory-EDA Emerging Strategy

Memory-EDA exhibits an original strategy. For all the datasets with complex interactions between variables ($I = I_{imp}$ and $I = I_{ic}$) for which **Memory-EDA** obtains the best results at the end of its learning process, we observe that its sampling strategy oscillates between very distant areas of the search space. This oscillation is made possible thanks to the update memory step allowing to alternate the sampling of the population of new individuals with different distributions during the same run.

To illustrate this phenomenon, we compute at each iteration for each individual x of the population the distance $D = \min(d_H(x, x^*), d_H(x, \bar{x}^*))$, where $d_H(x, x^*)$ and $d_H(x, \bar{x}^*)$ are the Hamming distances between x and x^* (respectively $\bar{x}^*$). x^* is the best solution found during each run and $\bar{x}^*$, the complementary solution of x^*.[2] This average distance obtained by the individuals of

---

[2] We compute the minimum between these two distances, because the QUBO instances generated by PUBO$_i$ are symmetric, i.e. for all $x \in \mathcal{X}$, $f(x) = f(\bar{x})$.

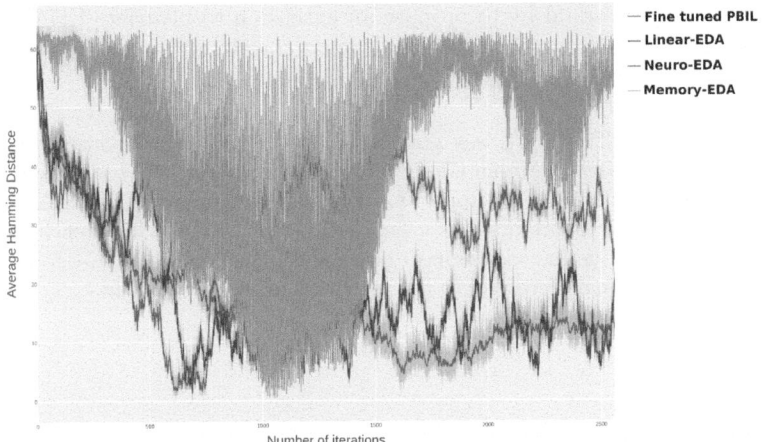

Fig. 2. Average Hamming distance between the sampled individuals of the population and x^* at each iteration. (Color figure online)

the population of different EDAs during the resolution of the first instance of the test set generated with parameters $n = 128$, $m = 0.20$ and $I = I_{imp}$ is displayed in Fig. 2.[3] **Memory-EDA** (in green) oscillates during the search and the average distances are in average much more important throughout the search compared to the other EDAs. We conclude that this strategy better explores the search space and can easily escape from local optima. **PBIL** (Red line) exhibits a less explorative behaviour since most of the individuals sampled during the run remain at a distance at most 20 from x^*. **Linear-EDA** (blue line) and **Neuro-EDA** (mauve line) seem to better explore the search space than **PBIL**.

6 Conclusion

We proposed a meta-learning framework to discover univariate EDAs for pseudo-boolean optimization problems. We highlighted that a neural network strategy using a larger memory that just a single probability value for each variable can be competitive with existing EDAs of the literature for different types of problem instances. In future work, we would like to investigate the impact of more complex neural network architectures for the design of new univariate strategies, such as convolutional graph neural networks, which could potentially take into account the interaction graph between variables induced by the Q matrix in white-box scenarios. Learning different generators to sample the individuals in the population in a differentiated way at each iteration could also be of interest.

---

[3] In this graph, we display the trajectory obtained for the resolution of a single instance, to better illustrate the dynamics of a single run, rather than an averaged run. We have checked that this is representative of different runs that can be obtained with other random seeds for this type of instances.

Another possibility would be to propose an extension to bivariate EDAs, which could be achieved, for example, by introducing correlated noise when sampling the different variables with a generative neural network.

Acknowledgment. This work was granted access to the HPC resources of IDRIS (Grant No. AD010611887R2) from GENCI. The authors would like to thank the Pays de la Loire region for its financial support for the Deep Meta project (Etoiles Montantes en Pays de la Loire). The authors also acknowledge ANR - FRANCE (French National Research Agency) for its financial support of the COMBO project (PRC - AAPG 2023 - Axe E.2 - CE23). We are grateful to the reviewers for their comments.

References

1. Baluja, S.: Population-based incremental learning: A method for integrating genetic search based function optimization and competitive learning (1994)
2. Chen, T., Tang, K., Chen, G., Yao, X.: Analysis of computational time of simple estimation of distribution algorithms. IEEE Trans. Evol. Comput. **14**(1), 1–22 (2010)
3. Doerr, B., Dufay, M.: General univariate estimation-of-distribution algorithms. In: International Conference on Parallel Problem Solving from Nature, pp. 470–484. Springer (2022)
4. Doerr, B., Krejca, M.S.: The univariate marginal distribution algorithm copes well with deception and epistasis. In: Paquete, L., Zarges, C. (eds.) Evolutionary Computation in Combinatorial Optimization EvoCOP 2020. Lecture Notes in Computer Science, vol. 12102, pp. 51–66. Springer (2020)
5. Doerr, B., Krejca, M.S.: Bivariate estimation-of-distribution algorithms can find an exponential number of optima. Theor. Comput. Sci. **971**, 114074 (2023)
6. Echegoyen, C., Lozano, J.A., Santana, R., Larrañaga, P.: Exact bayesian network learning in estimation of distribution algorithms. In: Proceedings of the IEEE Congress on Evolutionary Computation, CEC, pp. 1051–1058. IEEE (2007). https://doi.org/10.1109/CEC.2007.4424586
7. Friedrich, T., Kötzing, T., Krejca, M.S.: EDAs cannot be balanced and stable. In: Friedrich, T., Neumann, F., Sutton, A.M. (eds.) Proceedings of the 2016 on Genetic and Evolutionary Computation Conference, Denver, CO, USA, July 20 - 24, 2016, pp. 1139–1146. ACM (2016)
8. Gaon, M., Brafman, R.: Reinforcement learning with non-markovian rewards. In: Proceedings of the AAAI Conference on Artificial Intelligence. vol. 34, pp. 3980–3987 (2020)
9. Glover, F., Kochenberger, G., Hennig, R., Du, Y.: Quantum bridge analytics I: a tutorial on formulating and using QUBO models. Ann. Oper. Res. 1–43 (2022)
10. Glover, F., Laguna, M.: Tabu Search, pp. 2093–2229. Springer US, Boston, MA (1998)
11. Goudet, O., Goëffon, A., Saubion, F., Vérel, S.: Emergence of strategies for univariate estimation-of-distribution algorithms with evolved neural networks. In: Li, X., Handl, J. (eds.) Proceedings of the Genetic and Evolutionary Computation Conference Companion, GECCO 2024, Melbourne, VIC, Australia, July 14-18, 2024, pp. 195–198. ACM (2024). https://doi.org/10.1145/3638530.3654319

12. Hansen, N., Akimoto, Y., Baudis, P.: CMA-ES/pycma on Github. Zenodo (2019). https://doi.org/10.5281/zenodo.2559634
13. Hansen, N., Arnold, D.V., Auger, A.: Evolution strategies. Springer handbook of computational intelligence, pp. 871–898 (2015)
14. Hansen, N., Müller, S.D., Koumoutsakos, P.: Reducing the time complexity of the derandomized evolution strategy with covariance matrix adaptation (CMA-ES). Evol. Comput. **11**(1), 1–18 (2003)
15. Hansen, N., Ostermeier, A.: Completely derandomized self-adaptation in evolution strategies. Evol. Comput. **9**(2), 159–195 (2001)
16. Harik, G., Lobo, F., Goldberg, D.: The compact genetic algorithm. IEEE Trans. Evol. Comput. **3**(4), 287–297 (1999). https://doi.org/10.1109/4235.797971
17. Heidrich-Meisner, V., Igel, C.: Neuroevolution strategies for episodic reinforcement learning. J. Algorithms **64**(4), 152–168 (2009)
18. Jedidia, F.B., Doerr, B., Krejca, M.S.: Estimation-of-distribution algorithms for multi-valued decision variables. Theor. Comput. Sci. **1003** 114622 (2024). https://doi.org/10.1016/J.TCS.2024.114622
19. Jedrzejewski-Szmek, Z., Abrahao, K.P., Jedrzejewska-Szmek, J., Lovinger, D.M., Blackwell, K.T.: Parameter optimization using covariance matrix adaptation - evolutionary strategy (CMA-ES), an approach to investigate differences in channel properties between neuron subtypes. Front. Neuroinformatics **12**, 47 (2018)
20. Juels, A., Baluja, S., Sinclair, A.: The equilibrium genetic algorithm and the role of crossover. Unpublished manuscript (1993)
21. Krejca, M.S., Witt, C.: Theory of estimation-of-distribution algorithms. In: Doerr, B., Neumann, F. (eds.) Theory of Evolutionary Computation - Recent Developments in Discrete Optimization, pp. 405–442. Springer, Natural Computing Series (2020)
22. Lange, R.T., et al.: Discovering attention-based genetic algorithms via meta-black-box optimization. In: Silva, S., Paquete, L. (eds.) Proceedings of the Genetic and Evolutionary Computation Conference, GECCO 2023, Lisbon, Portugal, July 15-19, 2023, pp. 929–937. ACM (2023)
23. Larranaga, P., Bielza, C.: Estimation of distribution algorithms in machine learning: A survey. IEEE Transactions on Evolutionary Computation (2023)
24. Larrañaga, P., Lozano, J.A. (eds.): Estimation of Distribution Algorithms. Springer, Genetic Algorithms and Evolutionary Computation (2002)
25. Lodewijks, B.: Mapping NP-hard and NP-complete optimisation problems to quadratic unconstrained binary optimisation problems. arXiv preprint arXiv:1911.08043 (2019)
26. Mühlenbein, H., Paass, G.: From recombination of genes to the estimation of distributions i. binary parameters. In: Voigt, H., Ebeling, W., Rechenberg, I., Schwefel, H. (eds.) Parallel Problem Solving from Nature. Lecture Notes in Computer Science, vol. 1141, pp. 178–187. Springer (1996)
27. Pelikan, M., Hauschild, M., Lobo, F.G.: Estimation of distribution algorithms. In: Kacprzyk, J., Pedrycz, W. (eds.) Springer Handbook of Computational Intelligence, pp. 899–928. Springer Handbooks, Springer (2015)
28. Pelikan, M., Muehlenbein, H.: The bivariate marginal distribution algorithm. In: Roy, R., Furuhashi, T., Chawdhry, P.K. (eds.) Advances in Soft Computing, pp. 521–535. Springer, London, London (1999)
29. Segovia-Dominguez, I., Valdez, S.I., Aguirre, A.H.: A boltzmann multivariate estimation of distribution algorithm for continuous optimization. In: Rosa, A.C., Guervós, J.J.M., Filipe, J. (eds.) ECTA 2014, pp. 251–258. SciTePress (2014). https://doi.org/10.5220/0005079902510258

30. Shapiro, J.L.: Drift and scaling in estimation of distribution algorithms. Evol. Comput. **13**(1), 99–123 (2005)
31. Tari, S., Verel, S., Omidvar, M.: Pubo$_i$: A tunable benchmark with variable importance. In: Cáceres, L.P., Verel, S. (eds.) Evolutionary Computation in Combinatorial Optimization - 22nd European Conference, EvoCOP 2022, Held as Part of EvoStar 2022, Madrid, Spain, April 20-22, 2022, Proceedings. Lecture Notes in Computer Science, vol. 13222, pp. 175–190. Springer (2022)

A Selective Vehicle Routing Problem for the Bloodmobile System

Aldy Gunawan[1](✉), Samuel Alan Darmasaputra[2], Sy Hoang Do[2], and Vincent F. Yu[2]

[1] School of Computing and Information Systems, Singapore Management University, Singapore, Singapore
aldygunawan@smu.edu.sg
[2] Department of Industrial Management, National Taiwan University of Science and Technology, Taipei, Taiwan
{m10901858,d11101801,vincent}@mail.ntust.edu.tw

Abstract. Mobile blood collection has the advantage of greater reach compared to blood drives at fixed donation sites and is preferable for individuals with limited time or means of transportation. Bloodmobiles are widely used in healthcare logistics to increase the number of donors and donation frequency and to better match blood demand with collection. Bloodmobiles are stationed at predetermined locations, while shuttles are assigned to visit these locations to collect the donated blood. This problem is formulated as the Selective Vehicle Routing Problem under the Bloodmobile System (SVRP-BM). This research extends the Selective Vehicle Routing Problem with Integrated Tours problem (SVRPwIT) by considering: (i) multiple shuttles, (ii) multiple blood types, (iii) multiple trips for the bloodmobiles, and (iv) the visiting availability of the donation sites. The proposed adaptive large neighborhood search (ALNS) algorithm, with a simulated annealing (SA) acceptance criterion, is tested on generated instances adopted from a real-life case of the Surabaya Red Cross. SVRP-BM provides a strategic solution to Surabaya's blood shortage by optimizing blood collection.

1 Introduction

Blood transfusion is one of medical science's branches with a significant and undeniable role in providing services to patients. According to the World Health Organization, every country struggles to collect enough blood from safe donors to meet national needs [1]. In 2017, the global blood supply was estimated at 272 million units, while demand reached approximately 303 million units, resulting in a 30-million-unit deficit [2]. This supply challenge is also evident in Surabaya City, where the Surabaya Red Cross faced a blood shortage. In 2021, the city's blood demand of 84,963 blood bags exceeded the supply of 80,498 blood bags. Since blood cannot be synthetically produced, donation activities are the sole source of this vital resource. Moreover, transfusion-dependent patients, such as those with beta-thalassemia and sickle cell disease, require regular blood transfusions, emphasizing the need for a stable blood supply [3]. Hospitals must adopt strict guidelines and mitigation strategies to effectively manage blood transfusion orders [4].

Addressing these challenges requires innovative strategies for managing the blood supply. Donation campaigns via mobile blood collection drives play a pivotal role in maintaining a steady supply [5]. Studies on bloodmobile systems highlight the importance of optimizing collection and distribution routes. Gunpinar et al. [6] developed an MILP model considering variable visit durations and different bloodmobile types. Rezaei et al. [7] explored routing bloodmobiles under crisis conditions, where helicopters are used to collect blood at the end of the day and deliver it to crisis-stricken cities.

Sahinyazan et al. [8] proposed the Selective Vehicle Routing Problem with Integrated Tours (SVRPwIT) model and designed the routing plan for both bloodmobiles and a shuttle to collect donated blood, assuming that a bloodmobile can receive a single visit from a shuttle at the end of the day. Rabbani et al. [9] extended this work by requiring that a shuttle must visit a bloodmobile within a fixed time window. They presented a fuzzy mathematical model with the goal of maximizing the potential amount of blood collected while minimizing operational costs. They applied a simulated annealing algorithm to solve the problem. Building on these approaches, this research proposes the Selective Vehicle Routing Problem under the Bloodmobile System (SVRP-BM), which is an extension of the SVRPwIT model [8]. SVRP-BM aims to optimize routing plans for both bloodmobiles and shuttles over a planning period. The key contributions are as follows.

- Extend the SVRPwIT model by considering (i) multiple shuttles, (ii) multiple blood types, (iii) multiple trips for the bloodmobiles, and (iv) visiting availability of the donation sites to propose SVRP-BM.
- Formulate a mixed integer programming (MIP) model and develop an adaptive large neighborhood search (ALNS) with simulated annealing (SA) acceptance criteria.

The rest of the paper runs as follows. Section 2 describes the proposed network and formulates the problem as a mixed-integer linear programming (MILP) model. Section 3 designs the proposed ALNS algorithm. Section 4 presents the computational results. Section 5 provides final conclusions and future research directions.

2 Model Development

2.1 Problem Description

The Selective Vehicle Routing Problem under the Bloodmobile System (SVRP-BM) is an extension of the SVRPwIT model, which allows multi-trips for bloodmobiles, multiple types of blood, the use of multiple shuttles, and visiting availability of each donation site. In this model, bloodmobiles collect blood from donation sites each day, while shuttles are deployed to pick up blood from selected stops. Bloodmobiles can return to the blood center (BC) and resume their routes the next day, as multi-trip operations are allowed. Donation sites are planned based on the blood potential at each site and demand at BC. At the start of each day, bloodmobiles depart from BC to visit one donation site, considering its availability. Each site can be visited up to three times and not necessarily on consecutive days. If a bloodmobile moves to a new site or returns to BC, then it travels at night and stays until the next morning's donation activity. Shuttles collect blood from bloodmobiles at the end of the day, except when the bloodmobile is scheduled to return to BC the following day. This system reduces the need for daily returns to BC.

Figure 1 shows an example of an SVRP-BM network including BC, 3 bloodmobiles, 2 shuttles, 10 donation sites, and 3 days of planning horizon. On the first day, the yellow, green, and red bloodmobiles depart from BC to donation sites 1, 5, and 7, respectively. After arriving, they prepare for and conduct blood donation activities. Once the donations are collected, shuttles depart from BC to pick up the donated blood from the yellow and green bloodmobiles. However, no shuttle visits the red bloodmobile since it is scheduled to return to BC after its donation activity. At night, the yellow bloodmobile moves to donation site 2, the green bloodmobile remains at donation site 5 for ongoing activities, and the red bloodmobile returns to BC. On day 2, the shuttles again collect blood from the green and yellow bloodmobiles. At the end of day 2, all bloodmobiles return to BC, marking the conclusion of the planning horizon (considered as day 3 in the model - not shown in the figure).

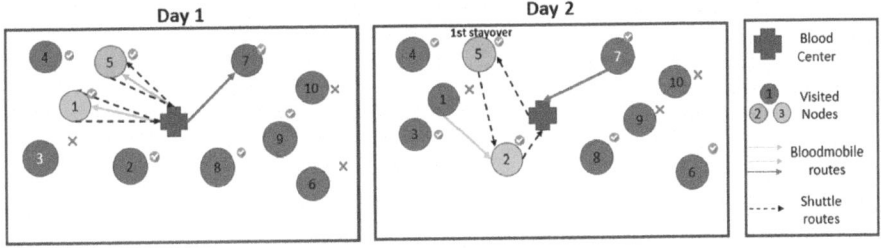

Fig. 1. An example of the SVRP-BM network

2.2 Mathematical Model

Let $\{0\}$ denote BC, $N = \{1, .., n_S\}$ is the set of n_S donation sites, $T = \{1, .., n_T\}$ is the set of n_T types of blood, and $D = \{1, ..., n_D\}$ is the set of n_D day of planning period. Each donation $i \in N$ has a potential amount of blood for each type $t \in T$ denoted by b_i^t and the available schedule on day $d \in D$ is represented by $a_i^d \in \{0, 1\}$, which means site i is available for visit on day d or not. A fleet of n_K bloodmobiles ($K = \{1, ..., n_K\}$) and n_{Sh} shuttles ($S = \{1, ..., n_{Sh}\}$) is available in BC to collect the desired amount of blood type $t \in T$ B_t^* in planning periods. The maximum capacity of shuttles is Q_{Sh}.

The SVRP-BM model can be defined by a directed graph $G = (V, A)$, where V is the set of all nodes, and A is the set of all arcs. In more detail, $V = \{0\} \cup N$ and $A = \{(i,j) : i,j \in V, i \neq j\}$. Each arc $(i,j) \in A$ is associated with travel cost c_{ij}. To solve the SVRP-BM problem, a mixed-integer linear programming (MILP) problem is formulated with decision variables as follows. There are three binary variables: x_{ij}^d equals 1 if the bloodmobile travels between $i,j \in V$ on day $d \in D$ and otherwise 0; y_{ij}^{sd} equals 1 if shuttle $s \in S$ traverses arc $(i,j) \in A$ on day $d \in D$ and otherwise 0; and z_i^d equals 1 if node $i \in N$ requires a shuttle on day $d \in D$. The continuous variable u_i^s represents the visiting order of node $i \in V$ in a tour of shuttle $s \in S$. Integer variable B_d determines the available bloodmobiles on day $d \in D$.

$$Minimize \sum_{(i,j)\in A} \sum_{d\in D} c_{ij} x_{ij}^d + \sum_{(i,j)\in A} \sum_{d\in D} c_{ij} y_{ij}^d \quad (1)$$

Subject to:

$$\sum_{i\in V} x_{ij}^d = z_j^d + x_{j0}^{d+1} \qquad \forall j \in N, d \in \{1,...,n_D - 1\} \quad (2)$$

$$\sum_{d=1}^{n_D-1} (x_{j0}^{d+1} + z_j^d) \leq 3 \qquad \forall j \in N \quad (3)$$

$$\sum_{i\in V} x_{ij}^d = \sum_{i\in V} x_{ji}^{d+1} \qquad \forall j \in N, d \in \{1,...,n_D - 1\} \quad (4)$$

$$\sum_{j\in V} x_{0j}^1 \leq n_K \quad (5)$$

$$\sum_{i\in N} \sum_{j\in V} x_{ij}^1 = 0 \quad (6)$$

$$\sum_{i\in V} x_{ij}^d \leq a_j^d \qquad \forall j \in N, d \in D \quad (7)$$

$$B_1 = n_K \quad (8)$$

$$\sum_{j\in V} x_{0j}^d \leq B_d \qquad \forall d \in \{2,...,n_D\} \quad (9)$$

$$B_d = B_{d-1} + \sum_{i\in N} x_{i0}^{d-1} - \sum_{j\in N} x_{0j}^{d-1} \qquad \forall d \in \{2,...,n_D\} \quad (10)$$

$$\sum_{i\in V} y_{ij}^{sd} = \sum_{i\in V} y_{ji}^{sd} \qquad \forall j \in N, s \in S, d \in \{1,...,n_D - 1\} \quad (11)$$

$$\sum_{i\in V} \sum_{s\in S} y_{ij}^{sd} = z_j^d \qquad \forall j \in N, d \in \{1,...,n_D - 1\} \quad (12)$$

$$\sum_{i \in N} y_{0i}^{sd} \leq 1 \qquad \forall s \in S, d \in \{1, ..., n_D - 1\} \qquad (13)$$

$$\sum_{i \in N} y_{i0}^{sd} \leq 1 \qquad \forall s \in S, d \in \{1, ..., n_D - 1\} \qquad (14)$$

$$\sum_{i \in V} \sum_{j \in N} \sum_{t \in T} b_j^t y_{ij}^{sd} \leq Q_{Sh} \qquad \forall s \in S, d \in \{1, ..., n_D - 1\} \qquad (15)$$

$$\sum_{i \in V} \sum_{j \in V} \sum_{d \in D} b_j^t x_{ij}^d \geq B_t^* \qquad \forall t \in T \qquad (16)$$

$$u_i^s - u_j^s + 1 \leq (n_S - 1)(1 - y_{ij}^{sd}) \quad \forall i, j \in N, s \in S, d \in \in \{1, ..., n_D - 1\} \quad (17)$$

$$1 \leq u_i^s \leq n_S \qquad \forall i \in N, s \in S \qquad (18)$$

$$x_{ij}^d \in \{0, 1\} \qquad \forall i, j \in V, d \in D \qquad (19)$$

$$y_{ij}^{sd} \in \{0, 1\} \qquad \forall i, j \in V, s \in S, d \in D \qquad (20)$$

$$z_i^d \in \{0, 1\} \qquad \forall i \in N, d \in D \qquad (21)$$

$$u_i^s \geq 0 \qquad \forall i \in V, s \in S \qquad (22)$$

$$B_d \geq 0 \qquad \forall d \in D \qquad (23)$$

The objective function (1) minimizes total cost, including the travel cost of bloodmobiles and shuttles. Constraint (2) ensures that if a bloodmobile visits node j on day d, then node j should be visited by the shuttle on the same day or the bloodmobile immediately returns to the depot on the next day. Constraint (3) restricts a node to be visited at most three times during the planning period. Flow conservation is ensured by Constraint (4) whereby if a bloodmobile visits node j on day d, then it should leave node j on the next day. On the first day of the planning period, the number of bloodmobiles starting from BC cannot exceed the total number of bloodmobiles at BC, as guaranteed by Constraint (5). Constraint (6) forces that no trip starts from any node except BC on the first day. Constraint (7) ensures that node j can only be visited on day d if it is available on that day. The number of available bloodmobiles on day $d \in D$ is determined by Constraints (8)–(10). In more detail, Constraint (8) shows that the number of available bloodmobiles on the first day of the planning period equals the total number of bloodmobiles, and from the second day until the end the number of bloodmobiles starting from BC not exceed the number of available bloodmobiles on that day as illustrated by Constraint (9). The number of available bloodmobiles at BC on a certain day is estimated by Constraint (10). Constraint (11) ensures connectivity of the path of each shuttle, and Constraint (12) guarantees that if the node requires a shuttle, then the shuttle will come to take the collected blood. Shuttles have to start and end at BC as illustrated by Constraints (13) and (14). Constraint (15) restricts the blood collected by each shuttle must not exceed its capacity. Constraint (16) ensures the amount of collected blood achieves a desired amount of each blood type. Constraints

(17) and (18) are the sub-tour elimination constraints. The domain of variables appears in Constraints (19)–(23).

3 Adaptive Large Neighborhood Search (ALNS)

This section describes the proposed algorithm. We first introduce the solution representation, followed by ALNS.

3.1 Solution Representation

There are two solution structures for SVRP-BM: the route for bloodmobiles and the route for shuttles.

The Route for Bloodmobiles. A two-dimension array consisting of d days and m bloodmobiles (Fig. 2) is introduced. The length of each bloodmobile's route is flexible, but must not exceed the number of planning periods plus 2 (accounting for the start and end points at BC). This flexibility exists, because bloodmobiles are not required to operate until the end of the planning horizon; they can conclude their routes early if blood demand is fully met. The route sequence details the order of visited nodes $n \in V$ on the specific day of each bloodmobile.

Bloodmobile Route

Day -	0	1	2	3	.	.	d	
BM - 1	0	n_{11}	n_{12}	n_{13}	.	.	n_{1d}	0
BM - 2	0	n_{21}	n_{22}	n_{23}	.	.	n_{2d}	0
⋮								
BM - m	0	n_{m1}	n_{m2}	n_{m3}	.	.	n_{md}	0

Example of a BM route consisting of 3 BM, 7 days of planning horizon, and 10 donation sites

		1	2	3	4	5	6	7	
BM - 1	0	9	9	3	3	9	2	2	0
BM - 2	0	2	10	10	0	5	5	5	0
BM - 3	0	4	4	4	7	7	6	6	0

Fig. 2. Solution representation of bloodmobiles

The Route for Shuttles. The shuttle route solution is represented by: (i) a sequence of BC and donation sites $n \in N$ visited by bloodmobiles on a given day, (ii) k shuttles, with each shuttle route separated by a delimiter (0), and (iii) d working days. Figure 3 illustrates this solution. The sequence from left to right shows the order in which donation sites n, visited by bloodmobile m on day d, are serviced. A shuttle visits a donation site n if the bloodmobile assigned to that site does not return to BC on the following day.

Shuttle Route

— Shuttle - 1   — Shuttle - 2 — Shuttle - k —

Day - 1	0	n_{11}	n_{21}	0	.	0	.	n_{m1}	0
Day - 2	0	n_{12}	n_{22}	0	.	0	.	n_{m2}	0
⋮									
Day - d	0	n_{1d}	n_{2d}	0	.	0	.	n_{md}	0

Example of a shuttle route based on a given BM route (2 shuttles)

Day - 1	0	9	2	0	4	0
Day - 2	0	9	10	0	4	0
Day - 3	0	3	4	0	0	
Day - 4	0	3	7	0	0	
Day - 5	0	9	5	0	7	0
Day - 6	0	2	5	0	6	0
Day - 7	0	0	0			

Fig. 3. Solution representation of shuttles

3.2 Adaptive Large Neighborhood Search with Simulated Annealing Acceptance Criteria

This study presents a modified adaptive large neighborhood search (ALNS) algorithm that integrates the simulated annealing (SA) acceptance mechanism to enhance its optimization capabilities. The ALNS has shown its effectiveness in solving various transportation problems such as vehicle routing problems ([10,11], and [12]), electric vehicle routing problem ([13,14]) and orienteering problem ([15,16]). The modification by integrating the SA acceptance mechanism allows the algorithm to explore the solution space more effectively, avoiding local optima while aiming for a global optimum. Figure 4 shows the process of the proposed ALNS.

The algorithm is started by generating an initial solution for bloodmobile and shuttle routes, which is treated as the current solution (σ) and best solution (σ^*) solution. The objective value of a solution is denoted by $\Omega(\sigma)$. The initial solution for bloodmobiles is created by assigning the available nodes with the highest average blood potential to the bloodmobiles. For the shuttles, nodes are assigned if they have been visited by the bloodmobiles, the bloodmobiles do not return to the BC after visiting those nodes, and the shuttles have enough remaining capacity.

The current temperature T is set to initial temperature T_0; both the current iteration $iter$ and non-improve count $non-improving$ are set to 0. The algorithm then enters its dual-loop structure: the outer loop handles temperature reduction, while the inner loop focuses on iterative improvements at each temperature level. During each iteration of the inner loop, one of the destroy (i.e., random removal, worst removal, Shaw removal, or route removal) and repair (i.e., random insert, greedy insert, greedy insert with noise, regret-2, or regret-2 with noise) operators is applied to the current solution σ to generate a new neighborhood solution σ'. The destroy heuristics remove nodes from the bloodmobile route, while the repair heuristics reintegrate these nodes to form a new solution. In this study, the destroy and repair heuristics follow the approach of Ropke et al. [17]. The number of nodes removed is determined randomly within a defined range $[\eta_{min}, \eta_{max}]$. After applying these heuristics, the feasibility of σ' is always checked.

If the new bloodmobile solution σ' is feasible, then the algorithm moves to the second phase, where initial shuttle routes are generated based on the updated bloodmobile schedule; otherwise, we ignore the infeasible solution. This phase is followed by a shuttle route improvement stage, which optimizes shuttle routes for better efficiency and lower operational costs. Once σ' is fully constructed, the acceptance mechanism is started. If the new solution improves the best solution ($\Omega(\sigma') < \Omega(\sigma^*)$), then σ^* is updated, and the corresponding destroy and repair operators are rewarded with the highest score. If σ' improves the current solution ($\Omega(\sigma') < \Omega(\sigma)$), then it replaces σ, and operators receive slightly lower rewards. However, for the worst case, if σ' is worse than current solution, then it may still replace σ with a probability determined by the Boltzmann equation. This stochastic acceptance mechanism helps the algorithm avoid local optima. The inner loop terminates when either the maximum number of iterations Max_{iter} is reached or when the algorithm experiences $Max_{non-improve}$ consecutive iterations without improving σ^*. At this point, after N_{ALNS} iterations, operator weights and selection probabilities are updated. The outer loop then reduces the temperature by cooling rate α ($T = \alpha T$), and the process repeats until the temperature reaches T_f. Finally, the algorithm returns the best solution σ^* and its objective $\Omega(\sigma^*)$.

4 Numerical Experiments

Section 4.1 presents instances and parameter values of the proposed ALNS. Section 4.2 shows the results for different scenarios.

4.1 Parameter Settings

The test instances in this study are based on a case study from Surabaya City, Indonesia. A randomly selected BC in the city, equipped with 3 bloodmobiles

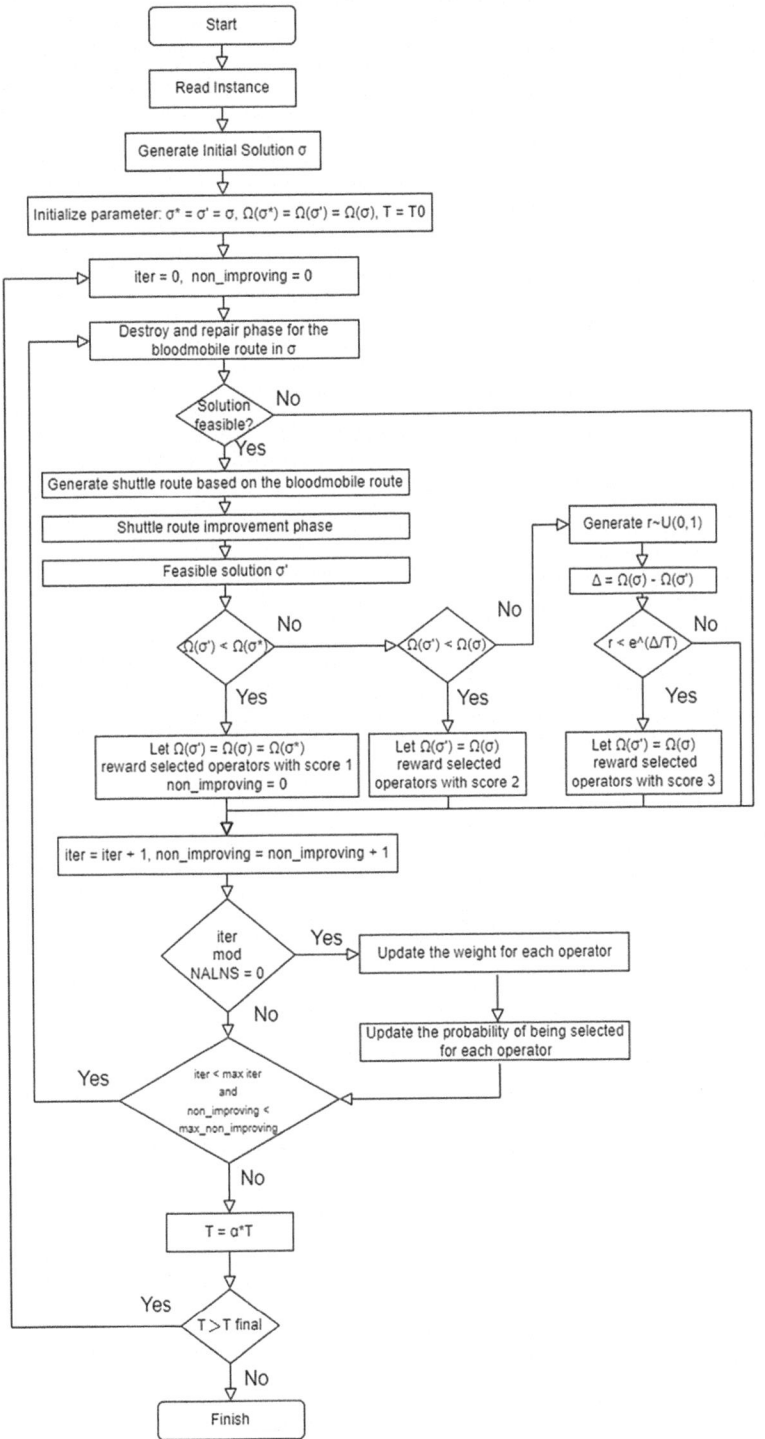

Fig. 4. Flow chart of the proposed ALNS

and 2 shuttles, serves as the basis for the instances. Three planning periods of 5, 6, and 7 days are considered, reflecting actual operational timeframes. The demand is categorized by the number of blood bags required from hospitals, for four types of blood, each accounting for a specific proportion of total demand. Table 1 outlines the parameters used in the test instances.

Table 1. Instances' parameters

Parameter	Values
Number of donation sites n_S	10, 20, 30, 40, 50
Number of days n_D	5, 6, 7
Number of blood types n_T	4 (O, A, B, AB)
Demand scenarios	Low (700 − 1100 blood bags) Medium (1100 − 1700 blood bags) High (1700 − 2200 blood bags)

Using these parameters, the model generates a total of 45 test instances, varying by demand scenario, planning period, and number of donation sites. This set-up ensures a comprehensive evaluation of the algorithm's performance under different conditions.

Those parameters (i.e., N_{ALNS}, T_0, T_{final}, α, $Max_{non-improve}$, and Max_{iter}) significantly impact ALNS performance. To optimize these parameters, 20 SVRP-BM instances are randomly selected for the tuning process with a one-factor-at-a-time (OFAT) method. For each instance, ALNS is executed 10 times to evaluate its performance under varying parameter settings. Table 2 presents a comprehensive overview of the candidate values, initial settings, and the final parameter values. The tuning sequence follows the order listed in the table, ensuring a systematic approach to parameter optimization.

Table 2. Tuned parameters for ALNS

Parameter	Candidate values	Initial value	Tuned value
N_{ALNS}	200, 250, 300, 350	250	250
T_0	55, 60, 65, 70	65	65
T_f	3, 4, 5, 6	5	3
α	0.96, 0.97, 0.98, 0.99	0.99	0.96
$Max_{non-improve}$	50, 100, 300, 500	100	300
Max_{iter}	20, 40, 60, 80	20	80

4.2 Experiment Results

This section evaluates the performance of the proposed ALNS algorithm by running it ten times for each instance in every scenario. To the best of our knowledge, no previous research exists for this specific problem; therefore, no benchmark solutions exist for comparison. As a result, the solutions obtained by the proposed ALNS are compared against those produced by CPLEX. The best and average solutions from the proposed algorithm are denoted as f_{best} and f_{avg}, respectively. In some instances, CPLEX can only provide feasible solutions rather than optimal ones - or in some other cases, no solution at all. The optimal solution provided by CPLEX is denoted as f^*, while f_{fe} represents the feasible solution. To assess the robustness of the proposed algorithm, the gap between the best and average solutions from ALNS is calculated using the formula $f_r(\%) = \frac{f_{avg}-f_{best}}{f_{avg}} \times 100$. Additionally, the gap between CPLEX and the proposed ALNS is calculated as follows: $gap(\%) = \frac{f_{best}-f^*}{f^*} \times 100$ or $gap(\%) = \frac{f_{best}-f_{fe}}{f_{fe}} \times 100$.

Low Demand Scenario. Table 3 shows the results of SVRP-BM under the low demand scenario. In this scenario, both CPLEX and the proposed ALNS consistently achieve optimal solutions. Notably, the proposed algorithm is capable of producing the optimal solution across all runs. In terms of computational time, the proposed ALNS takes longer at an average of 126.30 s compared to 36.97 s for CPLEX.

Medium Demand Scenario. Table 4 provides the detailed results of SVRP-BM under the medium demand scenario, comparing the performance of the proposed ALNS algorithm with the CPLEX solver. The results indicate that CPLEX is unable to find optimal solutions within the time limit for 5 out of 15 instances, offering only feasible solutions that are inferior to those obtained by ALNS, with an average percentage gap of -3.63%. Moreover, ALNS demonstrates a significant advantage in computational efficiency, solving all instances at an average of 908.46 s, compared to CPLEX's 8663.35 s. Additionally, ALNS still demonstrates robustness under a medium demand scenario with an average gap f_r of 0.33% between its best and average solutions.

High Demand Scenario. The results appear in Table 5. CPLEX fails to find the optimal solution within the time limit for 10 out of 15 instances, offering only feasible solutions. In these 10 instances, ALNS consistently outperforms CPLEX, achieving an average gap of -2.73%. ALNS maintains robustness, as evidenced by an average gap (f_r) of 0.54% between its best and average solutions. ALNS significantly reduces computational time, solving all instances on average at 922.94 s compared to CPLEX's 13369.39 s. This highlights ALNS's efficiency and effectiveness in handling complex, high-demand scenarios where CPLEX struggles. In conclusion, high- and medium-demand scenarios are harder to solve due to the complexity of the problems.

Table 3. Detailed results of SVRP-BM under low demand scenario

Instance	CPLEX			ALNS				gap(%)
	f^*	f_{fe}	$\bar{t}(sec)$	f_{best}	f_{avg}	$f_r(\%)$	$\bar{t}(sec)$	CPLEX
10N-5D-L	113.21	113.21	12.90	113.21	113.21	0.00	39.78	0.00
10N-6D-L	122.00	122.00	6.91	122.00	122.00	0.00	55.15	0.00
10N-7D-L	150.95	150.95	10.80	150.95	150.95	0.00	79.59	0.00
20N-5D-L	40.45	40.45	16.30	40.45	40.45	0.00	79.25	0.00
20N-6D-L	50.26	50.26	11.41	50.26	50.26	0.00	105.16	0.00
20N-7D-L	68.38	68.38	97.03	68.38	68.38	0.00	158.06	0.00
30N-5D-L	33.46	33.46	72.90	33.46	33.46	0.00	111.82	0.00
30N-6D-L	38.93	38.93	25.66	38.93	38.93	0.00	162.41	0.00
30N-7D-L	42.06	42.06	72.11	42.06	42.06	0.00	254.54	0.00
40N-5D-L	26.73	26.73	27.30	26.73	26.73	0.00	70.74	0.00
40N-6D-L	32.61	32.61	43.86	32.61	32.61	0.00	123.64	0.00
40N-7D-L	38.53	38.53	53.40	38.53	38.53	0.00	193.54	0.00
50N-5D-L	22.11	22.11	6.35	22.11	22.11	0.00	131.96	0.00
50N-6D-L	25.67	25.67	31.70	25.67	25.67	0.00	125.33	0.00
50N-7D-L	36.49	36.49	42.20	36.49	36.49	0.00	203.57	0.00
Average			36.97			0.00	126.30	0.00

Table 4. Detailed results of SVRP-BM under medium demand scenario

Instance	CPLEX			ALNS				gap(%)
	f^*	f_{fe}	$\bar{t}(sec)$	f_{best}	f_{avg}	$f_r(\%)$	$\bar{t}(sec)$	CPLEX
10N-5D-M	178.32	178.32	17.60	178.32	178.32	0.00	116.88	0.00
10N-6D-M	216.05	216.05	107.00	216.05	216.05	0.00	182.00	0.00
10N-7D-M	266.54	266.54	659.03	266.54	266.54	0.00	266.87	0.00
20N-5D-M	89.28	89.28	759.40	89.28	89.28	0.00	542.69	0.00
20N-6D-M	110.43	110.43	17873.70	110.43	110.43	0.00	444.33	0.00
20N-7D-M	___	153.94	18000.00	148.21	149.93	1.15	654.77	−3.44
30N-5D-M	54.34	54.34	1242.75	54.34	54.34	0.00	492.17	0.00
30N-6D-M	63.77	63.77	598.70	63.77	63.77	0.00	776.02	0.00
30N-7D-M	___	86.64	18000.00	82.09	83.21	1.35	1146.13	−5.25
40N-5D-M	47.69	47.69	2534.40	47.69	47.69	0.00	649.22	0.00
40N-6D-M	___	56.95	18000.00	54.49	54.85	0.67	1137.99	−4.32
40N-7D-M	___	69.22	18000.00	68.55	69.31	1.10	1747.78	−0.97
50N-5D-M	43.18	43.18	2169.80	43.18	43.18	0.00	1132.48	0.00
50N-6D-M	49.85	49.85	13987.90	49.85	49.85	0.00	1721.90	0.00
50N-7D-M	___	63.59	18000.00	60.93	61.33	0.65	2615.74	−4.18
Average			8663.35			0.33	908.46	−1.21

Table 5. Detailed results of SVRP-BM under high demand scenario

Instance	CPLEX			ALNS				gap(%)
	f^*	f_{fe}	$t(sec)$	f_{best}	f_{avg}	$f_r(\%)$	$t(sec)$	CPLEX
10N-5D-H	252.61	252.61	104.6	252.61	252.61	0.00	111.32	0.00
10N-6D-H	308.09	308.09	1191.20	308.09	308.09	0.00	160.39	0.00
10N-7D-H	---	378.22	18000.00	375.94	377.06	0.30	245.56	−0.60
20N-5D-H	163.81	163.81	625.61	163.81	163.81	0.00	618.99	0.00
20N-6D-H	---	182.93	18000.00	179.23	180.65	0.79	420.93	−2.02
20N-7D-H	---	263.11	18000.00	257.85	259.21	0.53	639.95	−2.00
30N-5D-H	113.04	113.04	10840.10	113.04	113.04	0.00	470.66	0.00
30N-6D-H	---	119.74	18000.00	114.86	115.34	0.42	732.91	−4.08
30N-7D-H	---	166.58	18000.00	160.95	161.27	0.20	1139.56	−3.38
40N-5D-H	---	97.23	18000.00	94.82	96.34	1.60	932.18	−2.48
40N-6D-H	---	98.55	18000.00	94.64	95.88	1.31	1066.94	−3.97
40N-7D-H	---	152.33	18000.00	147.76	149.44	1.14	1765.80	−3.00
50N-5D-H	74.64	74.64	7779.90	74.64	74.64	0.00	1168.32	0.00
50N-6D-H	---	82.39	18000.00	80.45	80.93	0.60	1709.26	−2.35
50N-7D-H	---	141.32	18000.00	136.42	138.06	1.20	2661.35	−3.47
Average			13369.39			0.54	922.94	−1.82

The results highlight several key factors that influence the objective values. First, higher demand levels lead to higher objective values, as more blood bags need to be collected and transported. This results in higher travel costs and more complex routes. Similarly, longer planning periods significantly impact objective values. While extended planning horizons provide greater flexibility for route optimization, they also require more resources, which drive up costs. Lastly, as the number of donation sites increases, the objective values tend to increase across all demand scenarios. This is due to the added travel distance and the increased complexity of coordinating bloodmobiles and shuttles to cover more locations effectively.

5 Conclusions

This research introduces a new model for the bloodmobile routing problem by extending the Selective Vehicle Routing Problem with Integrated Tours (SVR-PwIT), referred to as the Selective Vehicle Routing Problem under the Bloodmobile System (SVRP-BM). The SVRP-BM framework incorporates multiple shuttles, multi-trip scheduling for bloodmobiles, various blood types, and the visiting availability of donation sites. A Mixed-Integer Linear Programming (MILP) model is formulated, and a hybrid adaptive large neighborhood search (ALNS) algorithm with a simulated annealing (SA) acceptance criterion is proposed.

SVRP-BM instances are generated based on data from the Surabaya Red Cross, considering variations in number of donation sites, demand scenarios, and planning periods. The results show that the proposed ALNS outperforms the CPLEX solver. In addition, this model offers a strategic solution to address the problem of blood shortage in Surabaya by optimizing blood collection and distribution, ensuring timely delivery to meet demand.

For future research, incorporating real-world constraints at each blood donation site could enhance the model, such as time windows, unmet blood demand, etc. Considering the stochastic nature of blood donation potentials would also improve robustness, as donation quantities vary between times and locations. Finally, other robust parameter tuning approaches can be studied.

References

1. Organization, W.H., et al.: Towards 100% voluntary blood donation: a global framework for action (2010)
2. Samreen, S., Sales, I., Bawazeer, G., Wajid, S., Mahmoud, M.A., Aljohani, M.A.: Assessment of beliefs, behaviors, and opinions about blood donation in Telangana, India-a cross sectional community-based study. Front. Public Health **9**, 785568 (2021)
3. AlHamdan, N.A., AlMazrou, Y.Y., AlSwaidi, F.M., Choudhry, A.J.: Premarital screening for thalassemia and sickle cell disease in Saudi Arabia. Genet. Med. **9**(6), 372–377 (2007)
4. Ngo, A., Masel, D., Cahill, C., Blumberg, N., Refaai, M.A.: Blood banking and transfusion medicine challenges during the Covid-19 pandemic. Clin. Lab. Med. **40**(4), 587–601 (2020)
5. Melku, M., et al.: Knowledge, attitude and practice regarding blood donation among graduating undergraduate health science students at the university of Gondar, northwest Ethiopia. Ethiop. J. Health Sci. **28**(5) (2018)
6. Gunpinar, S., Centeno, G.: An integer programming approach to the bloodmobile routing problem. Transp. Res. Part E: Logist. Transp. Rev. **86**, 94–115 (2016)
7. Rezaei Kallaj, M., Abolghasemian, M., Moradi Pirbalouti, S., Sabk Ara, M., Pourghader Chobar, A.: Vehicle routing problem in relief supply under a crisis condition considering blood types. Math. Probl. Eng. **2021**(1), 7217182 (2021)
8. Şahinyazan, F.G., Kara, B.Y., Taner, M.R.: Selective vehicle routing for a mobile blood donation system. Eur. J. Oper. Res. **245**(1), 22–34 (2015)
9. Rabbani, M., Aghabegloo, M., Farrokhi-Asl, H.: Solving a bi-objective mathematical programming model for bloodmobiles location routing problem. Int. J. Ind. Eng. Comput. **8**(1), 19–32 (2017)
10. Liu, R., Jiang, Z.: A hybrid large-neighborhood search algorithm for the cumulative capacitated vehicle routing problem with time-window constraints. Appl. Soft Comput. **80**, 18–30 (2019)
11. Vincent, F.Y., Jodiawan, P., Gunawan, A.: An adaptive large neighborhood search for the green mixed fleet vehicle routing problem with realistic energy consumption and partial recharges. Appl. Soft Comput. **105**, 107251 (2021)
12. Vincent, F.Y., Jodiawan, P., Hou, M.L., Gunawan, A.: Design of a two-echelon freight distribution system in last-mile logistics considering covering locations and occasional drivers. Transp. Res. Part E: Logist. Transp. Rev. **154**, 102461 (2021)

13. Keskin, M., Çatay, B.: Partial recharge strategies for the electric vehicle routing problem with time windows. Transp. Res. Part C: Emerg. Technol. **65**, 111–127 (2016)
14. Schiffer, M., Walther, G.: An adaptive large neighborhood search for the location-routing problem with intra-route facilities. Transp. Sci. **52**(2), 331–352 (2018)
15. Santini, A.: An adaptive large neighbourhood search algorithm for the orienteering problem. Expert Syst. Appl. **123**, 154–167 (2019)
16. Hammami, F.: An efficient hybrid adaptive large neighborhood search method for the capacitated team orienteering problem. Expert Syst. Appl. **249**, 123561 (2024)
17. Ropke, S., Pisinger, D.: An adaptive large neighborhood search heuristic for the pickup and delivery problem with time windows. Transp. Sci. **40**(4), 455–472 (2006)

A Genetic Approach to the Operational Freight-on-Transit Problem

Corentin Juvigny[✉], Diego Delle Donne, and Laurent Alfandari

Department of Information Systems, Data Analytics and Operations, ESSEC
Business School, 3 Avenue Bernard Hirsch, 95021 Cergy, France
{juvigny,delledonne,alfandari}@essec.edu

Abstract. Last-mile delivery problems have become a subject of increasing academic study in recent years. This paper examines the *operational level Three-Tier Delivery Problem with Public Transportation* (3T-DPPT), which concerns the conveyance of customers' parcels from a warehouse, typically situated outside the city, to the customers in the city center, using public transport vehicles as an intermediate leg. The parcels are conveyed from the depot to public transport stops, then taken into the city center by public transportation vehicles, and finally delivered to the customers by freighters using green and lightweight means (or even walking). In this paper, we introduce a genetic algorithm (GA) to address the resolution of large-scale instances, which exact approaches in practice cannot tackle. We provide a detailed account of its encoding and the distinct genetic operators tailored for it. We undertake a comparative analysis of its performance vis-à-vis that of a compact mixed-integer linear programming formulation on a diverse array of instances of various sizes. The outcomes underscore the efficacy and robustness of the GA approach across different instance sizes, yielding solutions that are near the optimal ones in a relatively short span of time.

Keywords: Last-mile delivery · Freight-on-transit · Genetic algorithm · Mixed integer linear programming

1 Introduction

The development of e-commerce in recent years has caused logistical issues in retailers' supply chains to keep up with demand (Hossain et al. [10]). A particularly affected segment of this supply chain is the last-mile delivery (LMD) segment, which corresponds to the final stage of the distribution process, namely the delivery of goods from the last distribution center to the customer's doorstep. A set of distinctive characteristics distinguishes this segment of the supply chain. Indeed, this segment necessitates delivering a considerable number of small parcels to a vast customer base, each with a designated time window for receipt of their parcels. Moreover, given that the majority of customers reside in urban areas, the capacity of roadways is constrained. This is why a novel approach to last-mile delivery has emerged, capitalizing on the existing public transportation infrastructure. Such resources are typically underutilized throughout most of the day, as they are designed to meet the demands of peak-hour traffic. Different demonstration projects already exist across the world (Baron [1], Der Spiegel [24], Longhorn [16], Sustainable Bus [3], Deloison et al. [6]).

In this paper, we examine the three-echelon system, designated as 3T-DPPT, initially proposed by Mandal and Archetti [19]. In this system, the first echelon comprises trucks that deliver parcels from the last deposit area to city stops, which are often located outside the city limits. The second echelon involves public vehicles that carry dropped parcels into the city and discharge them at various stops. The third echelon encompasses the use of zero-emission vehicles, such as cargo bikes or autonomous robots, to retrieve parcels and deliver them to customer locations.

This paper makes a novel contribution to the field by proposing a new variant of the 3T-DPPT that does not consider freighter routes. Furthermore, it presents two resolution methods for this model: one based on linear programming and the other employing a genetic algorithm approach.

The structure of this paper is as follows. Firstly, a literature review is conducted. Secondly, the proposed new variant is detailed, along with its linear modeling. Thirdly, the genetic algorithm is described. Finally, both methods are assessed and compared on a set of instances, prior to the conclusion of the paper.

2 Literature Review

The idea of taking advantage of the spare capacity of public transport during non-pick hours has been studied in many papers. Kikuta et al. [15] carried out a pilot project to transport parcels in the metro instead of trucks and showed that this could reduce traffic jams and CO_2 emissions. Villa and Monzón [27], Zhou and Zhang [28] and Crepy [4] also studied the use of the metro for parcel delivery and showed its efficiency. Bruzzone et al. [2] presented an alternative using water-bus lines in the city of Venice, and He and Yang [8] analyzed the possibility of using urban buses in the Chinese city of Dalian, and showed that if the model accounts for excess carbon emissions and late-delivery penalties, their approach leads to reduce by 10% the cost and 13% the CO_2 emissions. Masson et al. [20] studied the problem of transferring goods from a fleet of buses to city freighters at bus stops. A variant of the pickup and delivery problem with transfers is used as a model for the corresponding optimization problem, which is solved using an adaptive large neighborhood search. The results are evaluated by calculating lower bounds using a column generation approach. Ghilas et al. [7] also used an adaptive large neighborhood search heuristic to solve a pickup and delivery problem with time windows and scheduled lines. They showed that they could handle up to 100 freight requests. Hörsting and Cleophas [9] employed a bi-objective model to minimize passenger inconvenience and delivery times for a system comprising one passenger tram line with shared capacity for parcel transport. The 3T-DPPT has been introduced by Mandal and Archetti [19].

They first proposed a compact formulation, which appears too large to be used in real-life instances. Thus, they divided the problem into three subproblems, each corresponding to an echelon of the problem. They introduced three heuristic algorithms, solving each subproblem in different orders. Delle Donne et al. [5] reformulated the previous compact formulation into an extended formulation, on which they applied a column generation approach to the first and third echelons. This new approach proved successful in finding solutions in instances where the previous one was not able to.

Metaheuristics are widely used approaches to solve large-scale, real-life-based applications (Talbi [25]), such as scheduling (Juvigny et al. [13]) or vehicle routing problems (Prodhon and Prins. [22]). A review of their applications in the Transit Route Network Design Problem was performed by Iliopoulou et al. [11]. Machado et al. [18] compared a metaheuristic approach with an exact ILP-based method and showed that both can be of interest as it can be easily predicted which of these algorithms should be employed regarding the instance characteristics. Sahli et al. [23] proposed a genetic algorithm for urban freight transport scheduling that uses passenger rail network capacities. In particular, they aimed to minimize the total waiting time of parcels at their departure stations. Their outcomes indicate that their approach significantly outperforms existing exact solvers. Vidal [26] proposed a hybrid genetic algorithm for the capacitated vehicle routing problem (CVRP) and showed that this approach stands as a leading metaheuristic, distinguished by its high solution quality, rapid convergence speed, and conceptual simplicity.

3 Problem Statement and Formulation

We propose and detail a new variant of the *Three-Tier Delivery Problem using Public Transportation* (3T-DPPT) first introduced by Mandal and Archetti [19]. Compared to the original version, we forbid indefinite waiting time of first-echelon trucks at transport stops, we consider only unitary size parcels, and we force third-echelon freighters to deliver a single customer (meaning they no longer make tours but only a round trip from the depot passing by the customer, as autonomous robots would do). We name this variant *Three-Tier Star Variant Delivery Problem using Public Transportation* (3T-SVDPPT). The motivations behind this variant are multifaceted. From practical considerations, freighters are often perceived as mechanized conveyors, delivery robots, or bicycles capable of transporting only a single parcel at a time. This aligns with the empirical observation that when freighter tours are permitted, the majority of these tours consist solely of roundtrips between a designated stop and a customer.

The objective of the problem is to deliver parcels from a unique Consolidation and Distribution Center (CDC) to a set of customers $\mathcal{C}$. We assume that every parcel has the same size, and each customer $i \in \mathcal{C}$ can only receive its good during a delivery time window $[\underline{T}_i, \overline{T}_i]$. All demands are known at the start of the problem. The delivery of a parcel is divided into three well-defined tiers.

In the first tier, a homogeneous fleet of trucks $\mathcal{D}$ is used to deliver parcels from the CDC, denoted as o, to a set of public transport stops $\mathcal{S}^{\text{IN}}$, called *drop-in stops*. Each truck has a carrying capacity of Q^1. Once dropped at a stop s, a parcel can only stay there for a time $\mathcal{W}_s$. Furthermore, it requires a service time T'_s for the truck to unload parcels at stop s, independently of the number of parcels unloaded. Hence, if a truck arrives at time t to the stop, the parcels it drops there can only be handled by public transport vehicles arriving at s between the times $t + T'_s$ and $t + T'_s + \mathcal{W}_s$. Considering $u, v \in \mathcal{S}^{\text{IN}} \cup \{o\}$, we note T_{uv} the time taken by delivery truck $d \in \mathcal{D}$ to travel from u to v and C^1_{uv} its associated cost. We assume that each truck tour is *elementary*, i.e., a single truck cannot visit each stop more than once, and *uninterrupted*, i.e., a truck leaves a stop as soon as it finishes unloading.

Table 1. Decision variables

Decision Variables	
Tier 1	
r_{isd}	equals 1 if the package for customer i is delivered by truck d to the drop-in stop s, 0 otherwise
w_{uvd}	equals 1 if a delivery truck d traverses arc (u, v), 0 otherwise
t_{ud}	time when delivery truck d leaves from node $u \in \mathcal{S}^{\text{IN}} \cup \{o\}$
Tier 2	
y^1_{isp}	equals 1 if the package for customer i is picked up by public vehicle p from drop-in stop s, 0 otherwise
y^2_{isp}	equals 1 if the package for customer i is dropped off by public vehicle p at drop-out stop s, 0 otherwise

In the second tier, public transport vehicles from a set $\mathcal{P}$ (for simplification, we henceforth assume it is a fleet of buses) carry the parcels from drop-in stops to drop-out stops. Each bus $p \in \mathcal{P}$ has a capacity Q^2_p and serves the stops $\mathcal{S}_p = (s^1_p, \ldots, s^{|\mathcal{S}_p|}_p)$. We note $\mathcal{S}^{\text{OUT}}$ the set of drop-out stops and $\mathcal{P}_s$ the set of buses stopping at stop s. We assume that $\mathcal{S}_p \subseteq \mathcal{S}^{\text{IN}} \cup \mathcal{S}^{\text{OUT}}$ and that they are distributed such that all the drop-in precede the drop-out stops. We denote t^s_p the arrival time of bus p at stop s (the time is scheduled and not subject to delay).

In the third tier, at each drop-out stop s, a set of freighters $\mathcal{F}_s$ waits to deliver the parcels to their final destination. The capacity of a stop s, i.e., the number of parcels that can be delivered from s, is Q^3_s. The delivery of a parcel can be started as soon as the parcel is dropped at a stop s after the service time T'_s and cannot be postponed more than $\mathcal{W}_s$ units of time. We note C^3_{is} the cost of delivering customer i's package. The time needed by a freighter to go from a drop-out stop $s \in \mathcal{S}^{\text{OUT}}$ to the customer's i location is known. Therefore, as freighters do not visit multiple customers, it is possible to pre-compute the subset $\mathcal{S}_i^{\text{OUT}}$ of drop-out stops which can be used to serve customer i, namely, those drop-out stops for which there exists at least one bus that can drop the parcel at a time which is compatible with the customer's time window (considering both the freighter travel time and the maximum waiting time $\mathcal{W}_s$ at the stop).

The objective of the 3T-SVDPPT is to find a distribution plan that guarantees the delivery of all parcels on time while minimizing the parcels' delivery costs. The latter is the sum of the route costs of the first and third-tier vehicles.

In order to model this problem using a Mixed-Integer Linear Programming (MILP) approach, it is necessary to introduce a set of decision variables for each problem tier. The decision variables of the initial tier shall model the truck tours, that is, determine which truck is responsible for delivering which parcel at which drop-in stop. The second tier shall indicate which bus loads the parcel and at which stop the latter is dropped off. As in the final tier, each freighter is to perform only a round trip and is to begin delivering as soon as the parcel is dropped off; thus, no decision is made by the freighter itself. Therefore, knowledge of the drop-out stops is sufficient to evaluate the cost of the final tier. This model is based on the one of Mandal and Archetti [19].

We detail the decision variables of the MILP in Table 1. A reminder of the sets and the parameters can be found in the appendix (Table 7) According to these notations, the complete modeling of the 3T-SVDPPT problem by a MILP is expressed as follows.

$$\sum_{s \in \mathcal{S}_{in}} \sum_{d \in \mathcal{D}} r_{isd} = 1 \quad \forall i \in \mathcal{C} \tag{1}$$

$$\sum_{i \in \mathcal{C}} \sum_{s \in \mathcal{S}^{\text{IN}}} r_{isd} \leq Q_d^1 \quad \forall d \in \mathcal{D} \tag{2}$$

$$\sum_{v \in \mathcal{S}_{in} \cup \{o\}} w_{ovd} = 1 \quad \forall d \in \mathcal{D} \tag{3}$$

$$\sum_{v \in \mathcal{S}_{in} \cup \{o\}} w_{vod} = 1 \quad \forall d \in \mathcal{D} \tag{4}$$

Constraints (1) ensure that each customer's package is assigned to one delivery truck that delivers it to a drop-in point. Constraints (2) model the truck maximum load capacity. Constraints (3) and (4) ensure that trucks start and end at the CDC.

$$\sum_{v \in \mathcal{S}_{in} \cup \{o\}, v \neq u} w_{uvd} = \sum_{v \in \mathcal{S}_{in} \cup \{o\}, v \neq u} w_{vud} \quad \forall u \in \mathcal{S}^{\text{IN}}, d \in \mathcal{D} \tag{5}$$

$$\sum_{v \in \mathcal{S}_{in} \cup \{o\}, v \neq u} w_{uvd} \geq \frac{1}{|C|} \sum_{i \in \mathcal{C}} r_{iud} \quad \forall u \in \mathcal{S}^{\text{IN}}, d \in \mathcal{D} \tag{6}$$

Constraints (5) are the flow conservation constraints. Constraints (6) guarantee that if a package is assigned to a drop-in via a delivery truck, the latter must visit the stop.

$$\sum_{d \in \mathcal{D}} r_{isd} = \sum_{p \in \mathcal{P}_s} y^1_{isp} \qquad \forall i \in \mathcal{C}, s \in \mathcal{S}^{\text{IN}} \tag{7}$$

$$t_{vd} \geq t_{ud} + T_{uv} + T'_v - M(1 - w_{uvd}) \qquad \forall u \in \mathcal{S}^{\text{IN}} \cup \{o\}, v \in \mathcal{S}^{\text{IN}}, u \neq v, d \in \mathcal{D} \tag{8}$$

$$t_{vd} \leq t_{ud} + T_{uv} + T'_v + M(1 - w_{uvd}) \qquad \forall u \in \mathcal{S}^{\text{IN}} \cup \{o\}, v \in \mathcal{S}^{\text{IN}}, u \neq v, d \in \mathcal{D} \tag{9}$$

Constraints (7) link the delivery trucks and the public buses, by forcing a public vehicle to pick up a customer's package left at a drop-in by a truck. Constraints (8) and (9) control the date time for each delivery truck as it visits the drop-in stop, and forces them to leave directly after dropping.

$$t_{sd} \leq \sum_{p \in \mathcal{P}_s} T_{sp} y^1_{isp} + M(1 - r_{isd}) \qquad \forall i \in \mathcal{C}, s \in \mathcal{S}^{\text{IN}}, d \in \mathcal{D} \tag{10}$$

$$\sum_{p \in \mathcal{P}_s} T_{sp} y^1_{isp} - t_{sd} \leq W_s + M(1 - r_{isd}) \qquad \forall i \in \mathcal{C}, s \in \mathcal{S}^{\text{IN}}, d \in \mathcal{D} \tag{11}$$

Constraints (10) ensure that a truck must arrive at a drop-in before the scheduled time of the public vehicle if it transports a package that must be picked up by the vehicle. Constraints (11) give a limit to the amount of time a package can wait at a drop-in stop.

$$\sum_{p \in \mathcal{P}} \sum_{s \in \mathcal{S}_p \cap \mathcal{S}_{in}} y^1_{isp} = 1 \qquad \forall i \in \mathcal{C} \tag{12}$$

$$\sum_{p \in \mathcal{P}} \sum_{s \in \mathcal{S}_p \cap \mathcal{S}^{out}_i} y^2_{isp} = 1 \qquad \forall i \in \mathcal{C} \tag{13}$$

$$\sum_{s \in \mathcal{S}_p \cap \mathcal{S}_{in}} y^1_{isp} = \sum_{s \in \mathcal{S}_p \cap \mathcal{S}^{out}_i} y^2_{isp} \qquad \forall i \in \mathcal{C}, p \in \mathcal{P} \tag{14}$$

$$\sum_{s \in \mathcal{S}_p \cap \mathcal{S}_{in}} \sum_{i \in \mathcal{C}_s} y^1_{isp} \leq Q^2_p \qquad \forall p \in \mathcal{P} \tag{15}$$

$$\sum_{i \in \mathcal{C}_s} \sum_{p \in \mathcal{P}_s} y^2_{isp} \leq Q^3_s \qquad \forall s \in \mathcal{S}^{\text{OUT}} \tag{16}$$

Constraints (12) and (13) ensure that one package is only loaded up (resp. discharged) in one drop-in (resp. drop-out). Constraints (14) guarantee that the vehicle picking up a package is the one dropping it off. Constraints (15) prevent public vehicle overloading. Constraints (16) limit the number of customers served from a drop-out stop (it can be viewed as the number of freighters available at this stop).

$$y_{isp}^2 = 0 \quad \forall i \in \mathcal{C}, \forall s \in \mathcal{S}^{\text{OUT}}, \forall p \in \mathcal{P} \setminus \mathcal{P}_{is} \quad (17)$$

where $\forall i \in \mathcal{C}, \forall s \in \mathcal{S}^{\text{OUT}}$, $\mathcal{P}_{is}$ is the set of buses that can deliver the parcel of customer i to stop s according to the time window of the customer (considering also the maximum waiting time $\mathcal{W}_s$ of a parcel at stop s). Constraints (17) forbid a customer from being delivered from an invalid drop-out stop.

$$\min \sum_{d \in \mathcal{D}} \sum_{u \in \mathcal{S}^{\text{IN}} \cup \{o\}} \sum_{\substack{v \in \mathcal{S}^{\text{IN}} \cup \{o\} \\ u \neq v}} C_{uv}^1 w_{uvd} + \sum_{i \in \mathcal{C}} \sum_{s \in \mathcal{S}^{\text{OUT}}} \sum_{p \in \mathcal{P}_s} C_{is}^3 y_{isp}^2 \quad (18)$$

Finally, the objective function (18) of the MILP is to minimize the sum of both the truck tours and the freighters' round trips.

4 A Genetic Algorithm

This section presents the genetic approach to compute an upper bound of the 3T-SVDPPT. Note that this approach can be easily modified to fit the previously proposed variants.

This section's outline is as follows: First, we present the general idea behind our approach. Then, we detail the genetic encoding representing a solution, followed by the genetic operators tailored for the latter. After this, we describe the greedy algorithm employed to compute the first leg.

4.1 General Idea

This genetic approach hinges on the idea that the problem can be divided into three parts. The first, henceforth denoted as the first leg, consists of finding the best routes for the trucks to deliver the customers' parcels from the depot to the different drop-in stops where they are loaded into public transport. The second henceforth dubbed the second leg, consists of assigning a parcel to a bus. The last leg is, therefore, to assign a drop-out stop to each parcel.

Considering that, we propose the following approach. A genetic algorithm is run by individuals representing valid solutions for the second and third legs. The first leg is computed only when the individual needs to be evaluated, and a greedy algorithm performs this operation. More precisely, the encoding of an individual encompasses the information for each customer on which bus and drop-out stop its parcel is assigned. Therefore, the greedy algorithm must determine which truck delivers the parcel and at which drop-in stop, considering the time constraints induced by the previous choice of buses. The capacity constraints of buses and drop-out stops are checked by the genetic algorithm, while those for trucks are assessed through the implementation of a greedy algorithm.

4.2 Encoding

For each customer, we pre-compute the possible pair of (buses/drop-out) stops that are compatible with the customer's time window. These tuples are stored in a structure called PS_{OUT}, depicted by Fig. 1. An individual is made up of a string ζ of which the value of the ith element corresponds to the index of a tuple in $PS_{\text{OUT}}[i]$, i.e. a valid pair (bus/drop-out) of the customer i. Thus, this string's size equals the number of customers. Note that to be valid, the inequality $0 \leq \zeta[i] < |PS_{\text{OUT}}[i]|$ must be satisfied.

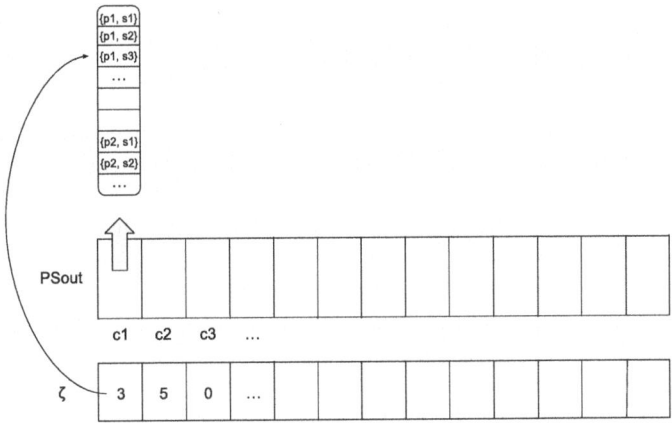

Fig. 1. Genetic Encoding Scheme

4.3 Genetic Operators

In a GA, a mating pool comprising individuals of the initial population (in which each individual is a valid solution to the studied problem) is formed by a selection operator. Crossover and mutation operators are then applied to them to breed a new population, the offspring. Subsequently, the latter and the initial population are merged following a replacement strategy to form a new population, and so forth until a stopping criterion is satisfied.

Initial Population. The initial population is generated randomly such that each chromosome refers to a valid tuple in PS_{OUT}, i.e. holds:

$$\forall i \in \mathcal{C}, 0 \leq \zeta[i] < |PS_{\text{OUT}}[i]|$$

Selection. A deterministic tournament selection (Miller and Goldberg [21]) is employed to select the individuals retained in the mating pool. This process exerts selective pressure upon a population of k competitors chosen randomly in the initial population. Then, the individual with the best fitness is integrated into the mating pool. This process is repeated until the mating pool's size reaches the original population's size. This average mating pool is therefore higher than the average population fitness. The selection in question is not applied to the entire population; rather, it is applied to a restricted population in which each individual has a reproductive probability $\rho_r \in [0, 1]$ to be included.

Mutation. The mutation operator is a k-mutations that works as follows. Considering an individual i, it selects k chromosomes, $k \leq |\mathcal{C}|/2$. Then, the values of the selected chromosomes are randomly redrawn to new possible ones.

Crossover. Two crossover operators are employed. The first is a swap operator. Let's consider two individuals α_1 and α_2. The swap operator selects an index k, then swaps the values α_1^k and α_2^k. The second operator is a TwoPoint crossover. The latter draws two indices j_1 and j_2 such that $0 \leq j_1 \leq j_2 < |\mathcal{C}|$. Then, for all indices k such that $j_1 \leq k \leq j_2$, the values α_1^k and α_2^k are swapped.

Replacement Strategy. Different replacement strategies have been considered. After tuning, we chose to use a global stochastic replacement strategy. The latter first adds parents to the offspring, then truncates using a global stochastic tournament with sort. A score is attributed to each individual corresponding to the number of duels it wins in the tournament, duels disputed against randomly chosen opponents. Then, we keep the p individuals with the best scores, where p is the initial population size. Such an approach provides softer selective pressure than pure truncate.

Terminate Criteria. The GA terminates when one of the following criteria is fulfilled; either more than 5000 generations are bred, or the time limit of half an hour is reached, or the fitness of the best individual does not improve for 50 consecutive generations (this criterion is not taken into account before the 250th generations, it can only stop the GA after the 300th generations).

Evaluation of an Individual. The cost of an individual corresponds to the cost of the solution it encodes. Therefore, it comprises the cost of the first leg's truck tours and the third leg's freighter rotations. While the latter can easily be retrieved from PS_{OUT}, the former must be calculated with the information furnished by the encoding. Two options are conceivable; the first would be to find the optimal truck tours by running a MILP that can be obtained from the MILP presented in Sect. 3. However, because of the complexity of it, we propose a heuristic approach based on a greedy approach that is further detailed in the next paragraph.

4.4 Greedy Approach for the First Leg

For the purposes of this discussion, let us consider a valid individual. Its chromosomes dictate the bus that will handle its parcel for each customer. Consequently, to assess the individual's cost, it is necessary to determine the location at which the parcel is to be unloaded for each customer and construct a route for the transportation of the parcels from the warehouse to the designated drop-in stops. This problem is NP-hard. In light of the aforementioned considerations, we put forth a heuristic approach that concurrently constructs the tour and determines the drop-in locations. The heuristic operates as follows.

Firstly, the costless drop-in stop assigned to each parcel is calculated. Then, the parcels are sorted in an unprocessed list in ascending order according to the delivery time required at their selected drop-in stop. After that, while there are still unprocessed parcels, the first remaining customer is selected and assigned to one of the available trucks in the depot. Subsequently, the latter commences its itinerary by proceeding to the customer's nearest drop-in stop from the depot. The time elapsed and overall cost are updated at each process stage. After that, it is determined whether any additional parcels can be delivered to the stop at the time of the truck's arrival. If this is the case, the items are added to the truckload, provided that the vehicle's maximum capacity is not exceeded. Then, suppose the vehicle is not at its maximum capacity. In that case, an examination is conducted among the remaining packages to ascertain which is the least costly and most readily accessible, and which drop-in stop it corresponds to. Once determined, this parcel is regarded as having been loaded into the truck from the depot, and the truck moves to the corresponding stop. The process is repeated until either the truck is full or no further parcels can be delivered within the required time windows and travel times. Subsequently, the truck returns to the depot and a new tour is generated. It should be noted that this new truck is independent of the former one; it does not need to wait for the previous truck to commence its tour. If the maximum number of trucks is reached while parcels remain to be assigned, the individual is considered invalid, i.e., encoding an infeasible solution.

5 Experiments

5.1 Protocol

The proposed algorithms were coded in C++20 using the framework ParadisEO [14] for the GA and the solver CPLEX 22.1.1 for the MILP. All runs were conducted on a computer with an Apple(R) M3 Pro and 18 GB of RAM. We used iRace [17] to tune the parameters of the GA. The tuned values are reported in Table 2.

We compare the computation times of each of the proposed methods, i.e., the CPLEX implementation of the MILP and the GA, in instances[1] of various sizes, as well as the gain of the GA over the MILP approach (given by formula (19)).

$$\delta := \frac{cost_{MILP} - cost_{GA}}{cost_{MILP}} \quad (19)$$

A positive value means the GA outperforms the MILP in the instance studied. An infinite symbol (∞) implies that the MILP did not find an integral solution,

---
[1] the dataset can be retrieved here: Juvigny et al. [12].

Table 2. Tuning parameters

Parameter	Value
Size of population	5000
Tournament selection size	2
Reproductive probability	0.85
Mutation probability	0.8
Crossover probability	0.75
Cross operator selection	0.75

while the GA did. The characteristics of each set of instances are reported in Table 3. All instances of each set (i.e., each column in this table) share the same properties. Computation times are reported in seconds. As the GA approach employs randomness, we launched 30 runs per instance. Hence, computation times and costs of the GA are the average times and costs over these 30 runs.

Table 3. Characteristics of instances

	B_1	B_2	B_3	B_4	C_1	C_2	C_3	C_4	C_5
$|\mathcal{C}|$	10	10	15	15	20	25	30	40	50
$|\mathcal{P}|$	15	30	30	45	45	45	60	60	60
$|\mathcal{S}|$	6	8	12	30	23	23	28	28	28

We consider two different time limits for the MILP runs. The *long-runs* have a time limit of 1800 seconds, while *short-runs* have a different time limit for each set of instances, which is equal to the average time of GA in the instance of this set. This allows us to assess the quality of the solutions yielded by the MILP in different contexts (and to better compare these values against those of the GA approach). This choice is driven by an operational perspective, in which a decision must be returned quickly.

5.2 Results

Table 4 shows the results obtained in the smaller sets of instances, namely B_1, B_2, and B_3. For each instance from each set, we report the average gain of the GA over the MILP approach (as stated in Eq. (19) and denoted here simply as δ) and the time taken by the latter (both for short and long runs of the MILP). The last two columns of the table report the average time and standard deviation of the solutions of the GA over the 30 runs on each instance. Table 5 and Table 6 depict the same information but for medium and large instances.

Table 4. Results on small instances. The first column shows the gap in objective value δ (in %) and computation time (in seconds) of the MILP long-run vs the GA, while the second displays them for MILP short-runs

Instances		MILP-long		MILP-short		GA	
		δ	time	δ	time	time	stdev(σ)
B_1	1	0.00	0.28	0.00	0.28	13.57	0.00
	2	0.00	0.88	0.00	0.88	13.79	0.00
	3	0.00	0.52	0.00	0.52	13.43	0.00
	4	−16.03	0.73	−16.03	0.73	13.30	0.00
	5	0.00	0.52	0.00	0.52	13.63	0.00
B_2	1	−12.76	1.01	−12.76	1.01	11.29	0.00
	2	−12.12	0.17	−12.12	0.17	13.82	0.00
	3	0.00	0.40	0.00	0.40	10.35	0.00
	4	−0.42	0.49	−0.42	0.49	15.96	0.00
	5	−5.75	0.41	−5.75	0.41	14.01	2.31
B_3	1	−4.50	1.52	−4.50	1.52	13.77	0.00
	2	−6.14	7.21	−6.14	7.21	20.43	0.00
	3	−5.73	6.89	−5.73	6.89	20.65	0.00
	4	−1.88	4.83	−1.88	4.83	20.58	0.00
	5	−0.17	5.70	−0.17	5.70	18.32	0.00
B_4	1	−19.89	1800	**2.15**	25	21.43	0.88
	2	**3.43**	1800	**4.34**	25	22.51	0.00
	3	−24.36	1800	**23.31**	25	21.25	1.46
	4	**9.99**	1800	**13.46**	25	31.15	0.00

A bold value means the GA outperforms the MILP

As evidenced in Table 4, in small instances, the MILP always beats the GA metaheuristic, as the former finds the optimal value in a very short amount of time. However, the bigger the instance, the better the GA solution is, compared to the MILP one (see Table 5). On most of the bigger instances, the MILP is not able to return any valid integral solution, while the GA always delivers these (see Table 6, instance sets C_4 and C_5). For the subset of instances where the MILP does find a feasible solution, the GA performance is on average only 3.36% lower than the long-run MILP (the worst behavior being mainly for small instances), but with a huge gap in computation times. Indeed, the GA always returns a solution in less than one minute. When limited to this amount of time, the MILP struggles to find a good solution, as the short-run outcomes expose. The GA outperforms the MILP by 7.77% in that case, if one excludes instances on which the MILP does not find a solution.

Table 5. Results on medium instances. The first column shows the gap in objective value δ (in %) and computation time (in seconds) of the MILP long-run vs the GA, while the second displays them for MILP short-runs

Instances		MILP-long		MILP-short		GA	
		δ	time	δ	time	time	stdev(σ)
C_1	0	−1.92	1800	−1.53	25	23.05	0.00
	1	−1.32	1800	**7.18**	25	20.19	0.28
	2	−8.25	1800	−7.05	25	21.38	0.33
	3	−19.18	1800	−2.49	25	27.46	0.04
	4	−3.90	1800	**3.72**	25	25.53	1.39
	5	−3.06	1800	**2.62**	25	22.92	2.76
	6	−11.29	1800	−4.41	25	24.30	1.47
	7	−3.91	1800	**1.17**	25	24.51	0.26
	8	−9.56	1800	−2.37	25	29.41	0.00
	9	0.00	1800	**3.62**	25	23.86	0.00
C_2	0	**0.76**	1800	**3.41**	30	32.43	0.48
	1	−1.03	1800	**14.02**	30	28.95	0.72
	2	−2.13	1800	**4.20**	30	28.67	0.72
	3	−4.48	1800	**9.39**	30	28.83	0.28
	4	−3.75	1800	**0.46**	30	28.88	1.51
	5	−5.82	1800	**3.03**	30	27.64	3.41
	6	−5.67	1800	**17.22**	30	30.84	2.44
	7	−2.63	1800	**4.35**	30	32.69	1.66
	8	−1.76	1800	−9.37	30	31.03	2.64
	9	**2.57**	1800	**3.05**	30	27.38	0.00
C_3	0	−1.32	1800	−0.33	40	40.40	2.24
	1	−1.32	1800	**22.76**	40	34.29	2.24
	2	**0.90**	1800	**10.20**	40	37.16	1.52
	3	**2.39**	1800	∞	40	35.86	1.03
	4	−0.38	1800	∞	40	35.17	1.01
	5	**0.11**	1800	∞	40	36.47	2.53
	6	−7.08	1800	∞	40	42.99	0.80
	7	−0.67	1800	**0.47**	40	38.65	0.37
	8	−2.92	1800	∞	40	35.12	0.61
	9	−0.25	1800	**3.44**	40	41.38	0.16

A bold value means the GA outperforms the MILP

Our results imply that when computation times are a critical issue (for example, if the model should be used daily by a stakeholder in need of fast decision-making), then the GA is the best option in all cases. Otherwise, the choice

Table 6. Results on large instances. The first column shows the gap in objective value δ (in %) and computation time (in seconds) of the MILP long-run vs the GA, while the second displays them for MILP short-runs.

Instances		MILP-long		MILP-short		GA	
		δ	time	δ	time	time	stdev(σ)
C_4	0	−4.17	1800	**22.81**	50	48.46	1.74
	1	∞	1800	∞	50	48.19	−
	2	∞	1800	∞	50	50.89	−
	3	∞	1800	∞	50	52.77	−
	4	−0.05	1800	**33.95**	50	51.76	2.33
	5	**1.20**	1800	**49.57**	50	47.84	1.53
	6	∞	1800	∞	50	52.69	−
	7	−3.72	1800	∞	50	52.53	1.07
	8	∞	1800	∞	50	48.34	−
	9	**6.61**	1800	**30.83**	50	53.53	2.85
C_5	0	**1.31**	1800	**6.56**	60	53.94	4.33
	1	∞	1800	∞	60	48.47	−
	2	∞	1800	∞	60	77.09	−
	3	∞	1800	∞	60	64.05	−
	4	∞	1800	∞	60	57.31	−
	5	∞	1800	∞	60	56.37	−
	6	∞	1800	∞	60	68.30	−
	7	∞	1800	∞	60	61.66	−
	8	∞	1800	∞	60	49.73	−
	9	∞	1800	∞	60	65.32	−

A bold value means the GA outperforms the MILP

between the two models should be made regarding the cardinality of customers, buses, and stops in the instance. The GA also shows a good robustness property, with an average standard deviation on the obtained solution of only 1.06%. However, in several instances, the GA standard deviation is low while the optimal solution is not found. This suggests that the GA can be prone to premature convergence. More developments will be carried out to increase diversification.

6 Conclusion

The objective of this study was to examine the potential of last-mile delivery services to leverage public transportation's spare capacity during off-peak hours, reducing the burden of delivery trucks on urban roadways. In particular, we proposed a new genetic algorithm and its associated encoding to solve a variant of the problem family, the 3T-SVDPPT. We compared it to a mixed-integer linear program. The latter ensures the optimality of the returned value (unless it reaches the time limit) while the former is much quicker. The results demonstrate that the GA provides solutions that are close to optimal while maintaining efficient computation times and a good robustness property. In subsequent research, our objective is to examine the potential advantages of integrating these two approaches. Indeed, the optimal solution identified by the GA can be provided as a preliminary solution to the MILP. Furthermore, a preliminary version of the MILP could be utilized as a substitute for the greedy evaluation of the initial tier in the GA, potentially leading to enhanced solutions. However, due to the considerable computational time required by this approach, it may be more advantageous to only replace the greedy algorithm at the conclusion of the GA, to enhance the truck tours of the best individual identified.

Acknowledgments and Disclosure of Funding. This work was partially funded by CY Initiative of Excellence, France (grant "Investissements d'Avenir" ANR-16-IDEX-0008").

Appendix

This appendix contains the sets and parameters (Table 7) that are incorporated into the MILP formulation.

Table 7. Sets & Parameters

Sets	
o	the CDC
$\mathcal{C}$	set of customers
$\mathcal{D}$	set of deliver trucks
$\mathcal{S}$	set of all stops
$\mathcal{S}^{\text{IN}}$	set of *drop-in* stops
$\mathcal{S}^{\text{OUT}}$	set of *drop-out* stops
$\mathcal{S}_i^{\text{IN}}$	set of *drop-in* stops from where customer i's package can be loaded
$\mathcal{S}_i^{\text{OUT}}$	set of *drop-out* stops from where customer i can be served
$\mathcal{P}$	set of public transportation vehicles (buses)
$\mathcal{P}_s$	set of vehicles that stop at stop s
Parameters	
C_{uv}^1	cost of traversing arc (u,v) by a delivery truck
C_{is}^3	cost of delivering customer i's package from stop s and going back to the stop
T_{sp}	time when public vehicle p is scheduled to reach stop s
Q_d^1	capacity of delivery truck $d \in \mathcal{D}$
Q_p^2	capacity of public vehicle $p \in \mathcal{P}$
Q_s^3	number of freighters at drop-out stop $s \in \mathcal{S}$
T_{uv}	time taken by delivery trucks to traverse arc (u,v)
$\overline{T}_i$	latest time of customer i to be served
$\underline{T}_i$	earliest time of customer i to be served
T'_s	service time required at stop s
$\mathcal{W}_s$	maximum time a package can be left at stop s

References

1. Baron, E.: Amazon looks to turn public buses into mobile delivery stations. Accessed **1**, 2021 (2019)
2. Bruzzone, F., Cavallaro, F., Nocera, S.: The integration of passenger and freight transport for first-last mile operations. Transp. Policy **100**, 31–48 (2021)
3. Bus, S.: What if trams were also to deliver goods? Regiokargo project launched in karlsruhe. Sustainable Bus (2021)
4. Crepy, M.M.E.A.: Leveraging public transit for robust last-mile distribution. Ph.D. thesis, Massachusetts Institute of Technology (2020)
5. Delle Donne, D., Santini, A., Archetti, C.: Integrating public transport in sustainable last-mile delivery: Column generation approaches. Eur. J. Oper. Res. (2025)
6. Deloison, T., et al.: The future of the last-mile ecosystem: Transition roadmaps for public-and private-sector players. World Econ. Forum (2020)
7. Ghilas, V., Demir, E., Van Woensel, T.: An adaptive large neighborhood search heuristic for the pickup and delivery problem with time windows and scheduled lines. Comput. Oper. Res. **72**, 12–30 (2016)
8. He, Y., Yang, Z.: Parcel delivery by collaborative use of truck fleets and bus-transit vehicles. Transp. J. **57**(4), 399–428 (2018)
9. Hörsting, L., Cleophas, C.: Scheduling shared passenger and freight transport on a fixed infrastructure. Eur. J. Oper. Res. **306**(3), 1158–1169 (2023)
10. Hossain, M.B., Wicaksono, T., Nor, K.M., Dunay, A., Illes, C.B.: E-commerce adoption of small and medium-sized enterprises during COVID-19 pandemic: Evidence from south Asian countries. J. Asian Finance, Econ. Bus. **9**(1), 291–298 (2022)
11. Iliopoulou, C., Kepaptsoglou, K., Vlahogianni, E.: Metaheuristics for the transit route network design problem: a review and comparative analysis. Public Transport **11**, 487–521 (2019)
12. Juvigny, C.: Replication Data for: Operational Freight-on-Transit problem (2025). https://doi.org/10.7910/DVN/SB4IMT
13. Juvigny, C., Baste, J., Lozenguez, G., Doniec, A., Jourdan, L.: A hybrid genetic approach for bi-level flexible job shops arising from selective deconstruction. In: 2023 IEEE Congress on Evolutionary Computation (CEC), pp. 1–8. IEEE (2023)
14. Keijzer, M., Merelo, J.J., Romero, G., Schoenauer, M.: Evolving objects: a general purpose evolutionary computation library. Artif. Evol. **2310**, 829–888 (2002). http://www.lri.fr/~marc/EO/EO-EA01.ps.gz
15. Kikuta, J., Ito, T., Tomiyama, I., Yamamoto, S., Yamada, T.: New subway-integrated city logistics Szystem. Procedia. Soc. Behav. Sci. **39**, 476–489 (2012)
16. Longhorn, D.: Passenger trains converted to deliver parcels to city centres. rail business daily. Rail Business Daily (2021)
17. López-Ibáñez, M., Dubois-Lacoste, J., Pérez Cáceres, L., Birattari, M., Stützle, T.: The irace package: iterated racing for automatic algorithm configuration. Oper. Res. Perspect. **3**, 43–58 (2016). https://doi.org/10.1016/j.orp.2016.09.002
18. Machado, B., Pimentel, C., de Sousa, A.: Integration planning of freight deliveries into passenger bus networks: exact and heuristic algorithms. Transp. Res. Part A: Policy Pract. **171**, 103645 (2023)
19. Mandal, M.P., Archetti, C.: A decomposition approach to last mile delivery using public transportation systems. arXiv preprint arXiv:2306.04219 (2023)
20. Masson, R., Trentini, A., Lehuédé, F., Malhéné, N., Péton, O., Tlahig, H.: Optimization of a city logistics transportation system with mixed passengers and goods. EURO J. Transp. Logist. **6**(1), 81–109 (2017)

21. Miller, B.L., Goldberg, D.E., et al.: Genetic algorithms, tournament selection, and the effects of noise. Complex Syst. **9**(3), 193–212 (1995)
22. Prodhon, C., Prins, C.: Metaheuristics for Vehicle Routing Problems, pp. 407–437. Springer International Publishing, Cham (2016).https://doi.org/10.1007/978-3-319-45403-0_15
23. Sahli, A., Behiri, W., Belmokhtar-Berraf, S., Chu, C.: An effective and robust genetic algorithm for urban freight transport scheduling using passenger rail network. Comput. Ind. Eng. **173**, 108645 (2022)
24. Spiegel, D.: Scheuer will paketauslieferung per u-bahn testen. Accessed (2020)
25. Talbi, E.: Metaheuristics: from design to implementation. John Wiley Sons Google Schola **2**, 268–308 (2009)
26. Vidal, T.: Hybrid genetic search for the CVRP: open-source implementation and swap* neighborhood. Comput. Oper. Res. **140**, 105643 (2022)
27. Villa, R., Monzón, A.: A metro-based system as sustainable alternative for urban logistics in the era of e-commerce. Sustainability **13**(8), 4479 (2021)
28. Zhou, F., Zhang, J.: Freight transport mode based on public transport: Taking parcel delivery by subway as an example. In: Sixth International Conference on Transportation Engineering, pp. 745–754. American Society of Civil Engineers Reston, VA (2019)

LON/D — Sub-problem Landscape Analysis in Decomposition-Based Multi-objective Optimization

Arnaud Liefooghe[1](✉), Gabriela Ochoa[2], and Sébastien Verel[1]

[1] LISIC, Université du Littoral Côte d'Opale, Calais, France
{arnaud.liefooghe,sebastien.verel}@univ-littoral.fr
[2] University of Stirling, Scotland, UK
gabriela.ochoa@stir.ac.uk

Abstract. We explore the underlying difficulties of sub-problems arising from decomposition in multi-objective optimization. Decomposition algorithms, such as MOEA/D, split the original multi-objective problem into a set of single-objective sub-problems using a scalarizing function. A weighting coefficient vector defines each sub-problem. We examine the relative difficulty of these sub-problems based on their weight vector and the chosen scalar function—either weighted sum or weighted Tchebycheff. Our approach involves creating a landscape for each sub-problem and analyzing its local optima network (LON). We contribute by jointly visualizing the LONs of sub-problems, defining LON features for decomposition, and examining their interaction with problem properties and their impact on algorithm performance. An extensive experimental analysis of bi-objective NK-landscapes reveals that landscape properties depend not only on the weight vector and scalar function but also on the objectives' intrinsic difficulty and their degree of conflict. These factors directly affect the relative performance of MOEA/D for each sub-problem. Among the landscape features explored, the size of each sub-problem's global optimum basin of attraction showed the strongest impact on the performance of decomposition-based multi-objective optimization.

Keywords: Multi-objective combinatorial optimization · Decomposition · MOEA/D · Landscape analysis · Local optima network · ρmnk-landscapes

1 Introduction

Decomposition has emerged as an important paradigm for approximating the Pareto set in evolutionary multi-objective optimization [13,23,26]. It breaks down the multi-objective optimization problem into a number of scalar (single-objective) sub-problems, each targeting a different region of the Pareto front. These sub-problems are solved simultaneously and cooperatively. Sub-problems are defined by a scalarizing function, with each sub-problem having a particular weight vector that reflects the relative importance of the objectives. Decomposition relies on two concurrent assumptions: (1) that a diverse set of weight vectors translates into good coverage of the Pareto front (diversity), and (2) that the search process can find a high-quality solution for each sub-problem (convergence). We challenge this second assumption by examining the landscape of each individual sub-problem defined by decomposition.

This paper aims to clarify the interplay between decomposition and sub-problem solving. We specifically examine how the scalarizing function and weight vector affect the search difficulty of sub-problems in decomposition-based algorithms —like MOEA/D [26]— for multi-objective (combinatorial) optimization. Our contributions are as follows:

(1) We define a local optima network (LON) model for decomposition (called LON/D) to characterize the landscape of sub-problems (Sect. 3).
(2) We critically assess the visualization of LON/D on selected example problems under different layouts, highlighting their key characteristics (Sect. 3).
(3) We introduce numerical features extracted from the LON/D and explore how they change with respect to sub-problem definitions (Sect. 4).
(4) We experiment with MOEA/D and relate its performance to LON/D features, identifying key factors that explain successful solving for each individual sub-problem (Sect. 5).

We focus on small problems (NK-landscapes) to provide an unbiased picture of the landscape through full enumeration. By limiting our study to two objectives, we enable easier visualization, thus enhancing interpretability and explainability.

Several studies pointed out the varying levels of intrinsic difficulty among sub-problems; see, e.g., [12,21,27]. However, these studies primarily adopt a solving perspective, focusing on improving search efficiency by adjusting resource allocation to sub-problems. As a result, it is still unclear why these differences in difficulty occur in terms of fundamental landscape properties induced by scalarizing functions and their settings. An early investigation on the distribution of (scalar) local optima for multi-objective traveling salesperson problems shows global convexity and Pareto front coverage through scalarization parameters [4]. Later studies found that scalarizing function contour lines and settings affect search behavior and attraction points when solving sub-problems [8]. In [6,7], decomposition is used to define sub-problem landscapes, but scalar landscape features are then aggregated to characterize the global multi-objective landscape rather than individual sub-problems. These aforementioned research papers neither use LONs nor provide landscape visualization support. A notable exception is [22], which examines the LON of individual objectives and their connection to the Pareto set, though it does not explore the landscapes of scalar sub-problems.

2 Background

2.1 Multi-objective Optimization

A multi-objective optimization problem aims to maximize an m-dimensional objective function vector $F\colon X \to Z$. Each solution $x \in X$ maps to a vector $z = F(x)$ in the objective space $Z \subseteq \mathbb{R}^m$. Given two objective vectors $z, z' \in Z$, z is dominated by z' if there exists a $j \in \{1, \ldots, m\}$ such that $z_j < z'_j$, and $z_i \leqslant z'_i$ for all $i \in \{1, \ldots, m\}$. A solution $x \in X$ is dominated by $x' \in X$ if $F(x)$ is dominated by $F(x')$. An objective vector $z^\star \in Z$ is *non-dominated* if no other $z \in Z$ dominates it. A *Pareto optimal* (po) solution $x^\star \in X$ is one where $F(x^\star)$ is non-dominated. The *Pareto set* comprises all Pareto optimal solutions, while the *Pareto front* is the mapping of this set in the objective space.

2.2 Decomposition

A multi-objective optimization problem can be *decomposed* into multiple single-objective sub-problems, each targeting different regions of the Pareto front. MOEA/D has emerged as one of the most popular multi-objective approaches based on decomposition [26]. Each sub-problem is defined using a specific weight vector, which determines the relative importance of objectives and is used within a parameterized scalarizing function. Various scalarizing functions can be used, with the weighted sum (WS) and weighted Tchebycheff (WT)[1] functions being common choices:

$$f_{\text{WS}}(x \mid w) = \sum_{i \in \{1, \ldots, m\}} w_i \cdot F_i(x) \qquad (1)$$

$$f_{\text{WT}}(x \mid w) = -\max_{i \in \{1, \ldots, m\}} w_i \cdot (z_i^\star - F_i(x)) \qquad (2)$$

where $x \in X$ is a solution, $w = (w_1, \ldots, w_m)$ is a weighting coefficient vector with $w_i \geqslant 0$ for all $i \in \{1, \ldots, m\}$, and $z^\star = (z_1^\star, \ldots, z_m^\star)$ is a reference point. This reference point is typically set as the ideal point, such that $z_i^\star = \max_{x \in X} F_i(x)$, $i \in \{1, \ldots, m\}$ [16], or a good approximation of it. The scalarizing function $f\colon X \mapsto \mathbb{R}$ and the set of uniformly generated weight vectors $W = (w^1, \ldots, w^\mu)$ define the scalar sub-problems. One solution $x \in X$ with the optimal (maximal) scalar function value $f(x \mid w^i)$ is sought for each sub-problem $i \in \{1, \ldots, \mu\}$.

MOEA/D maintains one solution per sub-problem in its population $P = (x^1, \ldots, x^\mu)$. For each sub-problem $i \in \{1, \ldots, \mu\}$, a set $B(i)$ of t closest weight vectors defines cooperating sub-problems. The population evolves through iterative, cooperative optimization. At a given iteration dealing with sub-problem $i \in \{1, \ldots, \mu\}$, solutions from $B(i)$ are selected to create an offspring y by means of variation. The offspring replaces the current solution x^j in sub-problem $j \in B(i)$ if $f(y \mid w^j)$ improves upon $f(x^j \mid w^j)$. The algorithm cycles through sub-problems until a stopping condition is met.

---

[1] For simplicity, we use the *inverse* of the Tchebycheff function, such that both scalar functions are to be maximized.

2.3 Landscape Analysis and Local Optima Networks

A *landscape* is defined by a triplet $(X, \mathcal{N}, f)$ where: X is the solution space, $\mathcal{N} \colon X \mapsto 2^X$ is a neighborhood relation, and $f \colon X \mapsto \mathbb{R}$ is a fitness function. A higher-level abstraction, focusing on the distribution and connectivity pattern of local optima, is captured by the notion of *local optima networks* (LONs) [19].

A LON is defined as a directed, weighted graph $G = (N, E)$. Nodes N represent local optima, and an edge $e \in E$ exists between two nodes if there is a non-zero probability for the search process to transition from one node to another. The basin of attraction of a local optimum is the set of solutions that converge to it when applying a simple hill-climbing algorithm. The basin's size—the number of these solutions—could be used to represent the width of a LON node. We adopt the concept of escape edges [24]: An edge's weight represents the probability of moving from one local optimum's basin of attraction to another. This transition occurs when a perturbation is applied, followed by a local search.

3 LON/D and Visualization of Sub-problem Landscapes

In the context of decomposition, we simply define a (single-objective) landscape for each (scalar) sub-problem, with the fitness function being the scalarizing function parameterized by the corresponding sub-problem's weight vector. To examine the sub-problem landscapes, we adopt the LON model. Drawing inspiration from MOEA/D, we name our approach LON/D to highlight its adaptation in the context of decomposition. In this section, we introduce the considered benchmark problems and then visualize the LON/D for selected settings.

3.1 Benchmark Problems

We consider ρmnk-landscapes [25] as multi-objective multi-modal combinatorial benchmark problems with objectives correlation. Candidate solutions are binary strings of length n. The neighborhood $\mathcal{N}$ (Sect. 2.3) is based on the conventional *1-bit-flip* operator: two solutions are neighbors if the Hamming distance between them is one. The objective function vector $F = (F_1, \ldots, F_i, \ldots, F_m)$ is defined as $F \colon \{0,1\}^n \to [0,1]^m$ such that each objective F_i is to be maximized. The objective value $F_i(x)$ of a solution $x = (x_1, \ldots, x_j, \ldots, x_n)$ is the average value of the individual contributions associated with each variable x_j. The contribution of x_j depends on its own value and the values of $k < n$ variables other than x_j, chosen uniformly at random. By increasing k, landscapes can be gradually tuned from smooth to rugged [11]. The contribution values follow a multivariate uniform distribution such that $\rho > \frac{-1}{m-1}$ defines the correlation among the objectives. The positive (resp. negative) correlation ρ decreases (resp. increases) the degree of conflict between the objective values. ρmnk-landscapes show different characteristics and degrees of difficulty for multi-objective algorithms [14], including decomposition approaches [3,15,21].

This paper focuses on small two-objective problems, allowing for both the exact construction of LONs through exhaustive enumeration and the visualization of solutions in the objective space. We generate 90 ρmnk-landscapes following the parameters listed in Table 1. For each parameter setting, 10 instances are randomly generated. This allows us to investigate landscapes ranging from smooth to rugged, with conflicting, uncorrelated, or (positively) correlated objectives.

Table 1. Considered benchmark problems and their parameters.

description	values
number of variables	$n = 16$
number of interactions	$k \in \{1, 2, 4\}$
number of objectives	$m = 2$
objectives correlation	$\rho \in \{-0.4, 0.0, 0.4\}$

3.2 LON/D Visual Inspection

Visualization plays a crucial role in understanding complex networks. Using node-edge diagrams, where nodes appear as points and edges as lines, we can visually explore network structures, identify patterns, and gain insight into their behavior. Key characteristics can be emphasized through visual attributes such as color, size, and shape. For visualization purposes, this section starts with five evenly spaced weight vectors $w = (w_1, w_2)$ such that $w_1 \in \{0, 0.25, 0.5, 0.75, 1\}$ and $w_2 = 1 - w_1$. The reference point in WT is set as the (exact) ideal point.

Figures 1 and 2 show LON/D visualizations for example problems in the objective space, offering a natural graph layout to display the LON of each sub-problem. Figure 1 depicts landscapes with increasing variable interactions (sub-figures (a) to (c)) and independent objectives, while Fig. 2 does so for landscapes with conflicting (sub-figure (a)) and correlated objectives (sub-figure (b)) under fixed (moderate) variable interactions. Each sub-figure displays the LON/D for five weight vectors (from left to right) and the two scalarizing functions: WS (top row) and WT (bottom row). The plots contain two sets of data:

(1) The LON/D model, where nodes represent scalar local optima (slo) whose size is proportional to their basin sizes. Nodes are colored following the node legend: red for scalar global optima (sgo), orange for Pareto optimal solutions (po), and blue for others.
(2) The fully enumerated set of points, identified by the **exact** legend, showing all Pareto optimal (po) and Pareto local optimal (plo) solutions, as well as the ideal point. We recall that a plo is not dominated by any of its neighbors [20].

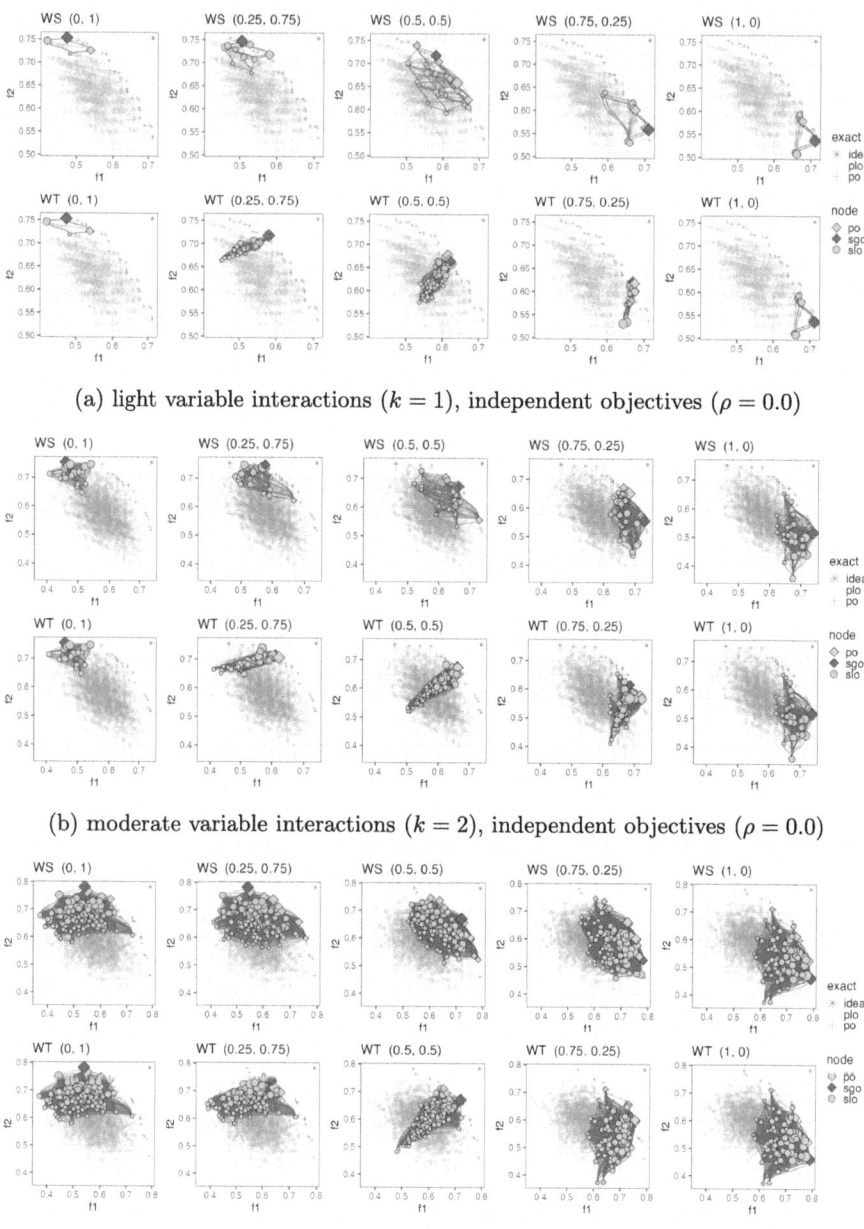

Fig. 1. LON/D examples for light, moderate and strong variable interactions. For each combination of weight vector and scalarizing function (top row: WS, bottom row: WT), two sets of data are displayed: the LON/D model (node legend) and fully enumerated solutions (exact legend). The legends use the following abbreviations: po for Pareto optimal, plo for Pareto local optimal, sgo for scalar global optimal, and slo for scalar local optimal solutions.

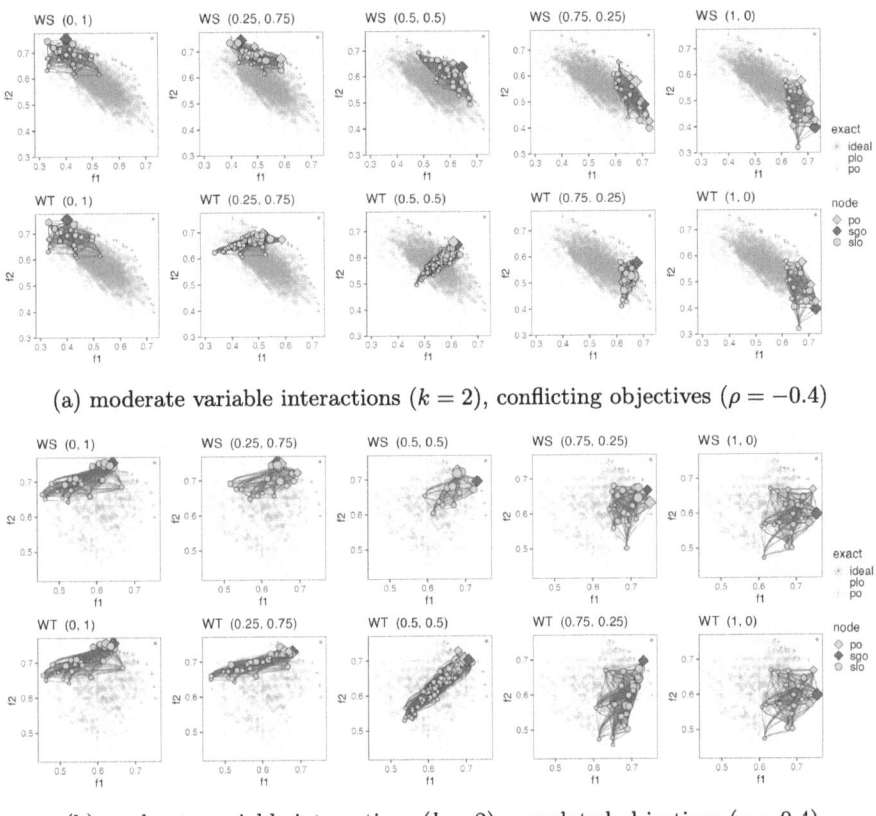

Fig. 2. LON/D examples for conflicting and positively correlated objectives. See Fig. 1 for legend explanations.

Firstly, we observe that the last weight vector $w = (1,0)$ corresponds to the standard (single-objective) LON for the first objective, while the first vector $w = (0,1)$ corresponds to the LON for the second objective. The key distinction lies in the use of the (ideal) reference point for WT, as detailed in Eq. (2). This is confirmed in each sub-figure, where the extreme LON/D models (leftmost and rightmost plots) are identical for both scalarizing functions.

Secondly, a visual inspection of the LON/D models reveals a key insight: each slo found in the sub-problems created by decomposition corresponds to a plo. This finding implies that scalar local optima from weighted sum (WS) and weighted Tchebycheff (WT) have no dominating neighbors. To further our understanding, it would be worth exploring how the proportion of plo covered by sub-problems' slo varies depending on the number of weights and the types of scalarizing functions. We leave this open for future research.

The main observations from the LON/D models' response to variable interactions in Fig. 1 are as follows. The nodes for the sequence of weight vectors (from left to right) cover a progression of bi-objective space regions. As anticipated, the node count increases in conjunction with increasing variable interactions. For a given level of variable interactions, the WT scalarizing function (bottom row) tends to produce more nodes (i.e. more slo) for intermediate weight vectors. However, this larger set of nodes is condensed into a narrower range of bi-objective values compared to the WS LON/D models (top row). Consequently, the nodes in WS LON/D models display a wider coverage of the objective space.

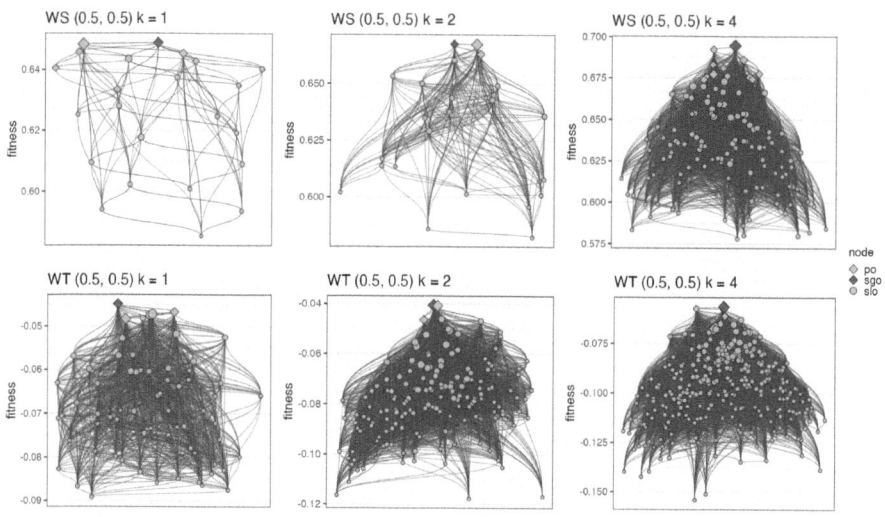

Fig. 3. LON/D examples with fitness layout for light, moderate and strong variable interactions. A single weight vector (0.5, 0.5) is shown for both scalarizing functions (top row: WS, bottom row: WT). See Fig. 1 for legend explanations.

To better illustrate the LON/D differences, we report in Fig. 3 some models from Fig. 1 for the intermediate weight vector $w = (0.5, 0.5)$ using an alternative graph layout. The y-axis shows the (single-objective) scalar fitness, while the x-axis uses a force-directed graph layout known as stress majorization [9]. Note that this layout could be applied to problems with more than two objectives. This visualization better highlights two crucial points: (1) the number of nodes increases substantially as the level of variable interactions (k) rises, and (2) the WT models (bottom row) have significantly more nodes and edges than the WS models (top row). Additionally, the plots highlight some Pareto optimal solutions (orange diamonds) that are sub-optimal under the scalarizing function.

Examining the impact of objectives correlation in Fig. 2, we observe that the nodes' distribution (in terms of objective values) mirrors the shape of the fully enumerated set for both (a) conflicting and (b) correlated objectives. As the objective correlation increases, the number of nodes decreases. Similar to Fig. 1, the LON/D models under WS (top row) display nodes covering a wider range of objective values.

4 LON/D Features vs. Sub-problem Settings

This section introduces nine features to characterize the LON/D, summarized in Table 2. Our selection includes established features from single-objective LON analysis, together with new features relevant to multi-objective optimization and decomposition. We then analyze how the features vary based on the problem and decomposition settings.

Table 2. Description of LON/D features.

feature	description
num-nodes	number of compressed nodes (i.e. local optima plateaus)
density-edge	density of edges
prop-pos	proportion of Pareto optimal nodes
strength-pos	normalized incoming strength of Pareto optimal nodes
strength-global	normalized incoming strength of scalar global optimal node
avg-rank	average rank of nodes with respect to non-dominated sorting
avg-fit-dev	average node deviation from best scalar fitness
global-basin-size	basin size of scalar global optimal node
global-basin-largest	whether the scalar global optimal node has the largest basin

4.1 Definition of Features

The features previously used in single-objective LON analysis [17,18] are: the number of nodes (num-nodes), compressing neighboring nodes with the same fitness value, i.e., plateaus; the ratio of actual edges to the maximum possible number of edges in the graph (density-edge); the normalized weighted incoming degree (strength) of scalar global optima (strength-global); the average difference between the fitness of local optima and the global optimum (avg-fit-dev); the size (number of solutions) of the basin of attraction containing the global optimum, or sum of basin sizes for multiple global optima (global-basin-size); and a Boolean indicating whether the largest basin falls into a global optimum (global-basin-largest). The newly proposed features relevant to multi-objective optimization are: the ratio of Pareto optimal nodes to the total number of nodes (prop-pos); the normalized incoming weighted degree (strength) of Pareto optimal nodes (strength-pos); and the average rank of nodes, where rank follows non-dominated sorting [10] (avg-rank).

4.2 Results and Discussion

Due to space restrictions, we analyze six of the nine features in detail below. We selected these based on their strong correlation with performance, which will be further elaborated in Sect. 5. To provide a more comprehensive view of the LON/D features across sub-problems, we now examine *nine* weight vectors, $w_1 \in \{0, 0.125, 0.25, 0.375, 0.5, 0.625, 0.75, 0.875, 1\}$ and $w_2 = 1 - w_1$. This new set includes all the weights from the previous set along with additional ones.

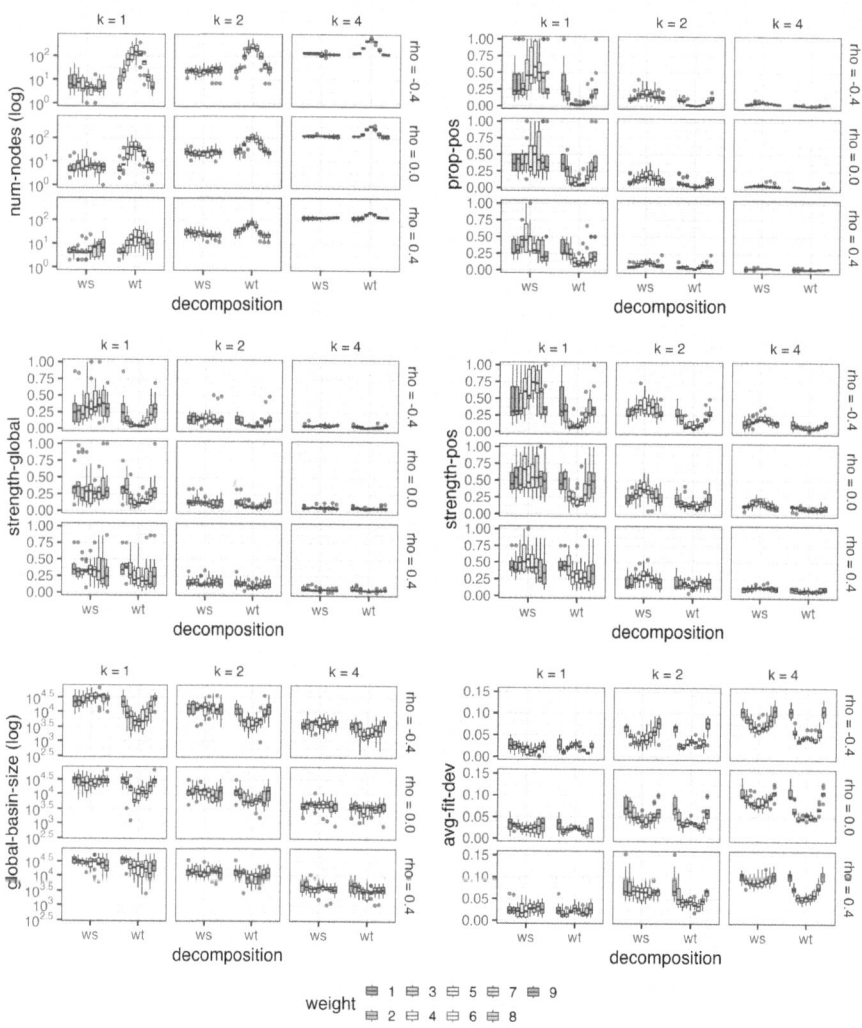

Fig. 4. Distribution of LON/D features with respect to sub-problem landscapes.

Figure 4 reports the distribution of the considered features across all landscape parameters (with 10 random instances per benchmark setting), the two scalarization functions and nine weighting coefficient vectors. We observe clear differences between the WS and WT scalarizing functions and the three levels of variable interactions (k). The influence of objective correlation (ρ), however, is less pronounced. The number of nodes (num-nodes) is given on a logarithmic scale to better showcase the wide range of values. It increases with the level of variable interactions and is notably higher for WT decomposition, displaying an inverted V-shape that peaks for intermediate weight vectors. WT, being a non-linear function, naturally induces more local optima than WS. Surprisingly, we observe a higher number of scalar local optima for intermediate weight vectors. This suggests that the landscape of individual objectives is easier to navigate than their weighted aggregation under WT. The pattern in the number of scalar local optima across different weight vectors highlights the complex nature of WT decomposition. The next four metrics—prop-pos, strength-pos, strength-global, and global-basin-size—exhibit a similar pattern. They decrease sharply as variable interactions increase and are significantly higher for WS decomposition. Intuitively, larger values for these features indicate "easier" landscapes. This observation reinforces that small k-values and that WS decomposition tend to lead to easier landscapes. This creates an inverted V-shape for WS, peaking at intermediate weights. In contrast, the WT decomposition plots display a U-shape, with lower values for intermediate weight vectors. Interestingly, the landscape at intermediate weights tends to be easier for WS and harder for WT, as captured by these LON/D features. Finally, the average deviation from best scalar fitness (avg-fit-dev) increases with the level of variable interactions. Both scalarizing functions exhibit similar values. Across weight vectors, the general trend shows a U-shape, with lower values for intermediate vectors. However, WT decomposition often displays an inverted M-shape, spiking up at extreme and intermediate weight vectors. The relative quality of scalar local optima thus tends to decrease for these weights. However, we note that the range of values is also influenced by the Pareto front shape, which is known to be globally convex for ρmnk-landscapes [25].

5 LON/D Features vs. Sub-problem Solving

In this section, we experiment with MOEA/D using both WS and WT, and evaluate its performance for each sub-problem defined by decomposition. We then analyze the correlation and association between sub-problem solving and LON/D features using correlation analysis and decision trees.

5.1 Experimental Setup

We run two MOEA/D variants—one using the WS scalarizing function and another using WT—on the same 90 ρmnk-landscapes investigated so far. Each algorithm runs independently 30 times per instance. We set the stopping condition to 3 276 calls to the evaluation function, which represents 5% of the search space. For variation, we employ uniform crossover and standard bit-flip mutation with a rate of $1/n$. The number of collaborating sub-problems is $t = 10$, and the population size is $\mu = 65$. Why 65 individuals? Simply because this reasonable setting also conveniently includes all the weight vectors we have examined so far in our sub-problem landscape analysis. It allows us to measure MOEA/D's performance on the entire population and, crucially, to focus on the same 9 weight vectors we considered previously.

5.2 Algorithm Performance

We begin by measuring the overall algorithm performance, using the entire MOEA/D population, in terms of hypervolume [28]. Specifically, we calculate the relative hypervolume deviation from the exact Pareto front. Let hv be the hypervolume covered by the population. The relative hypervolume deviation is $(hv^\star - hv)/hv^\star$, where $hv^\star$ is the optimal hypervolume. A lower value indicates better performance. We set the hypervolume reference point to the lower bound of the objective space. This is reported in Fig. 5 (left). For both scalarizing functions, we observe that the performance of MOEA/D tends to decline as the number of variable interactions (k) increases and the degree of conflict among objectives ($-\rho$) intensifies. When comparing WS and WT, we observe that WT consistently provides a better Pareto front approximation, except when both k and ρ are large, where there is no significant difference.

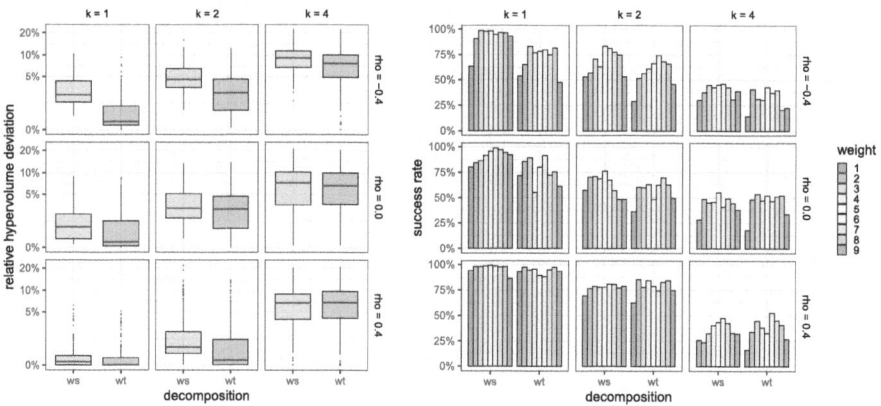

Fig. 5. Performance of MOEA/D using the WS and WT scalarizing functions, shown globally in terms of hypervolume (left) and for 9 individual sub-problems in terms of success rate (right).

Let us now delve deeper into performance analysis and examine how MOEA/D performs on each sub-problem. On the right-hand side of Fig. 5, we report the success rate of MOEA/D variants in identifying a scalar global optimum for each of the same nine evenly spaced weight vectors. For a given instance, we simply calculate the proportion of times MOEA/D was able to hit the sgo for each considered sub-problem. A global optimum in the scalar problem (sgo) always corresponds to a Pareto optimal point in the original multi-objective problem (po) [16]. As expected, we observe that the success rate typically decreases as the number of variable interactions (k) increases. However, there is no clear trend regarding the correlation between objectives (ρ). When comparing WS with WT, the former tends to achieve scalar global optima with a slightly higher success. This finding contrasts with the hypervolume results shown on the left-hand side of the figure. While WT offers superior properties by potentially identifying Pareto optimal points in concave regions of the Pareto front [16], this advantage comes at the cost of a more complex fitness function. This underscores an intricate interplay between solving individual sub-problems and how solutions are distributed in the objective space for each scalarizing function. We analyze below the correlation between LON/D features and the success rate over sub-problems.

5.3 Impact of LON/D Features on Algorithm Performance

Figure 6 shows the Spearman's rank correlation coefficient between each LON/D feature and the success rate for individual sub-problems. We examine the complete set of sub-problems—90 instances with 9 weight vectors each—yielding 810 sub-problem landscapes. The correlation is given for each scalarizing function.

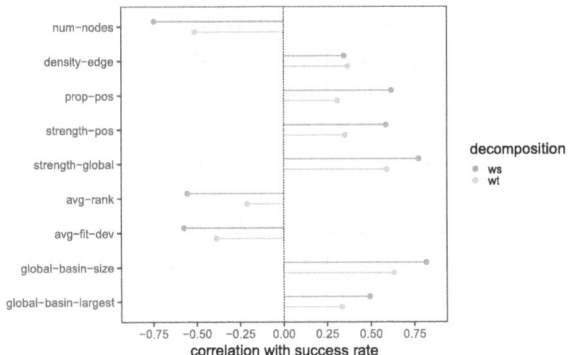

Fig. 6. LON/D features vs. success rate correlation.

Overall, we observe similar trends for both WS and WT. The success rate increases as the number of nodes (num-nodes), their average rank (avg-rank), and their average fitness deviation from the best (avg-fit-dev) decrease. Conversely, the search becomes more effective with increases in edge density (density-edge), proportion of Pareto-optimal nodes (prop-pos), incoming strength of Pareto (strength-pos) and scalar global (strength-global) optima, and basin size of scalar global optima (global-basin-size). Success also improves when the largest basin leads to a scalar global optimum (global-basin-largest). These trends align with our intuitions about landscape difficulty. Although the trends are similar across all features, the absolute correlation is consistently higher for WS than for WT. We attribute this to the complex interactions of success across sub-problems for WT. For both scalarizing functions, global-basin-size shows the strongest correlation with success rate (0.82 for WS, 0.63 for WT). This indicates that the larger the basin size of scalar global optima, the higher the probability of reaching it. The next most correlated features are strength-global (0.77 for WS, 0.59 for WT) and num-nodes (−0.75 for WS, −0.51 for WT). These correlations suggest that subproblem solving becomes more efficient not only when there are fewer scalar local optima, but also when the incoming degree towards sgo in the LON/D is higher—in other words, when global optima are well-connected to other nodes.

We conclude with a simple, interpretable regression tree that predicts success rate using the established CART approach [5]. We combine LON/D features and benchmark parameters as predictors. Figure 7 shows the decision trees obtained for WS (left) and WT (right). The numbers beneath each node indicate the count and proportion of sub-problems covered. We contend that they provide a valuable tool for explaining algorithm performance within the context of our analysis.

Despite limiting the tree's maximum depth to three levels for better interpretability, the error rate remains below 0.05 for both models. The R^2 is 0.67 for WS and 0.51 for WT, indicating that such a simple model explains 67% and 51% of the variance in success rates, respectively. While the tree levels align with the feature importance conveyed by our correlation analysis, their combined effect is more evident. For both scalarizing functions, the primary decision depends on the basin size of scalar global optima (global-basin-size). Interestingly, the proportion of Pareto optimal nodes (prop-pos) appears on both sides. However, a larger value, when combined with other features, now typically leads to a lower success rate. This suggests that when the basin size of scalar global optima is relatively small, Pareto optimal solutions act as sub-optimal attractors for the scalar sub-problem—echoing our previous observations from Fig. 3.

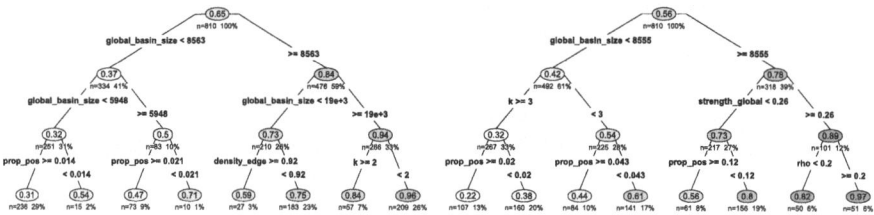

Fig. 7. Decision tree for the success rate by sub-problem obtained by MOEA/D using WS (left) and WT (right).

6 Conclusions and Open Issues

Decomposition offers a natural bridge between single-objective optimization tools, such as landscape analysis, and the multi-objective optimization realm.

This approach allows us to study and visualize multi-objective landscapes holistically while uncovering specific challenges in various regions of the objective space. As a result, we gain deeper insights into the complexities of multi-objective optimization problems. In this paper, we leveraged LONs to analyze the landscape induced by sub-problems defined through decomposition in multi-objective optimization. Our approach, LON/D, serves as both a landscape visualization technique and a framework with features that capture various aspects of problem difficulty. Our experiments demonstrated that LON/D features effectively explain algorithm performance in terms of sub-problem solving.

There are several ways in which our analysis could be extended. Firstly, we could broaden our scope to include large-size, many-objective optimization problems, encompassing those from continuous optimization. While visualizing and characterizing large-scale multi-objective landscapes remains largely unexplored, LON/D allows us to leverage established sampling techniques from single-objective optimization. Secondly, we could investigate other decomposition settings, such as penalty-based boundary intersection [26]. Finally, we intend to examine problems with *heterogeneous* objectives [2] in terms of landscape difficulty; i.e., with a different k value per objective in multi-objective NK-landscapes [1,7]. In terms of algorithm performance, our analysis left two important aspects unexplored for MOEA/D: (1) the value of sub-problems' cooperation, where we could assess performance with independent sub-problem solving to quantify the benefits of cooperation on the success rate for each sub-problem; and (2) the influence of extreme sub-problems on the reference point setting, where we could initialize it at the (exact) ideal point to evaluate its effects across all sub-problems.

Disclosure of Interests. The authors have no competing interests to declare that are relevant to the content of this article.

References

1. Aguirre, H.E., Tanaka, K.: Working principles, behavior, and performance of MOEAs on MNK-landscapes. Eur. J. Oper. Res. **181**(3), 1670–1690 (2007). https://doi.org/10.1016/j.ejor.2006.08.004
2. Allmendinger, R., Knowles, J.: Heterogeneous Objectives: State-of-the-Art and Future Research, pp. 317–335. Springer (2023). https://doi.org/10.1007/978-3-031-25263-1_12
3. Basseur, M., Liefooghe, A., Tari, S.: MOW-P: A simple yet efficient partial neighborhood walk for multiobjective optimization. In: IEEE Congress on Evolutionary Computation (CEC 2024), pp. 1–8 (2024). https://doi.org/10.1109/CEC60901.2024.10611767
4. Borges, P., Hansen, M.: A basis for future successes in multiobjective combinatorial optimization. Tech. Rep. IMM-REP-1998-8, Institute of Mathematical Modelling, Technical University of Denmark, Lyngby, Denmark (1998)
5. Breiman, L., Friedman, J., Stone, C.J., Olshen, R.A.: Classification and Regression Trees. Taylor & Francis, Andover, UK (1984)

6. Cosson, R., Derbel, B., Liefooghe, A., Aguirre, H.E., Tanaka, K., Zhang, Q.: Decomposition-based multi-objective landscape features and automated algorithm selection. In: 21st European Conference on Evolutionary Computation in Combinatorial Optimization (EvoCOP 2021). Lecture Notes in Computer Science, vol. 12692, pp. 34–50 (2021). https://doi.org/10.1007/978-3-030-72904-2_3
7. Cosson, R., Santana, R., Derbel, B., Liefooghe, A.: On bi-objective combinatorial optimization with heterogeneous objectives. Eur. J. Oper. Res. **319**(1), 89–101 (2024). https://doi.org/10.1016/J.EJOR.2024.06.029
8. Derbel, B., Brockhoff, D., Liefooghe, A., Verel, S.: On the impact of multiobjective scalarizing functions. In: Parallel Problem Solving from Nature (PPSN 2014). Lecture Notes in Computer Science, vol. 8672, pp. 548–558 (2014). https://doi.org/10.1007/978-3-319-10762-2_54
9. Gansner, E.R., Koren, Y., North, S.: Graph drawing by stress majorization. In: Graph Drawing, pp. 239–250. Springer, Berlin, Heidelberg (2005)
10. Goldberg, D.E.: Genetic Algorithms in Search. Optimization and Machine Learning. Addison-Wesley, Boston, MA (1989)
11. Kauffman, S.A.: The Origins of Order. Oxford University Press (1993)
12. Lavinas, Y.C., Ladeira, M., Aranha, C.: Faster convergence in multiobjective optimization algorithms based on decomposition. Evol. Comput. **30**(3), 355–380 (2022). https://doi.org/10.1162/EVCO_A_00306
13. Li, K.: Decomposition multi-objective evolutionary optimization: From state-of-the-art to future opportunities (2021). https://arxiv.org/abs/2108.09588
14. Liefooghe, A., Daolio, F., Derbel, B., Verel, S., Aguirre, H.E., Tanaka, K.: Landscape-aware performance prediction for evolutionary multi-objective optimization. IEEE Trans. Evol. Comput. **24**(6), 1063–1077 (2020)
15. Marquet, G., Derbel, B., Liefooghe, A., Talbi, E.G.: Shake them all! - Rethinking selection and replacement in MOEA/D. In: Parallel Problem Solving from Nature (PPSN 2014). Lecture Notes in Computer Science, vol. 8672, pp. 641–651 (2014). https://doi.org/10.1007/978-3-319-10762-2_63
16. Miettinen, K.: Nonlinear Multiobjective Optimization. Kluwer Academic Publishers, Boston, MA (1999)
17. Ochoa, G., Veerapen, N.: Mapping the global structure of TSP fitness landscapes. J. Heuristics **24**(3), 265–294 (2018). https://doi.org/10.1007/s10732-017-9334-0
18. Ochoa, G., Veerapen, N., Daolio, F., Tomassini, M.: Understanding phase transitions with local optima networks: number partitioning as a case study. In: European Conference on Evolutionary Computation in Combinatorial Optimization (EvoCOP 2017). Lecture Notes in Computer Science, vol. 10197, pp. 233–248 (2017)
19. Ochoa, G., Verel, S., Daolio, F., Tomassini, M.: Local optima networks: a new model of combinatorial fitness landscapes. In: Recent Advances in the Theory and Application of Fitness Landscapes. Emergence, Complexity and Computation, vol. 6. Springer (2014). https://doi.org/10.1007/978-3-642-41888-4_9
20. Paquete, L., Schiavinotto, T., Stützle, T.: On local optima in multiobjective combinatorial optimization problems. Ann. Oper. Res. **156**, 83–97 (2007). https://doi.org/10.1007/s10479-007-0230-0
21. Pruvost, G., Derbel, B., Liefooghe, A., Li, K., Zhang, Q.: On the combined impact of population size and sub-problem selection in MOEA/D. In: 20th European Conference on Evolutionary Computation in Combinatorial Optimization (EvoCOP 2020). Lecture Notes in Computer Science, vol. 12102, pp. 131–147 (2020)
22. Tanaka, S., Takadama, K., Sato, H.: Impacts of single-objective landscapes on multi-objective optimization. In: IEEE Congress on Evolutionary Computation (CEC 2022), pp. 1–8 (2022). https://doi.org/10.1109/CEC55065.2022.9870226

23. Trivedi, A., Srinivasan, D., Sanyal, K., Ghosh, A.: A survey of multiobjective evolutionary algorithms based on decomposition. IEEE Trans. Evol. Comput. **21**(3), 440–462 (2016)
24. Verel, S., Daolio, F., Ochoa, G., Tomassini, M.: Local optima networks with escape edges. In: 10th International Conference on Artificial Evolution (EA 2011). Lecture Notes in Computer Science, vol. 7401, pp. 49–60. Springer (2011). https://doi.org/10.1007/978-3-642-35533-2_5
25. Verel, S., Liefooghe, A., Jourdan, L., Dhaenens, C.: On the structure of multiobjective combinatorial search space: MNK-landscapes with correlated objectives. Eur. J. Oper. Res. **227**(2), 331–342 (2013). https://doi.org/10.1016/j.ejor.2012.12.019
26. Zhang, Q., Li, H.: MOEA/D: a multiobjective evolutionary algorithm based on decomposition. IEEE Trans. Evol. Comput. **11**(6), 712–731 (2007). https://doi.org/10.1109/TEVC.2007.892759
27. Zhou, A., Zhang, Q.: Are all the subproblems equally important? Resource allocation in decomposition-based multiobjective evolutionary algorithms. IEEE Trans. Evol. Comput. **20**(1), 52–64 (2016)
28. Zitzler, E., Thiele, L., Laumanns, M., Fonseca, C.M., Grunert da Fonseca, V.: Performance assessment of multiobjective optimizers: an analysis and review. IEEE Trans. Evol. Comp. **7**(2), 117–132 (2003). https://doi.org/10.1109/TEVC.2003.810758

Visualizing Pseudo-Boolean Functions: Feature Selection and Regularization for Machine Learning

Corentin Masson[1,2,3](✉), Xavier F. C. Sánchez-Díaz[1], and Ole Jakob Mengshoel[1]

[1] Norwegian University of Science and Technology, Trondheim, Norway
{xavier.sanchezdz,ole.j.mengshoel}@ntnu.no
[2] University of Grenoble Alpes, LEGI, CNRS, Grenoble, France
[3] University of Grenoble Alpes, LIG, CNRS, Grenoble, France
corentin.masson@univ-grenoble-alpes.fr

Abstract. The concept of evolutionary fitness landscapes, along with their visualizations, was introduced in biology by Sewall Wright in the 1930 s. The study of fitness landscapes is also important in artificial intelligence, evolutionary computation, and machine learning. In this paper, we discuss the difficulty of visualizing pseudo-Boolean fitness landscapes in the context of feature selection for machine learning. Visualization techniques for fitness landscapes, specifically hinged bitstring maps and local optima networks, are used to derive information from the landscapes. Specifically, we consider the problem of feature selection for machine learning with random forests in the setting of several real-world datasets as a case study. Using these techniques, we highlight the transformation on the multimodal structure of the feature selection fitness landscapes. This work improves the understanding of visualization for feature selection in machine learning, and promises to lead to improved feature selection and visualization methods.

Keywords: Fitness landscapes · Machine learning · Feature selection · Regularization · Pseudo-Boolean functions · Visualization

1 Introduction

Context. Efficiently visualizing theoretical and practical problems is a great challenge in computer science [5,39,40]. In many cases, these problems are high-dimensional. Examples of such high-dimensional spaces include combinatorial pseudo-Boolean functions, which can be found in many areas of AI and computer science. The human mind needs low-dimensional graphics, typically two- or three-dimensional plots, to easily see and analyze data.

© The Author(s), under exclusive license to Springer Nature Switzerland AG 2025
M. S. Krejca and M. Wagner (Eds.): EvoCOP 2025, LNCS 15610, pp. 150–166, 2025.
https://doi.org/10.1007/978-3-031-86849-8_10

Challenge. Unfortunately, mapping from high- spaces to low-dimensional Euclidean spaces ($\mathbb{R}^2$ or $\mathbb{R}^3$) suitable for visualization is non-trivial. Combinatorial functions, including pseudo-Boolean functions, are especially challenging when it comes to creating a readable two- or three-dimensional visualization for non-trivial arities n. A dramatic example of this behavior is the neighborhood structure of a Boolean lattice, which is often completely lost in such visualizations.

At the same time, solving real-world optimization problems often involves search in a Boolean lattice. Specifically, many problems in AI are essentially about optimizing pseudo-Boolean functions: computing models in propositional logic [12,36]; computing the most probable explanations in Bayesian networks [19–21]; maximizing machine learning (ML) accuracy by selecting the right features (feature selection); [9,14,15,41,47]; finding, from a dense measurement (sound, image, video), the best sparse components (sparse signal recovery) [26,30]; and unsupervised sentence summarization [35].

In this paper, we study complex pseudo-Boolean functions that come from real-world problems. The pseudo-Boolean functions are induced from real-world datasets of varying dimensions and domains using random forest classifiers. These classifiers have proved to perform well on high-dimensional problems and have been successfully used in various fields [2,18,31]. Consequently, we study how to visualize the feature selection problem using random forest classifiers [3].

Contributions. We make the following contributions:

- Through novel mathematical analysis, we show that there are fundamental limitations in the visualization of pseudo-Boolean functions in 2D Euclidean space. As a consequence, one needs to make difficult compromises to visualize these landscapes.
- In response to the above-mentioned fundamental compromises, we study a visualization approach that efficiently packs the exponentially sized Boolean lattice into 2D Euclidean space. We present and discuss this visualization method, known as hinged bitstring maps (HBMs), in detail and consider both its pros and cons.
- As a case-study of pseudo-Boolean function landscapes, we study by means of HBMs the essential problem of feature selection in ML. Specifically, we induce random forest classifiers for relatively low-dimensional datasets where the number of features is varied exhaustively. Our HBM-based landscape analysis reveals several interesting points about these datasets and classifiers.

A few previous works discuss pseudo-Boolean fitness landscapes of feature selection for ML [16,23,24,34]. However, these works are somewhat different from this paper. Liefooghe et al. study different ML classifiers and data sets [16], and conclude that very different multimodal pseudo-Boolean function landscapes are induced when feature selection is used. However, they are not concerned with the visualization of the fitness landscape induced by these classifiers. Sánchez-Díaz et al. consider HBM visualizations, but have as their main focus the combined impact of feature selection and regularization on landscape multimodality [34]. They do not study the mathematical rationale for and limitations of HBMs. Mostert et al. study feature selection landscapes for the k-NN classifier, and also consider the use of local optima networks [23,24]. We are here using a complementary ML classifier, namely random forest, and we also study different datasets. A key finding of Liefooghe et al. is that the different combinations of ML classifiers and featured selected are quite different in terms of their fitness landscape characteristics, at least when analyzed using the Spearman rank correlation coefficient [16]. Previous works on feature selection landscapes highlight the importance of researching different classifiers, different datasets, and varying regularization [16,23,24,34], and we add to this research effort in this work.

2 Background

2.1 Notation and Optimization

Two different distance metrics will be used here. The first one is the traditional Euclidean distance. As we only use it in $\mathbb{R}^2$, it will be written as $d_{\mathbb{R}^2}$. We define $d_{\mathbb{R}^2}$ in terms of two points A and B:

$$\forall A, B \in \mathbb{R}^2, \ d_{\mathbb{R}^2}(A,B) = \sqrt{(x_A - x_B)^2 + (y_A - y_B)^2}. \tag{1}$$

The second distance metric we use is the Hamming distance d_H, defined in terms of n-dimensional bitstrings $\boldsymbol{b}$ and $\boldsymbol{b}'$ [11]:

$$\forall n \in \mathbb{N} \ \forall \boldsymbol{b}, \boldsymbol{b}' \in \mathbb{B}^n, \ d_H(\boldsymbol{b}, \boldsymbol{b}') = \sum_{i=1}^{n} |\boldsymbol{b}_i - \boldsymbol{b}'_i|. \tag{2}$$

The Hamming distance is simply the sum of the differences between each pair of bits in absolute value.

Multimodal function optimization is important in artificial intelligence [33, 45]. The term 'fitness' is often used when the objective function needs to be maximized (for example, the classification accuracy for an ML classifier), while 'energy' is used for minimization (e.g., classification error) [6,42].

Our search space is a set of bitstrings, $\mathbb{B}^n$, with $n \in \mathbb{N}$. We seek to optimize (without loss of generality, minimize) an energy function $h: \mathbb{B}^n \to \mathbb{R}$. A global optimum is then any $\boldsymbol{b}^* \in \mathbb{B}^n$ such that for all $\boldsymbol{b}' \in \mathbb{B}^n$, $h(\boldsymbol{b}^*) \leq h(\boldsymbol{b}')$.

Apart from finding a global optimum, we are interested in finding local optima, defined in terms of a neighborhood operator. We define the neighborhood $\mathcal{N}$ of a bitstring $\boldsymbol{b}$ as the set of bitstrings that can be obtained by changing exactly one bit of $\boldsymbol{b}$:

$$\forall n \in \mathbb{N} \ \forall \boldsymbol{b} \in \mathbb{B}^n, \ \mathcal{N}(\boldsymbol{b}) = \left\{ \boldsymbol{b}' \in \mathbb{B}^n \mid d_H(\boldsymbol{b}, \boldsymbol{b}') = 1 \right\}. \tag{3}$$

The set of local optima is then defined as the set of all bitstrings whose energy is lower or equal to the energy of any of their neighbors:

$$\mathcal{L} = \left\{ \boldsymbol{b}^+ \in \mathbb{B}^n \mid \forall \boldsymbol{b}' \in \mathcal{N}(\boldsymbol{b}^+), h(\boldsymbol{b}^+) \leq h(\boldsymbol{b}') \right\}. \tag{4}$$

We will also need to split bitstrings into halves for the purpose of conversion into real space. This will be written the following way:

$$\forall \boldsymbol{b} \in \mathbb{B}^{2n} \; \exists \boldsymbol{b}_1, \boldsymbol{b}_2 \in \mathbb{B}^n \mid \boldsymbol{b} = (\boldsymbol{b}_1 || \boldsymbol{b}_2). \tag{5}$$

For simplicity, we will only use an even numbers of bits, but the approach can easily be generalized.

2.2 Local Optima Networks (LONs)

A local optima network (LON) is a weighted graph providing a compact representation of an energy landscape [8,29]. LONs are of great interest for this work in which we deal with multimodal pseudo-Boolean function landscapes, but they have been used on different domains as well [16,28,38]

We first define basins of attraction, which are central in LONs [7]. The definition of a basin of attraction relies on a hill-climbing algorithm, $HC(\boldsymbol{b}, h)$. Given an initial bitstring $\boldsymbol{b}$ and an objective function h, HC returns a local optimum by taking the best neighbor (with the least energy) of the current bitstring at each iteration. For simplicity, we assume each bitstring has a unique best neighbor.

We can now define the basin of attraction (BoA) of a local optimum $\boldsymbol{b}_i^+ \in \mathbb{B}^n$—the set of bitstrings from which the hill-climbing algorithm returns $\boldsymbol{b}_i^+$:

$$\mathcal{BA}_i = \left\{ \boldsymbol{b} \in \mathbb{B}^n \mid HC(\boldsymbol{b}, h) = \boldsymbol{b}_i^+ \right\}. \tag{6}$$

A LON can be represented as a weighted graph of the form $G = (V, E)$. The set of vertices V is the set of local optima of the objective function. E is the set of *basin-transition edges*. Given two different local optima $\boldsymbol{b}_i^+, \boldsymbol{b}_j^+ \in \mathbb{B}^n$, an edge $e_{ij} \in E$ links their corresponding vertices if there are neighboring points in their respective BoAs. The weight of e_{ij} is the number of such pairs:

$$w(e_{ij}) = \left| \left\{ (\boldsymbol{b}_i, \boldsymbol{b}_j) \in \mathbb{B}^{2n} \mid \boldsymbol{b}_i \in \mathcal{BA}_i, \boldsymbol{b}_j \in \mathcal{BA}_j, \boldsymbol{b}_i \in \mathcal{N}(\boldsymbol{b}_j) \right\} \right|. \tag{7}$$

Other definitions can be used, such as *escape edges* [29], but basin-transition edges are most useful in our case. Section 4.2 gives an example of a LON for a toy problem.

2.3 Feature Selection

In feature selection, the goal is to optimize the performance of an ML algorithm given a task (e.g., classification or regression) and a dataset [10,27]. The objective function to optimize depends on the ML task, but can for instance be the accuracy of a classifier or a given error function for a regressor.

The feature selection search space is formed by the sets of features used to train the model. Each feature can be represented by a bit—set to 1 if the feature is used for learning and 0 if it is not. Given a set of features $\boldsymbol{b}$, we model feature selection as a pseudo-Boolean *energy* function to minimize:

$$h(\boldsymbol{b}) = h_E(T(\boldsymbol{b})) + \epsilon \cdot h_P(\boldsymbol{b}), \tag{8}$$

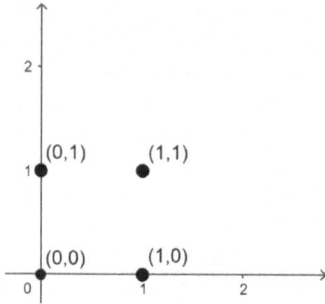

Fig. 1. Two-dimensional visualization in $\mathbb{R}^2$ of all elements in $\mathbb{B}^2$.

where $h_E(T(\boldsymbol{b}))$ is the classification error over a given dataset, using an ML model T. In this paper, we use a random forest classifier. We include in (8) a regularization (or penalty) term, $h_P(\boldsymbol{b})$, which penalizes according to the number of features used for training [4,9,46], i.e. $h_P(\boldsymbol{b}) = \sum b_i$. We control the degree of regularization using the ϵ parameter: putting more weight on minimizing the classification error with a small ϵ, or using as few features as possible if ϵ is large.

3 Analysis of Visualization

In $\mathbb{B}^n$, each bitstring has n neighbors with respect to the Hamming distance. When $n = 2$, we can plot the bitstrings in $\mathbb{R}^2$: using one axis for the first bit, and another one for the second bit. An example is illustrated in Fig. 1, using the x-axis for the first bit, and the y-axis for the second.

In this example, each pair of neighbors is separated by a distance of 1. However, there is already a distortion of the distance metrics from $\mathbb{B}^2$, as each pair of bitstrings whose Hamming distance is 2 is now separated by an Euclidean distance of $\sqrt{2} \approx 1.414$.

3.1 Visualizing $\mathbb{B}^2$

Let us focus on small bitstrings in $\mathbb{B}^2$. We have two metric spaces $\mathbb{B}^2$ and $\mathbb{R}^2$ with their respective distances d_H and $d_{\mathbb{R}^2}$. We are looking for an isometric embedding $f : \mathbb{B}^2 \to \mathbb{R}^2$, that is a map such that:

$$\forall b_1, b_2 \in \mathbb{B}^2, d_H(b_1, b_2) = d_{\mathbb{R}^2}(f(b_1), f(b_2)). \tag{9}$$

Even in this simple case, we show that the *distance preservation constraint*—transposing the Hamming distance into the Euclidean distance—prevents us to create an embedding into $\mathbb{R}^2$. The proof is as follows.

Lemma 1. *There is no isometric embedding from $\mathbb{B}^2$ into $\mathbb{R}^2$ that preserves the Hamming distance as the Euclidean distance.*

In the following, we note as $\mathcal{C}(M, r)$ the circle with center M and radius r.

Proof. Let $A, B, D, E \in \mathbb{R}^2$ be four points to which we want to map bitstrings from $\mathbb{B}^2$. Without loss of generality, A represents $(0,0) \in \mathbb{B}^2$ and takes $(0,0) \in \mathbb{R}^2$ as coordinates. Let B represent $(1,0) \in \mathbb{B}^2$. It could also take $(1,0) \in \mathbb{R}^2$ as coordinates—any point with a distance of 1 from A can be used, without loss of generality. Let D and E represent $(1,1)$ and $(0,1)$, respectively. From the required distances to A and B:

$$\begin{cases} d_{\mathbb{R}^2}(B, D)^2 = d_H((1,0), (1,1))^2 &= 1 \\ d_{\mathbb{R}^2}(A, D)^2 = d_H((0,0), (1,1))^2 &= 4, \end{cases} \tag{10}$$

we can now derive the coordinates of D:

$$\begin{cases} (x_D - 1)^2 + y_D^2 &= 1 \\ x_D^2 + y_D^2 &= 4. \end{cases} \tag{11}$$

By computing the difference between the two lines of Eq. 11, we obtain:

$$2x_D - 1 = 3 \tag{12}$$
$$(x_D, y_D) = (2, 0). \tag{13}$$

We have now shown that $A(0,0)$, $B(1,0)$ and $D(2,0)$ must be on a line. The inconsistency comes immediately when adding a fourth point E:

$$\begin{cases} d_{\mathbb{R}^2}(A, E)^2 = d_H((0,0), (0,1))^2 &= 1 \\ d_{\mathbb{R}^2}(D, E)^2 = d_H((1,1), (0,1))^2 &= 1. \end{cases} \tag{14}$$

We can now derive the coordinates of E:

$$\begin{cases} x_E^2 + y_E^2 &= 1 \\ (x_E - 2)^2 + y_E^2 &= 1. \end{cases} \tag{15}$$

By computing the difference between the two lines of Eq. 15, we obtain:

$$4x_E - 4 = 0 \tag{16}$$
$$(x_E, y_E) = (1, 0). \tag{17}$$

Then E and B coincide, which shows the impossibility to preserve the Hamming distance.

Geometrically, we sum up these equations by drawing the circles representing all possible neighbors of each point. Figure 2 shows the first and last steps of the construction of the visualization following the aforementioned choices. Points A and B are placed without loss of generality in Fig. 2A. In Fig. 2b, we add $\mathcal{C}(A, 2)$, which groups all possible neighbors of the neighbors of A (green), and $\mathcal{C}(B, 1)$, which groups all neighbors of B (blue). Then, point D (red) has to be at the intersection of these circles. Drawing $\mathcal{C}(D, 1)$, on which point E should be placed, illustrates this impossibility, as the only intersection of $\mathcal{C}(D, 1)$ with $\mathcal{C}(A, 1)$ is point B. □

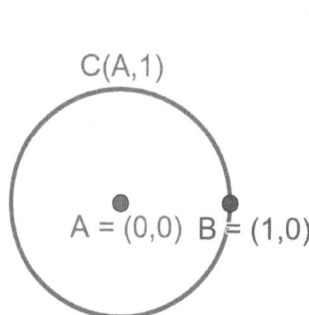
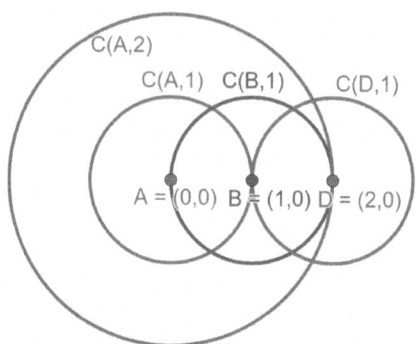

(a) Point A, its neighborhood $\mathcal{C}(A,1)$, and point B.

(b) Point D at the intersection of $\mathcal{C}(A,2)$ and $\mathcal{C}(B,1)$, and its neighborhood $\mathcal{C}(D,1)$.

Fig. 2. Tentative construction of a distance-preserving visualization of $\mathbb{B}^2$.

The *distance preservation constraint* needs to be relaxed, namely Eq. 9 will not hold for all pairs of bitstrings. Figure 1 shows a valid visualization of all bitstrings $\boldsymbol{b} \in \mathbb{B}^2$ in the case where Eq. 9 only holds for neighboring pairs in $\mathbb{B}^2$. Using this relaxation, it is also possible to represent bitstrings from $\mathbb{B}^3$. Table 1 lists the points used to represent $\mathbb{B}^3$ in $\mathbb{R}^2$ under this principle.

Points A and B are placed without loss of generality. Point D is then placed on $\mathcal{C}(A,1)$. We arbitrarily choose $(0,1)$ as its position, but any location on $\mathcal{C}(A,1)$ leads to a valid visualization. The same applies to point F, placed in $(-\frac{\sqrt{2}}{2}, -\frac{\sqrt{2}}{2})$. These choices produce a good visualization, but others are also possible.

Table 1. Points in $\mathbb{R}^2$ representing bitstrings in $\mathbb{B}^3$ after calculations (Eqs. 10 to 17).

Point	Coordinates		Point	Coordinates	
	$\mathbb{B}^3$	$\mathbb{R}^2$		$\mathbb{B}^3$	$\mathbb{R}^2$
A	$(0,0,0)$	$(0,0)$	F	$(1,0,0)$	$(-\frac{\sqrt{2}}{2}, -\frac{\sqrt{2}}{2})$
B	$(0,0,1)$	$(1,0)$	G	$(1,0,1)$	$(\frac{2-\sqrt{2}}{2}, -\frac{\sqrt{2}}{2})$
D	$(0,1,0)$	$(0,1)$	H	$(1,1,0)$	$(-\frac{\sqrt{2}}{2}, \frac{2-\sqrt{2}}{2})$
E	$(0,1,1)$	$(1,1)$	I	$(1,1,1)$	$(\frac{2-\sqrt{2}}{2}, \frac{2-\sqrt{2}}{2})$

Table 1 is then completed by solving the equations linking neighboring pairs. Resulting points are placed in Fig. 3. The most interesting visualization arises when adding edges between the pairs of neighboring points in $\mathbb{B}^3$, on the right-hand side: an example visualization is actually the perspective of a cube.

We have then a two-dimensional visualization of points in $\mathbb{B}^3$ with a constant distance of 1 between all pairs of neighbors, shown in Fig. 3. However, this

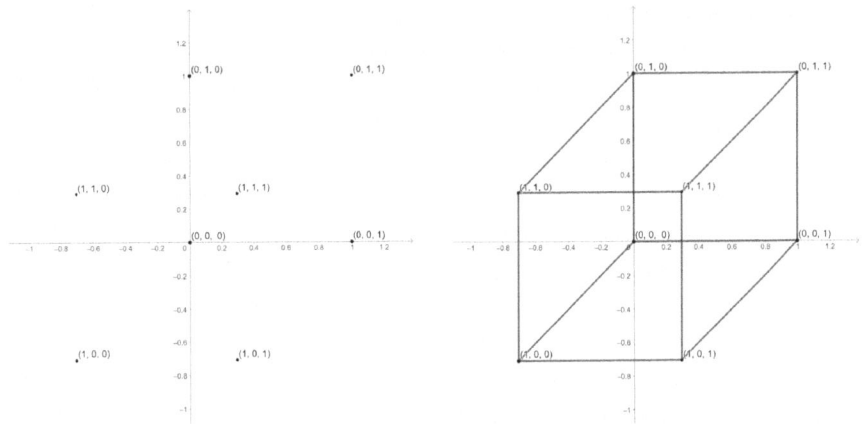

Fig. 3. Two-dimensional visualization in $\mathbb{R}^2$ of all elements in $\mathbb{B}^3$ with coordinates from Table 1, with and without edges between neighboring pairs

visualization is perhaps problematic, as some pairs of non-neighboring points are separated by distances smaller than 1. For instance, point A representing $(0, 0, 0)$ is nearer to points G $(1,0,1)$, H $(1,1,0)$ and I $(1,1,1)$ than it is to its neighbors.

4 Our Visualization Method

4.1 The Hinged Bitstring Map (HBM)

Mathematical work presented in Sect. 3 proves that the neighborhood structure of the bitstring landscapes we study may need to be sacrificed in order to obtain clear two-dimensional plots. To display all bitstrings in a compact figure, we use a technique proposed by Sánchez et al. [34], called the *hinged bitstring map* (HBM). A bitstring is split into two halves and each part is converted to its decimal representation. Then, each half is mapped to a point in $\mathbb{R}^2$: the first half is used as its x-coordinate while the second half is its y-coordinate. The conversion from bitstring to real space can then be formally written as:

$$\begin{cases} \mathbb{B}^n & \to \mathbb{R}^2 \\ \boldsymbol{b} = (\boldsymbol{b}_1 || \boldsymbol{b}_2) & \to (\mathrm{Dec}(\boldsymbol{b}_1), \mathrm{Dec}(\boldsymbol{b}_2)), \end{cases} \quad (18)$$

where $\mathrm{Dec}(\cdot)$ is the conversion from binary to decimal representation.

Because $\mathbb{B}$ is a discrete space, the scatter *plot geometry* (as defined in the layered grammar of graphics [43,44]) is an appropriate choice. The rest of the *plot aesthetics*—namely color, shape and size—can be used to represent other attributes of the data. We use color to show the energy of a given bitstring $\boldsymbol{b}$, and add an outline to highlight the optima $\boldsymbol{b}^*$ and $\boldsymbol{b}^+$. However, these elements can also be used to communicate other data attributes like fitness, connectedness, or feasibility of a solution, making the HBM a suitable choice for visualizing Boolean search spaces.

4.2 HBM Visualization of a Toy Problem

Now we showcase how the HBM works using a toy minimization problem:

$$\forall \boldsymbol{b} \in \mathbb{B}^6, \ h(\boldsymbol{b}) = \sin\left(2\operatorname{Dec}(\boldsymbol{b})\right), \tag{19}$$

which provides an easy-to-compute landscape with several local optima. Table 2 presents the coordinates in $\mathbb{R}^2$ obtained for elements of $\mathbb{B}^6$ as well as the energy h computed with Eq. 19.

Table 2. Construction of a visualization for a pseudo-Boolean toy problem (cf. Eq. 19). Local optima are shown in bold, and the global optimum is underlined.

Bitstring	x	y	Energy	Bitstring	x	y	Energy	Bitstring	x	y	Energy	Bitstring	x	y	Energy
000000	0	0	0.000	010000	2	0	0.551	100000	4	0	0.92	110000	6	0	0.984
000001	0	1	0.909	010001	2	1	0.529	100001	4	1	−0.027	110001	6	1	−0.573
000010	0	2	−0.757	**010010**	**2**	**2**	**−0.992**	100010	4	2	−0.898	110010	6	2	−0.506
000011	0	3	−0.279	010011	2	3	0.296	100011	4	3	0.774	110011	6	3	0.995
000100	0	4	0.989	010100	2	4	0.745	100100	4	4	0.254	110100	6	4	−0.322
000101	0	5	−0.544	**010101**	**2**	**5**	**−0.917**	**100101**	**4**	**5**	**−0.985**	110101	6	5	−0.727
000110	0	6	−0.537	010110	2	6	0.018	100110	4	6	0.566	110110	6	6	0.927
000111	0	7	0.991	010111	2	7	0.902	100111	4	7	0.514	110111	6	7	−0.044
001000	1	0	−0.288	011000	3	0	−0.768	**101000**	**5**	**0**	**−0.994**	111000	7	0	−0.89
001001	**1**	**1**	**−0.751**	011001	3	1	−0.262	101001	5	1	0.313	111001	7	1	0.785
001010	1	2	0.913	011010	3	2	0.987	101010	5	2	0.733	111010	7	2	0.237
001011	1	3	−0.009	011011	3	3	−0.559	101011	5	3	−0.923	**111011**	**7**	**3**	**−0.982**
001100	**1**	**4**	**−0.906**	011100	3	4	−0.522	101100	5	4	0.035	111100	7	4	0.581
001101	1	5	0.763	011101	3	5	0.993	101101	5	5	0.894	111101	7	5	0.499
001110	1	6	0.271	011110	3	6	−0.305	101110	5	6	−0.779	<u>**111110**</u>	<u>**7**</u>	<u>**6**</u>	<u>**−0.996**</u>
001111	**1**	**7**	**−0.988**	011111	3	7	−0.739	101111	5	7	−0.245	111111	7	7	0.33

Figure 4a shows the HBM visualization obtained on this toy problem. Each bitstring is represented by a point whose coordinates are as presented in Table 2. The color and size of a point show the energy $h(\boldsymbol{b})$ of the bitstring $\boldsymbol{b}$: the greener and the bigger, the better (i.e., with lower energy). The ten local optima are circled in blue. The global optimum, $h(111110) = -0.996$, is circled in red. These color choices will hold for all HBM plots in this paper.

In Fig. 4a, arrows have been drawn between bitstrings as a demonstration of how LONs are computed. Every bitstring is in the BoA of a local optimum. Arrows are drawn between each bitstring (except local optima) and its corresponding local optimum. This is only done here to demonstrate the construction of LONs, and not in other HBM plots.

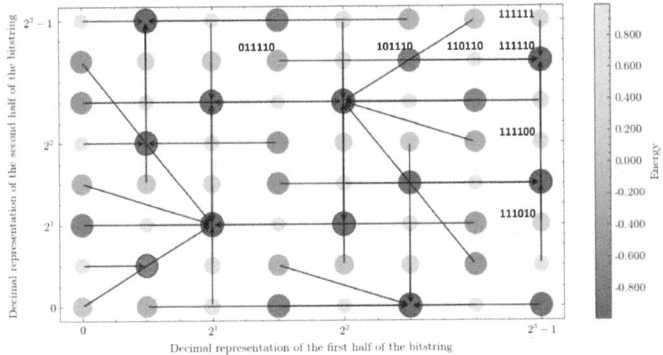

(a) Demonstration of the HBM plotting strategy with $2^6 = 64$ bitstrings. Colors reflect energy h of bitstrings. The global optimum has a BoA of size seven (including itself).

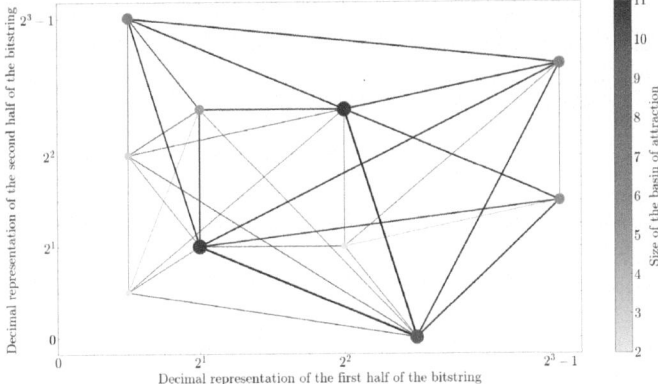

(b) Demonstration of LON construction following the HBM plot in 4a. The LON layout is performed according to the HBM layout. Colors reflect the size of the BoA.

Fig. 4. The toy problem from Eq. 19 visualized.

Consider the global optimum, $b^* = 111110$, towards the top-right corner of the HBM plot in Fig. 4a.[1] Altogether, we find seven bitstrings in the BoA of b^*. Then, when looking at the LON in Fig. 4b, one can see that the size and color of the global optimum correspond to a BoA of size seven. The analysis of the LON is enriched when plotting the LON over the HBM and presenting them side by side, as one can derive the position of a local optimum $b^+ \in \mathbb{B}^n$.

---

[1] At first sight, three arrows come into the node for $b^* = 111110$. Unfortunately, a limitation of this demonstration is that some arrows overlap. For instance, the two small pink bitstrings next to our global optimum (on the left and at the bottom) seem to have no arrow coming out of them: this can only be the case for local optima.

4.3 Pros and Cons of the HBM

The HBM provides a comprehensive view of pseudo-Boolean fitness landscapes. It displays effectively even relatively large spaces: interesting patterns are evidenced in Sect. 5.2 with the number of features going up to $n = 16$. It also supports comparison between similar spaces, for instance HBMs of a given feature selection problem under different transformations such as degree of regularization [34]. In Sect. 5.2, we show how HBMs give a deeper insight into the effect of regularization and a clearer view of the usefulness of the features.

Of course, HBMs also have some drawbacks, as the Boolean lattice gets distorted. The most noticeable example is the fact that some neighbors in combinatorial space are not neighbors in real space. However, all pairs of neighbors are either in the same row or in the same column, which can help locate them. An example of this is provided in Fig. 4a for $\boldsymbol{b}^* = 111110$.

Indeed, if two bitstrings are neighbors in bitstring space, it means that there is only one bit that is different. If this bit is in the first half of the bitstring, then they share the same second half, and they will have the same y-coordinate in real space. The same applies conversely—if their different bit is in the second half, they would have the same x-coordinate.

Table 3. Datasets used in this study with their respective number of features (n), number of examples (m) and number of classes.

Name	Subject Area	n	m	Number of Classes
Yeast [25]	Biology	8	1484	10
HTRU2 [17]	Physics and Chemistry	8	17898	2
Dry Bean [1]	Biology	16	13611	7
Bank Marketing [22]	Business	16	45211	2

5 Case Study: Feature Selection Under Regularization

To highlight the performance of HBMs, this work uses real-world feature selection problems. Feature selection is known as a complex multimodal pseudo-Boolean function optimization problem with high variability among different datasets, ML classifiers, and degree of regularization [16,23,34].

5.1 Datasets and Random Forest Classifiers

We use datasets from the UCI repository [13]. We focus on classification problems with a relatively small number of features n to be able to train a classifier using every subset of bitstring in $\mathbb{B}^n$. Exhaustive enumeration enables to precompute accuracy tables that can be used afterwards as an energy function h with a constant evaluation time.

There are 2^n subsets of features to enumerate exhaustively, so the running time is proportional to $\mathcal{O}(2^n m)$, where m is the number of examples in the dataset. With $n = 16$ and $m = 20000$, for instance, doing only one operation per example requires around 1.3 billion operations in total. The datasets chosen are presented in Table 3. We use two different number of features, $n = 8$ and $n = 16$. These problem sizes offer various challenges, both in terms of visualization and performance metrics. Further details about the datasets (accuracy tables and HBMs) can be found in the supplementary material.

We implemented the random forest classifier provided in the Scikit-learn package in Python [32]. The number of trees is set to 10, the function used to measure the quality of a split is the Shannon entropy [37], and the random seed is fixed to ensure reproducibility. All other parameters are default.[2]

Here we focused on computing an energy function that comes from a random forest classifier learned from and classifying real-world data. Thus, little hyperparameter tuning was done, as our hyperparameter choices produced good results on the datasets used. The number of trees (10) is quite low, but it enables to train a model and test it (with a train/test split of 70%/30%) for each combination of features in a reasonable time. The result of this work is contained in four accuracy tables, one for each dataset.[3] The tables supported as many experiments as needed with a constant evaluation time for the energy function.

Table 4. Description of accuracy tables' key features: number of global and local (non-global) optima. We include the bitstring representation of each global optimum and their accuracy.

Dataset	Number of optima		Accuracy	Bitstring
	Global	Local		
Yeast	1	2	0.580	11110011
HTRU2	1	20	0.981	10111110
Dry Bean	2	1154	0.930	0010011111101011
				0010101111101011
Bank marketing	1	389	0.903	1010001001111011

Table 4 presents the key features of the resulting accuracy tables. The number of local optima varies much between the datasets. The Yeast dataset is known to be difficult, in terms of achieving high accuracy, and our classifier performs on par with the baseline models cited by the UCI [25]. Interestingly, all globally optimal

---
[2] Hyperparameter values for the RandomForestClassifier of Scikit-learn: n_estimators=10, criterion='entropy', max_depth=None (no limits), max_features='sqrt', min_samples_split=2 (no limits), min_samples_leaf=1 (no limits).
[3] For space constraints, the Yeast and HTRU2 analyses are included in the supplementary material.

bitstrings contain zeros. This means that the best accuracy is not achieved by using all features in any of the dataset.

5.2 Hinged Bitstring Maps

The HBM landscape analysis simply starts with plotting the feature selection energy functions with the visualization technique presented in Sect. 4.2.

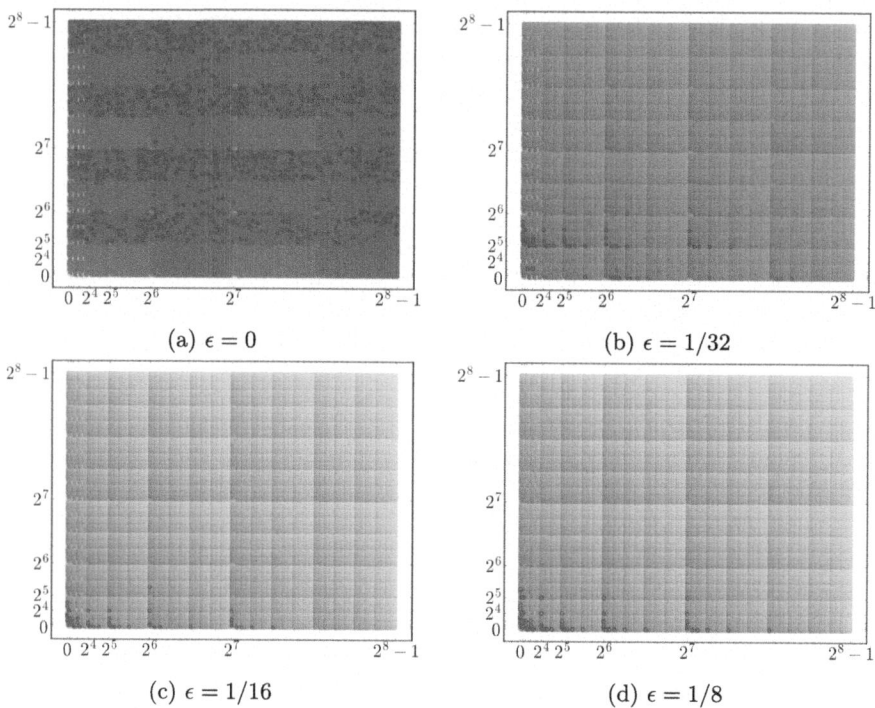

Fig. 5. Plot of the landscape defined by the accuracy of a Random Forest classifier on the Dry Bean dataset with four levels of regularization

Figures 5 and 6 are the plots for the Dry Bean and Bank Marketing datasets, respectively. As regularization increases, the top-right part of the HBM plots becomes more pink and worse according to h. In this part of the graph we find bitstrings with many ones, so they are more penalized by regularization. In contrast, the bottom-left parts of HBM plots become greener and better according to h when regularization increases; bitstrings contain few ones.

In Figs. 5 and 6 we see that local optima often lie on green 'strips' of bitstrings. This is especially true for $\epsilon = 0$ for Dry Bean and Bank Marketing, probably due to the fact that some features are essential to the low energy of the classifier. With $\epsilon > 0$ regularization, a 'grid' structure appears. Quarters of

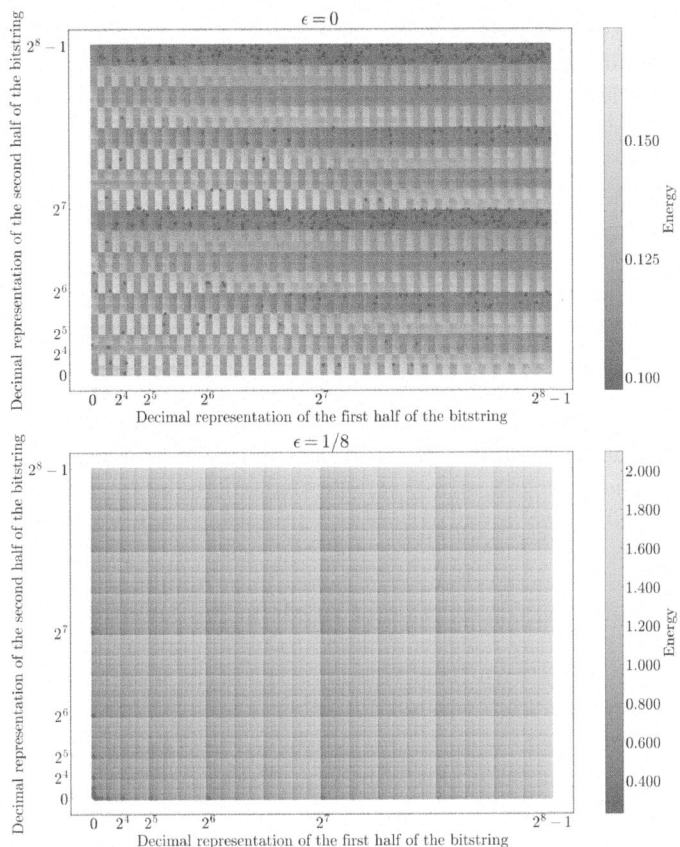

Fig. 6. HBM plots induced by the accuracy of a random forest classifier on the Bank Marketing dataset with two levels of regularization, $\epsilon = 0$ (top) and $\epsilon = 1/8$ (bottom).

green dots border quarters of pink dots due to the conversion made between binary and real spaces in the visualization. When a point in an HBM has its x-coordinate with $x = 2^k - 1$ for any $k \in \mathbb{N}$, its binary representation contains k ones. Its right neighbor, with x coordinate $x = 2^k$, is written with a single 1 in binary space. The drop in regularization penalty is significant between the two bitstrings, producing a clear vertical border between pink bitstrings with high penalty on the left, and green bitstrings with low penalty on the right. The same applies for the y-coordinates, producing horizontal lines. One can check that green lines correspond to 2^k points both in the x and y axis in Figs. 5 and 6 for $\epsilon > 0$ regularization.

One last interesting phenomenon can be noted for Bank Marketing with $\epsilon = 0$ regularization (cf. the top panel in Fig. 6): vertical strips of pink dots become strips of green dots when the y-coordinate changes. We theorize that this pattern emerges due to feature interaction—some combinations of features are useful for classification only if they are considered together, but not on their own. This explains why strips of given x-coordinates are green (lower energy h) for certain values of y, but pink (higher energy h) for others. Observing these interactions, as well as other transformations of the landscape (such as regularization) are the advantages of looking at the search space in its entirety.

6 Conclusion and Future Work

We have benefited from state-of-the-art visualization strategies such as local optima networks (LONs) and hinged bitstring maps (HBMs) to understand fitness landscapes. Our contributions include extension of recent work on visualization approaches for pseudo-Boolean function landscapes using feature selection problems, both with synthetic data and from publicly available repositories. Through mathematical analysis, we have shown fundamental limitations of visualization in pseudo-Boolean space, and highlighted the advantages that HBMs provide. Combining LONs and HBMs is a promising path for future work, as well as exploring other mappings for the *plot aesthetic* elements in HBMs to gain a better understanding of function landscapes. Additionally, it is of special interest to consider the construction of HBMs for instances with large n, and exploring the use of Gray code as opposed to using regular binary strings—to avoid having bitstrings with huge Hamming distances next to one another in HBMs.

References

1. Bean, D.: UCI machine learning repository (2020). https://doi.org/10.24432/C50S4B
2. Azar, A.T., Elshazly, H.I., Hassanien, A.E., Elkorany, A.M.: A random forest classifier for lymph diseases. Comput. Methods Programs Biomed. **113**(2), 465–473 (2014). https://doi.org/10.1016/j.cmpb.2013.11.004, https://www.sciencedirect.com/science/article/pii/S0169260713003751
3. Breiman, L.: Random forests. Mach. Learn. **45**, 5–32 (2001). https://doi.org/10.1023/A:1010950718922
4. Chen, J., et al.: Customizing graph neural networks using path reweighting. Inf. Sci. **674**, 120681 (2024)
5. Cossalter, M., Mengshoel, O.J., Selker, T.: Visualizing and understanding large-scale Bayesian networks. In: The AAAI-11 Workshop on Scalable Integration of Analytics and Visualization, pp. 12–21 (2011)
6. Eiben, A.E., Smith, J.E.: Introduction to Evolutionary Computing, Natural Computing Series, 2nd edn. Springer, Heidelberg (2015)
7. Elorza, A., Hernando, L., Mendiburu, A., Lozano, J.A.: Estimating attraction basin sizes of combinatorial optimization problems. Prog. Artif. Intell. **7**(4), 369–384 (2018)
8. Fieldsend, J.E.: Computationally efficient local optima network construction. In: Proceedings of the Genetic and Evolutionary Computation Conference Companion, pp. 1481–1488. GECCO 2018. Association for Computing Machinery, New York, USA (2018). https://doi.org/10.1145/3205651.3208263

9. Guyon, I., Elisseeff, A.: An introduction to variable and feature selection. JMLR **3**, 1157–1182 (2003)
10. Hall, M.A., Smith, L.A.: Feature selection for machine learning: comparing a correlation-based filter approach to the wrapper. In: Proceedings of FLAIRS, pp. 235–239 (1999)
11. Hamming, R.W.: Error detecting and error correcting codes. Bell Syst. Tech. J. **29**(2), 147–160 (1950). https://doi.org/10.1002/j.1538-7305.1950.tb00463.x
12. Hoos, H.H.: An adaptive noise mechanism for WalkSAT. In: Proceedings of AAAI, pp. 655–660 (2002)
13. Markelle, K., Longjohn, R., Nottingham, K.: UCI machine learning repository. https://archive.ics.uci.edu/
14. Kohavi, R., John, G.H.: Wrappers for feature subset selection. Artif. Intell. **97**(1–2), 273–324 (1997)
15. Leardi, R., Boggia, R., Terrile, M.: Genetic algorithms as a strategy for feature selection. J. Chemom. **6**(5), 267–281 (1992)
16. Liefooghe, A., Tanabe, R., Verel, S.: Contrasting the landscapes of feature selection under different machine learning models. In: Affenzeller, M., et al. (eds.) Parallel Problem Solving from Nature - PPSN XVIII, pp. 360–376. Springer Nature Switzerland, Cham (2024)
17. Lyon, R.: HTRU2. UCI machine learning repository (2017). https://doi.org/10.24432/C5DK6R
18. Masetic, Z., Subasi, A.: Congestive heart failure detection using random forest classifier. Comput. Methods Programs Biomed. **130**, 54–64 (2016). https://doi.org/10.1016/j.cmpb.2016.03.020, https://www.sciencedirect.com/science/article/pii/S0169260715303369
19. Mengshoel, O.J.: Understanding the role of noise in stochastic local search: analysis and experiments. Artif. Intell. **172**(8–9), 955–990 (2008)
20. Mengshoel, O.J., Roth, D., Wilkins, D.C.: Portfolios in stochastic local search: efficiently computing most probable explanations in Bayesian networks. J. Autom. Reason. **46**(2), 103–160 (2011)
21. Mengshoel, O.J., Wilkins, D.C., Roth, D.: Initialization and restart in stochastic local search: computing a most probable explanation in Bayesian networks. IEEE Trans. Knowl. Data Eng. **23**(2), 235–247 (2011)
22. Moro, S., Rita, P., Cortez, P.: Bank marketing. UCI machine learning repository (2012). https://doi.org/10.24432/C5K306
23. Mostert, W., Malan, K., Engelbrecht, A.: Filter versus wrapper feature selection based on problem landscape features. In: Proceedings of the Genetic and Evolutionary Computation Conference Companion, pp. 1489–1496 (2018)
24. Mostert, W., Malan, K.M., Ochoa, G., Engelbrecht, A.P.: Insights into the feature selection problem using local optima networks. In: Liefooghe, A., Paquete, L. (eds.) EvoCOP 2019. LNCS, vol. 11452, pp. 147–162. Springer, Cham (2019). https://doi.org/10.1007/978-3-030-16711-0_10
25. Nakai, K.: Yeast. UCI machine learning repository (1996). https://doi.org/10.24432/C5KG68
26. Needell, D., Tropp, J.A.: CoSaMP: iterative signal recovery from incomplete and inaccurate samples. Appl. Comput. Harmon. Anal. **26**(3), 301–321 (2009)
27. Ng, A.Y.: Feature selection, l 1 vs. l 2 regularization, and rotational invariance. In: Proceedings of the Twenty-First International Conference on Machine Learning, p. 78. ACM (2004)

28. Ochoa, G., Liefooghe, A., Verel, S.: Funnels in Multi-objective Fitness Landscapes. In: Affenzeller, M., et al. (eds.) Parallel Problem Solving from Nature – PPSN XVIII, pp. 343–359. Springer, Cham (2024). https://doi.org/10.1007/978-3-031-70055-2_21
29. Ochoa, G., Verel, S., Daolio, F., Tomassini, M.: Local Optima Networks: A New Model of Combinatorial Fitness Landscapes, pp. 233–262. Springer, Heidelberg (2014). https://doi.org/10.1007/978-3-642-41888-4_9
30. Pal, D.K., Mengshoel, O.J.: Stochastic CoSaMP: randomizing greedy pursuit for sparse signal recovery. In: Proceedings of ECML-PKDD, pp. 761–776 (2016)
31. Pal, M.: Random forest classifier for remote sensing classification. Int. J. Remote Sens. **26**(1), 217–222 (2005). https://doi.org/10.1080/01431160412331269698
32. Pedregosa, F., et al.: Scikit-learn: machine learning in Python. J. Mach. Learn. Res. **12**, 2825–2830 (2011)
33. Preuss, M.: Multimodal Optimization by Means of Evolutionary Algorithms, 1st edn. Springer Publishing Company, Incorporated (2015)
34. Sánchez-Díaz, X.F.C., Masson, C., Mengshoel, O.J.: Regularized feature selection landscapes: an empirical study of multimodality. In: Parallel Problem Solving from Nature – PPSN XVIII, pp. 409–426. Springer, Cham (2024)
35. Schumann, R., Mou, L., Lu, Y., Vechtomova, O., Markert, K.: Discrete optimization for unsupervised sentence summarization with word-level extraction. In: Proceedings of ACL, pp. 5032–5042 (2020)
36. Selman, B., Levesque, H., Mitchell, D.: A new method for solving hard satisfiability problems. In: Proceedings of the AAAI, pp. 440–446 (1992)
37. Shannon, C.E.: A mathematical theory of communication. Bell Syst. Tech. J. **27**(3), 379–423 (1948). https://doi.org/10.1002/j.1538-7305.1948.tb01338.x
38. Thomson, S.L., Ochoa, G.: On funnel depths and acceptance criteria in stochastic local search. In: GECCO '22, Proceedings of the Genetic and Evolutionary Computation Conference, pp. 287–295. Association for Computing Machinery, New York, NY, USA (2022). https://doi.org/10.1145/3512290.3528831
39. Tufte, E.R.: Visual Explanations: Images and Quantities. Graphics Press, Evidence and Narrative (1997)
40. Tufte, E.R.: The Visual Display of Quantitative Information (Second Edition). Graphics Press (2001)
41. Vafaie, H., De Jong, K.: Genetic algorithms as a tool for feature selection in machine learning (1997)
42. Vikhar, P.A.: Evolutionary algorithms: a critical review and its future prospects. In: 2016 International Conference on Global Trends in Signal Processing, Information Computing and Communication (ICGTSPICC), pp. 261–265 (2016). https://doi.org/10.1109/ICGTSPICC.2016.7955308
43. Wickham, H.: Practical tools for exploring data and models. Ph. D. thesis, Iowa State University (2008)
44. Wilkinson, L.: The Grammar of Graphics. Statistics and Computing, Springer-Verlag, New York, 2 edn. (2005). https://doi.org/10.1007/0-387-28695-0
45. Wong, K.C.: Evolutionary multimodal optimization: a short survey (2015)
46. Yang, J., Honavar, V.: Feature subset selection using a genetic algorithm. In: Liu, H., Motoda, H. (eds.) Feature Extraction, Construction and Selection: A Data Mining Perspective, pp. 117–136 (1998). https://doi.org/10.1007/978-1-4615-5725-8_8
47. Yu, T., Kveton, B., Mengshoel, O.J.: Thompson sampling for optimizing stochastic local search. In: ECML-PKDD, pp. 493–510 (2017)

Mixed-Binary Problems Optimized with Fast Discrete Solver

Timo Kötzing and Aishwarya Radhakrishnan(✉)

Hasso Plattner Institute, University of Potsdam, Potsdam, Germany
aishwarya.radhakrishnan@hpi.de

Abstract. Tackling optimization in mixed domains (continuous and discrete decision variables) has recently gained attention, causing the development of various extensions of continuous optimization algorithms. In order to more accurately address the combinatorial nature of the discrete aspects of the search space, we go a different way and combine two algorithms, one for each aspect of the search space. Focusing, in the discrete part, on binary search spaces, we use Population-Based Incremental Learning (PBIL) for discrete optimization. We combine this algorithm with the Covariance Matrix Adaptation Evolutionary Strategy (CMA-ES) to address the continuous part.

We compare our CMA-ES-PBIL with two leading variants of CMA-ES from the literature: CMA-ES with Margin (CMA-ESwM) and CMA-ES with Probability Distribution Model (CMA-ES-PDM). We conduct run time analysis on some separable and a non-separable benchmark functions and show that our hybrid algorithm significantly outperforms both CMA-ESwM and CMA-ES-PDM. Our results also show that the binary part is optimized much earlier than the continuous part, raising the question: Is run time evaluation the right measure to analyze mixed-binary algorithms?

1 Introduction

Mixed-integer problems are optimizations problems in which the search space contains both continuous and integer variables. Mixed-integer optimization algorithms face the additional challenge of handling a mixed search space, compared to purely continuous or discrete problems. A survey of analyses of mixed-integer problems optimized by evolutionary algorithms can be found in [2]. This survey also includes mixed-binary combinatorial optimization problems such as feature selection, portfolio selection, unit commitment problem, construction site layout planning, and integer programming problems. A further prominent problem using evolutionary algorithms to solve mixed-integer combinatorial problems is Neural Architecture Search (NAS), where the structure of a neural network (as well as the weights) is sought. A survey on evolutionary NAS [32] reviews over 200 research papers that employ various techniques to handle NAS.

As a first step towards understanding how simple hybrid mixed-binary algorithms, like the one we propose, perform on complex mixed-integer combinatorial problems, we analyze and compare the performance of mixed-binary algorithms on some existing mixed-binary benchmark functions.

For purely continuous search spaces, the Covariance Matrix Adaptation Evolutionary Strategy (CMA-ES) [15] has been widely used due to its ability to solve non-linear problems. The CMA-ES algorithm maintains a mean vector and a covariance matrix from which solutions are sampled at each iteration. The mean vector and the covariance matrix are updated at each iteration based on some best individuals in the current population. A comparison of 31 continuous algorithms, including CMA-ES, across various benchmark functions can be found in [3,13]. The CMA-ES algorithm has been adapted for mixed-integer problems by treating all variables as continuous, which are then discretized before function evaluations [10,12,18,21,25]. Even a modification of CMA-ES for NAS was proposed in [24].

In the literature, approaches to solve mixed-integer benchmark problems are mostly approached from continuous optimization algorithms [18,25]. The Covariance Matrix Adaptation Evolutionary Strategy with Margin (CMA-ESwM) [10] is one such extension of the CMA-ES that is widely used. A brief description of this algorithm can be found in Sect. 2.1. A comparison of various mixed-integer algorithms on the mixed-integer benchmark suite [26] is provided in [9].

For the discrete part of mixed-binary problems, we consider Population-Based Incremental Learning (PBIL) algorithm [4,5]. PBIL is an Estimation of Distribution Algorithm (EDA) that combines a simple genetic algorithm with competitive learning. At each iteration, it maintains a probability vector from which solutions are sampled. PBIL learns and updates the probability distributions based on a few best individuals in the current population, using a given learning rate. The PBIL algorithm has been used to optimize many binary optimization problems and has been modified to improve the performance or to adapt to specific problem setting; a brief survey about different versions of PBIL can be found in [8]. A brief description of PBIL can be found in Sect. 2.3.

Many combinatorial problems like KNAPSACK and VERTEX COVER problems, can be modeled as binary optimization problems. Therefore, we can use EDAs to optimize combinatorial problems. The competitive learning component of the PBIL makes it more effective than other EDAs and has been studied for solving combinatorial problems as well. The application of PBIL to combinatorial optimization is explored in [19]. A PBIL method for solving the traveling salesman problem and the serial colored traveling salesman problem is discussed in [16,20], respectively. PBIL has also been used to analyze dynamic optimization problems [29–31]. Additionally, PBIL has been used to solve mixed-integer linear programming models [7].

A theoretical study exploring the convergence of PBIL on linear and non-linear problems was conducted in [17]. In [6], a modification of PBIL to handle multi-valued search spaces was proposed and analyzed theoretically. This simple EDA has also been adapted to handle continuous search spaces, as explored empirically in [23].

The use of estimation of distribution algorithms has already been explored for some applied problems such as mixed-variable News vendor and design optimization problems [22,27]. Recently, in [21], a modified version of CMA-ES incorporating Probability Distribution Model (PDM) to handle discrete part of mixed-integer problems was proposed. A brief description of this algorithm, CMA-ES-PDM, can be found in Sect. 2.2.

In this paper, we propose a hybrid algorithm that combines the standard continuous optimization algorithm CMA-ES with PBIL. We call our hybrid algorithm CMA-ES-PBIL. The proposed algorithm uses CMA-ES to optimize the continuous part and PBIL to optimize the discrete part of the algorithm. We chose PBIL over other EDAs, such as the compact genetic algorithm (cGA) or the Univariate Marginal Distribution Algorithm (UMDA), for the following reasons. PBIL requires a population size, unlike cGA, which samples only two individual at a given iteration. This makes the hybridization of the continuous algorithm CMA-ES with PBIL straightforward. PBIL is more similar to CMA-ES than UMDA is, so for this first study we chose PBIL for the discrete part. For a complete description of CMA-ES-PBIL and to know more about its advantages over CMA-ESwM, refer to Sect. 3.

The paper is structured as follows. We start with the preliminaries in Sect. 2. In Sect. 3, we provide a detailed description of the proposed algorithm CMA-ES-PBIL. Section 4 presents an empirical run time comparison among CMA-ESwM, CMA-ES-PDM, and CMA-ES-PBIL. We further examine the performance of these algorithms specific to the binary part in Sect. 5. Sections 6 and 7 offer detailed analyses of the performance of CMA-ESwM and CMA-ES-PBIL on two different functions. Finally, we conclude in Sect. 8.

Upon acceptance of the paper, the code for all the experiments will be made available online. The python code for the experiments is build upon the GitHub code in [1], provided by the authors of the CMA-ESwM paper [10].

2 Preliminaries

In this section, we introduce basic notations, describe two mixed-binary algorithms, and define the benchmark functions used in the analyses.

Let $n, m \in \mathbb{N}$, $x = (x_1, \ldots, x_n) \in \mathbb{R}^n$, and $y = (y_1, \ldots, y_m) \in \mathbb{R}^m$. Then, $(x, y) = (x_1, \ldots, x_n, y_1, \ldots, y_m) \in \mathbb{R}^{n+m}$.

Let $n, \mu \in \mathbb{N}$ and $z_1, z_2, \ldots, z_\mu \in \mathbb{R}^n$. Then, $\frac{1}{\mu} \sum_{i=1}^{\mu} z_i$ denotes the coordinate-wise mean of the μ vectors.

Let $n \in \mathbb{N}$ and $a, b \in \mathbb{R}$. Then, $[a,b]^n = \{(x_1, \ldots, x_n) \in \mathbb{R}^n \mid \forall i \in [n], a \leq x_i \leq b\}$.

The benchmark functions we analyze are listed below. Let $N_{co}, N_{bi} \in \mathbb{N}$, $N = N_{co} + N_{bi}$, and $x \in \mathbb{R}^{N_{co}} \times \{0,1\}^{N_{bi}}$.

SPHEREONEMAX$(x) = \sum_{j=1}^{N_{co}} x_j^2 + N_{bi} - \sum_{j=N_{co}+1}^{N} x_j$.

SPHERELEADINGONES$(x) = \sum_{j=1}^{N_{co}} x_j^2 + N_{bi} - \sum_{j=N_{co}+1}^{N} \prod_{k=N_{co}+1}^{j} x_k$.

ELLIPSOIDONEMAX$(x) = \sum_{j=1}^{N_{co}} \left(1000^{\frac{j-1}{N_{co}-1}} x_j\right)^2 + N_{bi} - \sum_{j=N_{co}+1}^{N} x_j$.

$$\text{ELLIPSOIDLO}(x) = \sum_{j=1}^{N_{co}} \left(1000^{\frac{j-1}{N_{co}-1}} x_j\right)^2 + N_{bi} - \sum_{j=N_{co}+1}^{N} \prod_{k=N_{co}+1}^{j} x_k.$$

Throughout our analyses, we look at the optimization process of different mixed-integer algorithms, which tries to find the minimum of the functions mentioned above.

2.1 CMA-ESwM

We use the CMA-ESwM algorithm proposed in [10]. This algorithm is a modification of the standard CMA-ES algorithm [11,15]. Given $A^{(0)}$ as an identity matrix, $m^{(0)}, \sigma^{(0)}, C^{(0)}, \lambda$ as described in Table 1, and the default hyper parameters as mentioned in [10, Table 1], the CMA-ESwM performs the following steps at each iteration t. Here, we provide a brief outline of the algorithm; for a detailed description, please refer to [10].

Step 1. For $i \in \{1,\ldots,\lambda\}$, $y_i \sim \mathcal{N}(0, C^{(t)})$ and λ individuals are created as $x_i = m^{(t)} + \sigma^{(t)} y_i$.

Step 2. For $i \in \{1,\ldots,\lambda\}$, the affine-transformed solutions v_i are calculated as $v_i = m^{(t)} + \sigma^{(t)} A^{(t)} y_i$.

Step 3. For $i \in \{1,\ldots,\lambda\}$, the discretized v_i, i.e., $\bar{v}_i$ are evaluated by the objective function f. Then, the sets $\{x_{1:\lambda},\ldots,x_{\lambda:\lambda}\}$ and $\{y_{1:\lambda},\ldots,y_{\lambda:\lambda}\}$ are sorted such that the indices correspond to $f(\bar{v}_{1:\lambda}) \leq \cdots \leq f(\bar{v}_{\lambda:\lambda})$.

Step 4. Update $m^{(0)}, C^{(0)}$ and $\sigma^{(0)}$ based on the update rules given in [10,11].

Step 5. Modify $m^{(t+1)}$ and update $A^{(t)}$ according to Sects. 4.3 and 4.4 in [10].

2.2 CMA-ES-PDM

A modification of CMA-ES-IM with Integer Mutation [12], called CMA-ES-PDM (CMA-ES with Probability Distribution Models of integer variables) was proposed by Nguyen in [21]. This algorithm maintains a probability distribution model which gets updated based on the binary/integer part of the individuals collected, specifically for this purpose, over consecutive T^{period} iteration. The CMA-ES-PDM introduces following two main modification to the CMA-ES-IM.

- At each iteration, with probability β, the integer parts of the individuals are sampled from the probability distribution model.
- The probability distribution is updated based on the set of individuals, stack_x, collected over T^{period} consecutive iterations. The probability distribution model for the binary variables is created and updated in the following way; the probability that a given binary bit position is 1 is calculated as the mean of number of 1s in $\lfloor \rho |\text{stack}_x| \rfloor$ best individual in stack_x.

The first probability distribution model is created after collecting a set of $\max(20N, 1000)$ individuals, where N is the dimension of the search space. Other parameter values can be found in Table 1. For further explanation of the algorithm, refer to [21, Algorithm 2].

2.3 PBIL

The population-based incremental learning (PBIL) algorithm maintains a probability vector at each iteration. It starts with a probability of 1/2 for each bit position to be 1. At each iteration, λ individuals are sampled and sorted based on their fitness values. The μ best individuals are then selected, and the probability vector is updated according to the specified learning rate r, as described in Line 5 of Algorithm 1. If any values in the probability vector fall outside the interval $[\alpha, 1-\alpha]$, then that value is corrected to be in the suitable boundary (either α or $1-\alpha$) of this interval (lines $7-8$).

Algorithm 1: Population-Based Incremental Learning (PBIL) optimizing f, given $n, \lambda, \mu \in \mathbb{N}$ and $r, \alpha \in (0, 1)$.

1   $p_{n \times 1} \leftarrow (0.5, \ldots, 0.5)^T$;
2   $P \leftarrow \{x_i \in \{0,1\}^n \mid \forall i \in [\lambda], x_i \sim p\}$;
3   **while** *termination criteria not met* **do**
4   select top μ individuals $y_1 \ldots y_\mu$ in P based on f;
5      $p \leftarrow (1-r) \cdot p + r \cdot \frac{1}{\mu} \sum_{i=1}^{\mu} y_i$;
6   **for** *i in {1, ...n}* **do**
7   **if** $p_i > 1-\alpha$ **then** $p_i = 1-\alpha$;
8   **else if** $p_i < \alpha$ **then** $p_i = \alpha$;
9      $P \leftarrow \{x_i \mid \forall i \in [\lambda], x_i \sim p\}$;

---

3 CMA-ES-PBIL

We propose a hybrid algorithm for mixed-binary problems that combines two standard algorithms: CMA-ES and PBIL. This algorithm uses CMA-ES to optimize the continuous variables and PBIL to optimize the binary variables. The continuous and the binary part of the solutions are sampled separately from CMA-ES and PBIL, respectively. These parts are then combined and evaluated based on the fitness function that the algorithm aims to optimize.

At any given iteration t, the mean vector $m^{(t)}$ and the covariance matrix $C^{(t)}$ of the CMA-ES have dimensions $N_{co} \times 1$ and $N_{co} \times N_{co}$, respectively, while the probability vector $p_{pbil}^{(t)}$ corresponding to the PBIL has a dimension of $N_{bi} \times 1$. Given the parameter values for both CMA-ES and PBIL, the algorithm, which we call CMA-ES-PBIL, performs the following steps at each iteration t.

Step 1. For $i \in \{1, \ldots, \lambda\}$, $y_i \sim \mathcal{N}(0, C^{(t)}_{N_{co} \times N_{co}})$ and the continuous part of the λ individuals are created as $x_i = m^{(t)}_{N_{co} \times 1} + \sigma^{(t)} y_i$.

Step 2. For $i \in \{1, \ldots, \lambda\}$, the binary part of the individuals v_i are sampled from the distribution $p_{pbil}^{(t)}$.

Step 3. For $i \in \{1, \ldots, \lambda\}$, the mixed-binary individuals X_i are created by combining x_i and v_i, $X_i = (x_i, v_i)$.

Step 4. For $i \in \{1, \ldots, \lambda\}$, the individuals X_i are evaluated by the objective function f. Then, the sets $\{x_{1:\lambda}, \ldots, x_{\lambda:\lambda}\}$ and $\{v_{1:\lambda}, \ldots, v_{\lambda:\lambda}\}$ are sorted such that the indices correspond to $f(X_{1:\lambda}) \leq \cdots \leq f(X_{\lambda:\lambda})$.

Step 5. Update $m^{(0)}, C^{(0)}$ and $\sigma^{(0)}$ based on the update rules given in [10, 11] for the CMA-ES. Additionally, for the PBIL, update the probability vector $p_{pbil}^{(t)}$ based on the update rule specified in Algorithm 1 (line 5).

Table 1. Parameter values for CMA-ESwM, CMA-ES-PBIL and CMA-ES-PDM.

Parameters	CMA-ESwM	CMA-ES-PBIL	CMA-ES-PDM
λ	$4 + \lfloor 3\ln(N) \rfloor$	$4 + \lfloor 3\ln(N) \rfloor$	$4 + \lfloor 3\ln(N) \rfloor$
μ	$\lfloor \lambda/2 \rfloor$	$\lfloor \lambda/2 \rfloor$	$\lfloor \lambda/2 \rfloor$
Margin	$\alpha = \frac{1}{N_{bi} \cdot \lambda}$	$[\frac{1}{N_{bi} \cdot \lambda}, 1 - \frac{1}{N_{bi} \cdot \lambda}]$	$\alpha = \frac{1}{N_{bi} \cdot \lambda}$
$m^{(0)}$	u.a.r from $[1, 3]^N$	u.a.r from $[1, 3]^{N_{co}}$	u.a.r from $[1, 3]^N$
$C^{(0)}$	Identity Matrix $I_{N \times N}$	Identity Matrix $I_{N_{co} \times N_{co}}$	Identity Matrix $I_{N \times N}$
$\sigma^{(0)}$	1	1	1
Learning Rate r	–	0.15	–
ρ	–	–	0.1
β	–	–	0.9

The following two points are immediate advantages of the CMA-ES-PBIL.

- At each iteration CMA-ESwM maintains a covariance matrix of size $N \times N$ while CMA-ES-PBIL maintains a covariance matrix of size only $N_{co} \times N_{co}$. This speeds up a few calculations within the algorithm.
- The parameter μ of the CMA-ES and PBIL does not need to be the same value. Changing the parameters of PBIL does not directly affect the CMA-ES part. Therefore, parameters of the PBIL can be optimized to improve the performance in the binary part without necessarily compromising the performance of the CMA-ES in the continuous part.

Note that having covariance matrix only for the continuous variables comes at the cost of no longer correlating with discrete variables. This is not an issue for separable fitness functions. In Sect. 7, we analyze a non-separable function to show that this does not necessarily poses a problem for non-separable functions.

The parameter values for the CMA-ESwM, CMA-ES-PDM and CMA-ES-PBIL are listed in Table 1. Although parameters of PBIL could be tuned to give better performance, we consider the same parameter values for the CMA-ESwM and PBIL to ensure a level of fairness in our comparison analyses. The parameter λ represents the number of individuals in the population, while μ denotes the number of best individuals selected from the λ sampled individuals used for updating the CMAES and PBIL models. The parameters $m^{(0)}$, $C^{(0)}$ and $\sigma^{(0)}$ correspond to the mean, covariance matrix and the step size of the CMA-ES algorithm, respectively. The learning rate of PBIL is indicated by r. The parameters ρ and β refer to the proportion of stored individuals considered to update the probability distribution model and the probability of sampling the binary part from the probability distribution model in CMA-ES-PDM, respectively. Note that the margin parameter α is $\frac{1}{N_{bi}}$ and not $\frac{1}{N}$, as the margin is for the binary part to recover if it was initially guided in the wrong direction.

4 Empirical Run Time Analysis

In this section, we compare the CMA-ES-PBIL, CMA-ESwM, and CMA-ES-PDM based on the number of function evaluations each algorithm takes to find an individual with a fitness value of at most 10^{-10}. We start with an overall run time analysis before examining the detailed performance of these algorithms in optimizing the binary part.

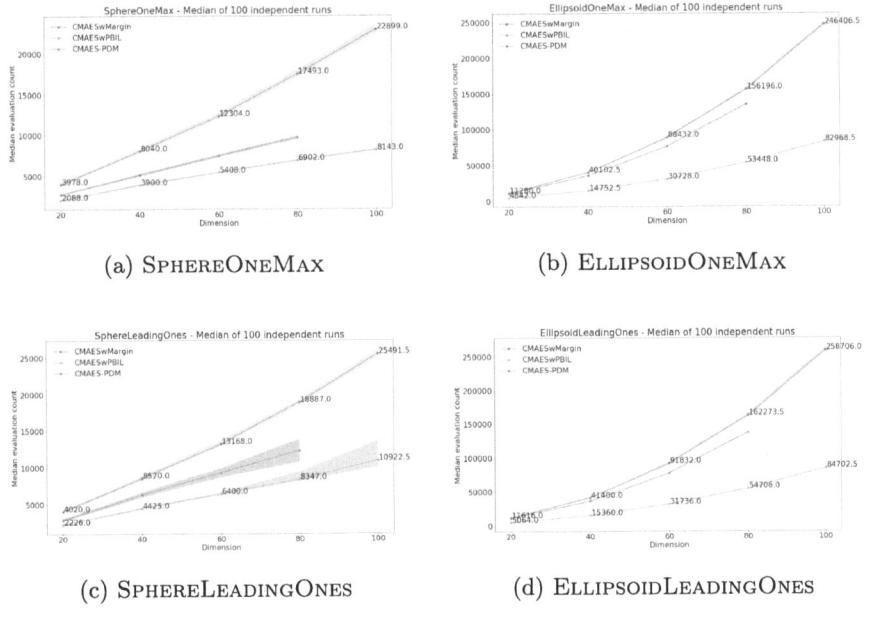

Fig. 1. Run time comparison of the CMA-ESwM, CMA-ES-PDM, and CMA-ES-PBIL. In each plot, the x-axis represents the total dimension $N = N_{co} + N_{bi}$, and the y-axis shows the median number of function evaluations.

All parameter values for the CMA-ESwM and CMA-ES-PBIL are provided in Table 1, and for CMA-ES-PDM the parameter values are as suggested by

Nguyen in [21]. We consider five different dimensions for our comparison, $N \in \{20, 40, 60, 80, 100\}$, and as commonly practiced, $N_{co} = N_{bi} = N/2$. For each dimension, we consider the median of 100 independent runs of each algorithm. The experiments for the CMA-ES-PDM were not re-run; instead, we report the values published in [21] for comparison, noting that values for $N = 100$ are not available.

A single run of an algorithm terminates when either the algorithm found an individual with a fitness value of at most 10^{-10}, or the minimum eigenvalue of $\delta^2 C$ is less than 10^{-30}. A run is considered successful if the algorithm finds an individual with fitness at most 10^{-10} when a termination criterion is met. We observe an empirical success rate of 1 for the CMA-ESwM and the CMA-ES-PBIL across all four objective functions considered. The CMA-ES-PDM, however, does not always succeed in the case of the SPHERELEADINGONES. More details about the success rate of the CMA-ES-PDM are available in [21, Table 2].

Each plot in Fig. 1 corresponds to one of the four benchmark functions used for the comparison. The x-axis represents the total dimension $N = N_{co} + N_{bi}$, and the y-axis shows the median number of function evaluations required to find an individual with a fitness value of at most 10^{-10}. The solid lines represent the median number of function evaluations each algorithm took to find an individual with a fitness value of at most 10^{-10}. The shaded region shows the area between the median of the 25th and 75th percentile. The orange and the green solid lines in Fig. 1 (b) and (d) do not appear to have the shaded area because of the y-axis scale is in thousands, while the corresponding 25th and 75th percentile values are in the hundreds.

In Fig. 1, we observe that the CMA-ES-PBIL outperforms the other two algorithms across all dimensions for all the four test functions considered. Since all three algorithms use the CMA-ES for the continuous part, the observed differences in run time are likely due to variations in their performance when optimizing the binary part. In the following sections we investigate this in detail.

5 Performance Analysis on Binary Variables

In this section, we examine the performance of the CMA-ESwM and CMA-ES-PBIL algorithms with respect to the binary variables. We evaluate how quickly each algorithm finds the optimum for binary variables and how long it takes for all the individuals in the population to reach this optimum.

For this analysis, at each iteration of the CMA-ESwM and CMA-ES-PBIL, we record the number of individuals among the λ individuals in the population that have a non-zero binary fitness. Each plot in Fig. 2 shows the median of these recorded values across 100 independent runs of the algorithms on each benchmark function. The plots in Fig. 2 represent results for dimensions $N = 40$; similar results for other dimensions are available in Appendix. There is no particular reason to choose $N = 40$, other than to save space for further analyses.

Since each algorithm run reaches a desired individual (with fitness value of at most 10^{-10}) at a different iterations, we find the minimum number of iterations,

l, among the 100 independent runs and display median number of individuals with non-zero binary fitness only up to l iterations. For $i \in \{1, \ldots, 100\}$, if t_i is the iteration at which the i-th independent run of the CMA-ES-PBIL found a desired individual, then $l = \min\{t_i \mid i \in \{1, \ldots, 100\}\}$. Let $u = \max\{t_i \mid i \in \{1, \ldots, 100\}\}$. The black dashed vertical lines represents the iteration l, while the black solid vertical line indicates the maximum iteration number u of the CMA-ES-PBIL over 100 independent runs.

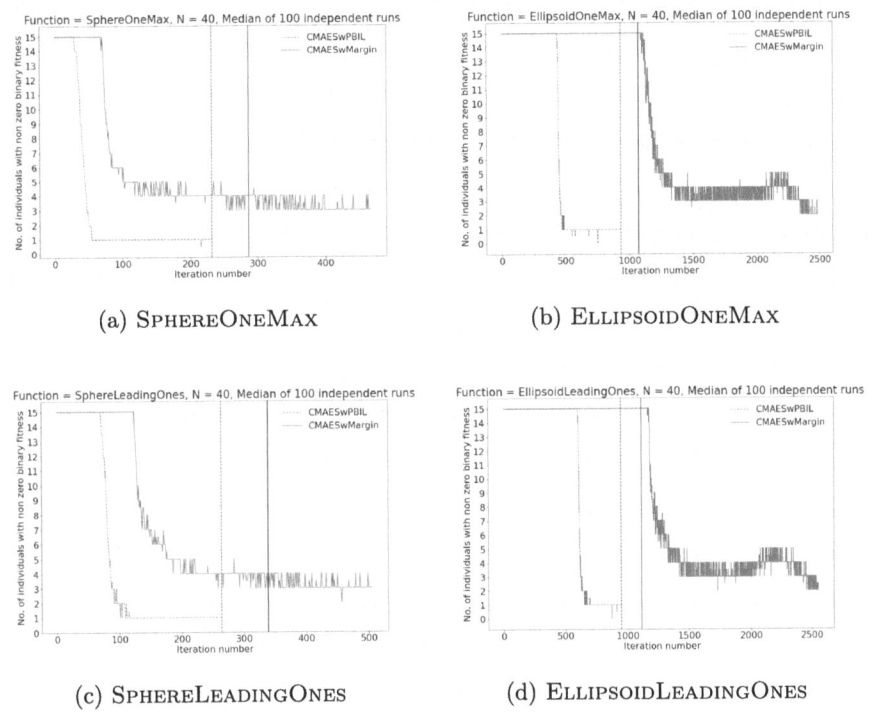

Fig. 2. Number of individuals with non-zero binary fitness at each iteration of the CMA-ESwM and CMA-ES-PBIL on all four benchmark function for dimension $N = 40$.

Note that the x-axis represents the iteration number, not the number of function evaluations. The y-axis indicates the number of individuals with non-zero binary fitness at a given iteration.

We observe that the CMA-ES-PBIL finds the optimum of the binary part within first half of its total iterations required to find an individual with a fitness of at most 10^{-10}. Also its probability distribution converges to the margin faster than the CMA-ESwM. Since both algorithms use CMA-ES for the continuous part, performance of the CMA-ES-PBIL is better for all the considered mixed-binary benchmark functions, because of its ability to optimize the binary part more efficiently and comparatively smaller covariance matrix.

The performance of the mixed-integer algorithms is often compared only based on the number of function evaluations required to find an individual with fitness of at most 10^{-10} [9,10,21,28]. This does not provide complete insight into the performance of the algorithms concerning the binary variables. Both CMA-ESwM and CMA-ES-PBIL rely on CMA-ES for optimizing the continuous variables. When comparing mixed-binary algorithms that share the same continuous optimization approach, it is important to include a performance measure specifically evaluating optimization in the binary part. Depending on the optimization goals, different comparison methods, like the one mentioned in this section, provide a better understanding of the optimization process of mixed-binary algorithms.

6 Analyses of SPHEREELLIPSOID Function

In this section, we introduce the SPHEREELLIPSOID function to analyze the behavior of CMA-ESwM and CMA-ES-PBIL on mixed-binary functions with ELLIPSOID or BINVAL-like binary part. ELLIPSOID and BINVAL are functions in which the coefficients are arranged in descending order of magnitude, with the leftmost bit having the highest coefficient.

SPHEREELLIPSOID$(x) = \sum_{j=1}^{N_{co}} x_j^2 + \sum_{j=N_{co}+1}^{N} 10^{\frac{-10}{N-j+1}}(1 - x_j)$.

For a given bit position j, instead of using the usual coefficient $1000^{\frac{j-1}{N_{bi}-1}}$, we have the coefficient described above to illustrate the following behavior of the CMA-ES-PBIL. Although the continuous part dominates fitness evaluations until the algorithm finds an individual with a fitness of at most $10^{-10/N_{bi}}$ in the continuous part, CMA-ES-PBIL still outperforms CMA-ESwM. This is noteworthy because, until the fitness value corresponding to the continuous part falls below $10^{-10/N_{bi}}$, PBIL need not get useful information from the fitness evaluations to update the probability distribution in the correct direction.

We conduct run time analysis (Fig. 3 (a)) to compare the performance of the CMA-ESwM and CMA-ES-PBIL on SPHEREELLIPSOID. We then extend our analyses to understand the optimization process of binary variables by looking at the probability vector, which contains information about the value of each binary bit position (Fig. 3 (b), (c) and (d)). Below, we provide a detailed explanation of the experiments conducted to understand the optimization processes of CMA-ESwM and CMA-ES-PBIL on SPHEREELLIPSOID.

In Fig. 3 (a), the x-axis represents the dimension and the y-axis indicates the number of function evaluations required to find an individual with fitness of at least 10^{-10}. The parameter values are consistent with those used for other benchmark functions and are listed in Table 1. Similar to the run time analysis in Sect. 4, the solid lines show the median of 100 independent runs, while the shaded regions represent the area between 25th and 75th percentiles. It can be observed that the CMA-ES-PBIL outperforms the CMA-ESwM in minimizing SPHEREELLIPSOID as well.

The Fig. 3 (b) gives insight into the optimization process of the CMA-ES-PBIL on the binary variables. Specifically, this plot shows how many iterations on average over 100 independent runs the CMA-ES-PBIL takes until the probability of being 1 value corresponding to each bit positions reaches a value of 0.9 or higher for the first time. The x-axis represents the binary bit position, and the y-axis indicates the first iteration at which the probability value exceeds 0.9. Each colored curve in the plot corresponds to a binary dimension in the set $\{10, 20, 30, 40, 50\}$.

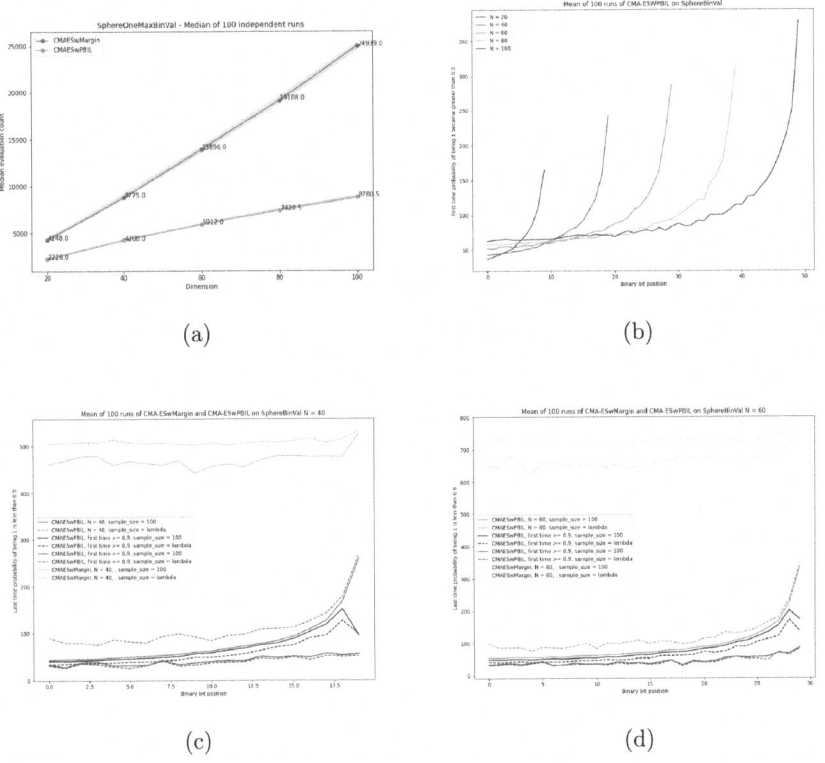

Fig. 3. Analyses of the CMA-ESwM and CMA-ES-PBIL on SphereEllipsoid function. The plot in (a) presents the run time analysis, while the remaining plots analyze the probability distribution calculated based on the population at each iteration. (Color figure online)

It can be observed that the optimization happens in such a way that the probability value corresponding to the bit position with highest coefficient value reaches a value greater than 0.9 faster compared to the bit position with lower coefficient values. This behavior is expected, as the fitness value is influenced by the bit positions with the highest coefficient value, and thus, the probability values corresponding to the other positions need not be guided in the right direction. This suggests that the CMA-ES-PBIL can recover efficiently even if initially the solutions are deviated from the optimum.

The CMA-ESwM does not maintain a probability vector for the binary variables at each iteration. In order to compare the performance of the CMA-ESwM with CMA-ES-PBIL in the binary part, we designed the following experiment. At each iteration t of the CMA-ESwM, we sampled 100 individuals based on the parameter values at that iteration and calculated a probability distribution based on these 100 individuals. Let $P_{samp}^t = \{z_i \in \mathbb{R}^{N_{bi}} \mid i \in \{1,\ldots,100\}\}$ be the set of all sampled individuals. Then, the probability vector based on these sampled individuals p_{samp}^t, which indicates the probability that a given bit position has a value of 1, is calculated as $p_{samp}^t = \frac{1}{100}\sum_{i=1}^{100} z_i$. We applied the same procedure for CMA-ES-PBIL.

For each binary bit position, we identify the iteration t at which this sampled probability value first exceeds 0.9, as well as the last iteration during which this value was below 0.9. The solid lines in Fig. 3 (c) and (d) are based on this sampled probability distribution. We also report the same for the probability distribution calculated based on the λ individuals in the population, indicated by the dotted lines in Fig. 3 (c) and (d). These plots are for dimension $N = 40$ and 60; additional plots for other dimensions ($N = 20, 80$ and 100) can be found in the Appendix.

In case of the CMA-ES-PBIL, except for the last few bit positions, once the probability value reaches at least 0.9 for the first time, it remains relatively stable, staying above 0.9 with little fluctuation. This behavior can be observed from the black and purple lines in Fig. 3 (c). In contrast, with CMA-ESwM, although the probability value of every bit reaches a value of at least 0.9 sooner than in CMA-ES-PBIL, it does not consistently maintain this distribution. For each bit position, the probability of it being 1 requires more iterations to consistently stay above 0.9 compared to the CMA-ES-PBIL. This behavior can be observed from the blue and orange lines in Fig. 3 (c). For each position, the iteration at which the probability values remain above 0.9 tends to occur later, often towards the end of the optimization process (orange line). In CMA-ES-PBIL, however, the probability reach and maintain values above 0.9 sequentially from the first bit to the last bit (purple lines).

This property of the CMA-ES-PBIL is beneficial when there are restrictions on the number of function evaluations. Since bit positions with higher coefficients are optimized first, CMA-ES-PBIL can yield better solutions compared to CMA-ESwM. This behavior of the CMA-ES-PBIL is observed because analyses beyond simple run time analysis were conducted. This instance further support our claim that deeper comparative analyses on the optimization of the binary part are beneficial, particularly when the continuous part of the comparing algorithms are handled by the same continuous optimization algorithm.

7 Analyses of ROSENBROCK Function

The benchmark functions we have considered until now are all separable functions. In this section, we look at a non-separable function, the well-known Rosenbrock function. We allow only binary values for the last half of the solution to make it a mixed-binary Rosenbrock function. We consider this mixed-binary ROSENBROCK function in order to analyze the performance of the CMA-ES-PBIL on non-separable functions and to observe differences in the performance of CMA-ESwM and CMA-ES-PBIL compared to the performance of the CMA-ES on continuous ROSENBROCK.

ROSENBROCK$(x) = \sum_{j=1}^{N-1}(100(x_j^2 - x_{j+1})^2 + (x_j - 1)^2)$.

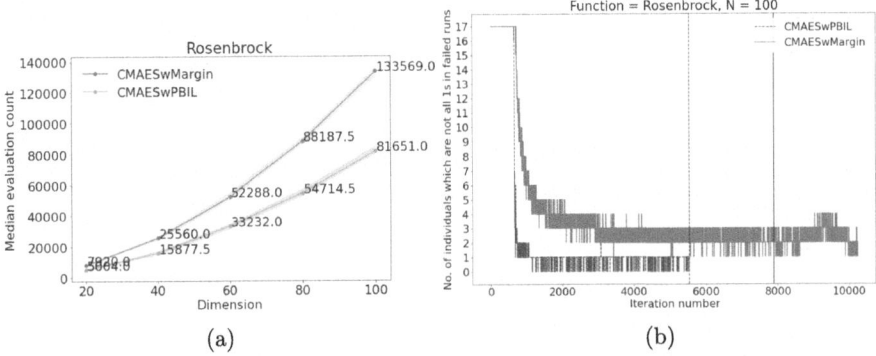

Fig. 4. Analyses of CMA-ESwM and CMA-ES-PBIL on ROSENBROCK function. The plot on the left (a) shows the run time analysis. The plot on the right (b) displays, for each iteration, the number of individuals in the population with a non-optimal binary component (not all 1s).

Note that the optimum of the ROSENBROCK function is the all 1 string. More properties of the ROSENBROCK function can be found in BBOB 2008 benchmark suite [3, Function 8] paper.

We conduct a run time experiment similar to the experimental setup mentioned in Sect. 4; we independently run both the CMA-ESwM and CMA-ES-PBIL algorithms 100 times with the parameters mentioned in Table 1. Unlike for other benchmark functions considered, neither of the algorithms consistently succeeds in finding an individual with fitness value of at most 10^{-10}. This behavior is expected, as similar results were observed when CMA-ES was used to optimize the continuous ROSENBROCK [9].

In Fig. 4 (a), each curve represents the median number of function evaluations required to find a desired individual across the successful runs of the algorithms. Recall that a run is considered successful if the algorithm finds an individual with a fitness value of at most 10^{-10}. In Table 2, we report the number of successful runs out of 100 independent runs for each algorithm across dimensions $N \in \{20, 40, 60, 80, 100\}$. For CMA-ESwM, the success rate tends to drop as dimensions increase, whereas for CMA-ES-PBIL, the success rate increases with higher dimensions. We further investigate whether these differences in the success rate are due to the better performance of the CMA-ES-PBIL on the binary part.

In Fig. 4 (b), we perform a similar analysis to that in Sect. 5, but focused on the unsuccessful runs. The x-axis shows the iteration number, and on the y-axis we have the number of individuals that are not all 1s in the binary part. Each curve represents the median value taken across failed runs. From this plot we observe that the decrease in the success rate of the CMA-ESwM is not because of its performance in the binary part. In most cases, both the CMA-ESwM and CMA-ES-PBIL found the optimum with respect to the binary part even though the run is unsuccessful. This implies that the failure of the algorithm is because the mean vector of the CMA-ES could not recover from the local optima, rather due to the difficulty in finding the optimum with respect to the binary part. We suspect that the increase in the success rate of the CMA-ES-PBIL for higher dimensions could be because of its covariance matrix size compared to the CMA-ESwM.

Table 2. Number of successful runs out of 100 independent runs for each dimension.

Algorithm	$N = 20$	$N = 40$	$N = 60$	$N = 80$	$N = 100$
CMA-ESwM	41	63	51	56	35
CMA-ES-PBIL	43	54	55	66	65

8 Conclusion

In this paper, we proposed a hybrid algorithm, CMA-ES-PBIL, for solving mixed-binary problems. We compare the total function evaluations required by CMA-ESwM, CMA-ES-PDM, and CMA-ES-PBIL to find a desired solution and showed that the CMA-ES-PBIL outperforms the other two algorithms across all the benchmark functions considered.

We observed that, during the optimization process, the algorithm takes at most half of the total run time to find the optimum in the binary part. Furthermore, when the binary part of the mixed-binary problem has decreasing coefficients, the algorithm finds the optimal value for each binary bit position in the same decreasing order as the coefficients. This behavior is useful when we have certain restrictions, such as limitations on the total budget for function evaluations. These observations suggests that run time analysis alone may not provide a complete understanding of the optimization process within the binary part of mixed-binary algorithm.

Finally, we analyzed a non-separable function and showed that CMA-ES-PBIL outperforms CMA-ESwM in this scenario as well. We observed that, although both CMA-ES-PBIL and CMA-ESwM occasionally fail to find a desired solution, such failures occur less frequently in higher dimensions for CMA-ES-PBIL. Additionally, we observed that, both in CMA-ESwM and CMA-ES-PBIL, failed runs are due to the continuous part rather than the binary part.

A natural extension of this work would be to explore multi-valued EDAs like r-PBIL mentioned in [6] and perform comparative analyses using the COCO platform on all available mixed-integer benchmark functions [14]. Note that we only proposed one potential hybridization. Further research is necessary to better understand the interplay between the continuous and binary parts, particularly for addressing practical mixed-integer combinatorial problems such as neural architecture search.

References

1. Cma-es with margin. https://github.com/EvoConJP/CMA-ES_with_Margin. Accessed 30 Jun 2024
2. Akay, B., Karaboga, D., Gorkemli, B., Kaya, E.: A survey on the artificial bee colony algorithm variants for binary, integer and mixed integer programming problems. Appl. Soft Comput. **106**, 107351 (2021). https://doi.org/10.1016/j.asoc.2021.107351
3. Auger, A., Finck, S., Hansen, N., Ros, R.: Bbob 2009: comparison tables of all algorithms on all noiseless functions. INRIA (2010)
4. Baluja, S.: Population-based incremental learning: a method for integrating genetic search based function optimization and competitive learning. Tech. Rep. CMU-CS-94-163, Carnegie Mellon University, Pittsburgh, PA (1994)
5. Baluja, S., Caruana, R.: Removing the genetics from the standard genetic algorithm. In: Prieditis, A., Russell, S. (eds.) Machine Learning Proceedings 1995, pp. 38–46. Morgan Kaufmann, San Francisco (CA) (1995). https://doi.org/10.1016/B978-1-55860-377-6.50014-1
6. Ben Jedidia, F., Doerr, B., Krejca, M.S.: Estimation-of-distribution algorithms for multi-valued decision variables. Theoret. Comput. Sci. **1003**, 114622 (2024). https://doi.org/10.1016/j.tcs.2024.114622
7. Fallah, M.K., Fazlali, M., Daneshtalab, M.: A symbiosis between population based incremental learning and LP-relaxation based parallel genetic algorithm for solving integer linear programming models. Computing , 1–19 (2021). https://doi.org/10.1007/s00607-021-01004-x
8. Folly, K.A.: A short survey on population-based incremental learning algorithm. In: 2019 IEEE Symposium Series on Computational Intelligence (SSCI), pp. 339–344 (2019). https://doi.org/10.1109/SSCI44817.2019.9002858
9. Hamano, R., Saito, S., Nomura, M., Shirakawa, S.: Benchmarking CMA-ES with margin on the bbob-mixint testbed. In: GECCO '22, Proceedings of the Genetic and Evolutionary Computation Conference Companion. Association for Computing Machinery, New York, NY, USA (2022). https://doi.org/10.1145/3520304.3534043
10. Hamano, R., Saito, S., Nomura, M., Shirakawa, S.: CMA-ES with margin: lower-bounding marginal probability for mixed-integer black-box optimization. In: GECCO '22, Proceedings of the Genetic and Evolutionary Computation Conference, pp. 639–647. Association for Computing Machinery, New York, USA (2022). https://doi.org/10.1145/3512290.3528827
11. Hansen, N.: The CMA evolution strategy: a tutorial. INRIA (2010). https://doi.org/10.48550/arXiv.1604.00772
12. Hansen, N.: A CMA-ES for mixed-integer nonlinear optimization (2011)

13. Hansen, N., Auger, A., Ros, R., Finck, S., Pošík, P.: Comparing results of 31 algorithms from the black-box optimization benchmarking bbob-2009. In: GECCO '10, Proceedings of the 12th Annual Conference Companion on Genetic and Evolutionary Computation, pp. 1689–1696. Association for Computing Machinery, New York, NY, USA (2010). https://doi.org/10.1145/1830761.1830790
14. Hansen, N., Auger, Anne annd Ros, R., Mersmann, O., Tušar, T., Brockhoff, D.: Coco: a platform for comparing continuous optimizers in a black-box setting. Optim. Methods Softw. **36**(1), 114–144 (2021). https://doi.org/10.1080/10556788.2020.1808977
15. Hansen, N., Müller, S.D., Koumoutsakos, P.: Reducing the time complexity of the derandomized evolution strategy with covariance matrix adaptation (CMA-ES). Evol. Comput. **11**(1), 1–18 (2003). https://doi.org/10.1162/106365603321828970
16. He, Z., Wei, C., Jin, B., Pei, W., Yang, L.: A new population-based incremental learning method for the traveling salesman problem. In: Proceedings of the 1999 Congress on Evolutionary Computation-CEC99 (Cat. No. 99TH8406), vol. 2, p. 1156 (1999). https://doi.org/10.1109/CEC.1999.782553
17. Höhfeld, M., Rudolph, G.: Towards a theory of population-based incremental learning. In: Proceedings of 1997 IEEE International Conference on Evolutionary Computation (ICEC '97), pp. 1–5 (1997). https://doi.org/10.1109/ICEC.1997.592258
18. Ikeda, K., Ono, I.: Natural evolution strategy for mixed-integer black-box optimization. In: GECCO '23, Proceedings of the Genetic and Evolutionary Computation Conference, pp. 831–838. Association for Computing Machinery, New York, NY, USA (2023). https://doi.org/10.1145/3583131.3590518
19. Li-Yong, Y., Bing-Yao, J.: Application and research of pbil algorithm on combinatorial optimization. In: 2010 International Conference on Artificial Intelligence and Computational Intelligence, vol. 2, pp. 562–565 (2010). https://doi.org/10.1109/AICI.2010.237
20. Meng, X., Li, J., Zhou, M., Dai, X., Dou, J.: Population-based incremental learning algorithm for a serial colored traveling salesman problem. IEEE Trans. Syst. Man Cybern. Syst. **48**(2), 277–288 (2018). https://doi.org/10.1109/TSMC.2016.2591267
21. Nguyen, D.M.: A combination of CMA-ES with probability distributions of integer variables for mixed-integer black-box optimization. In: GECCO '24 Proceedings of the Genetic and Evolutionary Computation Conference Companion, pp. 415–418. Companion, Association for Computing Machinery, New York, NY, USA (2024). https://doi.org/10.1145/3638530.3654220
22. Rao, S.S., Xiong, Y.: A hybrid genetic algorithm for mixed-discrete design optimization. J. Mech. Design **127**(6) (2004). https://doi.org/10.1115/1.1876436
23. Sebag, M., Ducoulombier, A.: Extending population-based incremental learning to continuous search spaces. In: Eiben, A.E., Bäck, T., Schoenauer, M., Schwefel, H.P. (eds.) Parallel Problem Solving from Nature — PPSN V, pp. 418–427. Springer Berlin Heidelberg, Berlin, Heidelberg (1998)
24. Sinha, N., Chen, K.W.: Neural architecture search using covariance matrix adaptation evolution strategy. Evol. Comput. **32**(2), 177–204 (2024). https://doi.org/10.1162/evco_a_00331
25. Thomaser, A., De Nobel, J., Vermetten, D., Ye, F., Bäck, T., Kononova, A.: When to be discrete: analyzing algorithm performance on discretized continuous problems. In: GECCO '23, Proceedings of the Genetic and Evolutionary Computation Conference, pp. 856–863. Association for Computing Machinery, New York, NY, USA (2023). https://doi.org/10.1145/3583131.3590410

26. Tušar, T., Brockhoff, D., Hansen, N.: Mixed-integer benchmark problems for single- and bi-objective optimization. In: GECCO 2019 -The Genetic and Evolutionary Computation Conference. Prague, Czech Republic (2019). https://inria.hal.science/hal-02067932
27. Wang, F., Li, Y., Zhou, A., Tang, K.: An estimation of distribution algorithm for mixed-variable newsvendor problems. IEEE Trans. Evol. Comput. **24**(3) (2020). https://doi.org/10.1109/TEVC.2019.2932624, cited by: 85
28. Watanabe, Y., Uchida, K., Hamano, R., Saito, S., Nomura, M., Shirakawa, S.: (1+1)-CMA-ES with margin for discrete and mixed-integer problems. In: GECCO '23, Proceedings of the Genetic and Evolutionary Computation Conference, pp. 882–890. Association for Computing Machinery, New York, NY, USA (2023). https://doi.org/10.1145/3583131.3590516
29. Yang, S., Yao, X.: Population-based incremental learning with associative memory for dynamic environments. IEEE Trans. Evol. Comput. **12**, 542–561 (2008)
30. Yao, X.: Dual population-based incremental learning for problem optimization in dynamic environments. In: Proceedings of the 7th Asia Pacific Symposium on Intelligent and Evolutionary Systems (2003)
31. Yao, X.: Experimental study on population-based incremental learning algorithms for dynamic optimization problems. Soft Comput. **9**, 815–834 (2005). https://doi.org/10.1007/s00500-004-0422-3
32. Zhou, X., Qin, A.K., Sun, Y., Tan, K.C.: A survey of advances in evolutionary neural architecture search. In: 2021 IEEE Congress on Evolutionary Computation (CEC), pp. 950–957 (2021). https://doi.org/10.1109/CEC45853.2021.9504890

Feature-Based Evolutionary Diversity Optimization of Discriminating Instances for Chance-Constrained Optimization Problems

Saba Sadeghi Ahouei[1](✉), Denis Antipov[2], Aneta Neumann[1], and Frank Neumann[1]

[1] Optimisation and Logistics, School of Computer and Mathematical Sciences, The University of Adelaide, Adelaide, SA, Australia
saba.sadeghiahouei@adelaide.edu.au
[2] LIP6, Sorbonne University, Paris, France

Abstract. Algorithm selection is crucial in the field of optimization, as no single algorithm performs perfectly across all types of optimization problems. Finding the best algorithm among a given set of algorithms for a given problem requires a detailed analysis of the problem's features. To do so, it is important to have a diverse set of benchmarking instances highlighting the difference in algorithms' performance. In this paper, we evolve diverse benchmarking instances for chance-constrained optimization problems that contain stochastic components characterized by their expected values and variances. These instances clearly differentiate the performance of two given algorithms, meaning they are easy to solve by one algorithm and hard to solve by the other. We introduce a $(\mu + 1)$ EA for feature-based diversity optimization to evolve such differentiating instances. We study the chance-constrained maximum coverage problem with stochastic weights on the vertices as an example of chance-constrained optimization problems. The experimental results demonstrate that our method successfully generates diverse instances based on different features while effectively distinguishing the performance between a pair of algorithms.

Keywords: Diversity Optimization · Benchmarking · Evolutionary Algorithms

1 Introduction

Finding an algorithm that performs perfectly on all optimization problems is impossible. Hence, with the vast number of algorithms and heuristics developed over the years, it is crucial to determine suitable algorithms for solving a specific problem. To do so, it is effective to assess the behavior of different algorithms in solving the problem regarding its characteristics. A diverse set of discriminating instances that accentuate the difference in the performance of algorithms helps to study their behavior, strengths, and weaknesses in solving a particular type of problem [4]. An instance is called *discriminating* for a pair of algorithms if it is easy to solve by one algorithm and hard to solve by the other.

In this paper, we propose a feature-based diversity approach to evolve diverse sets of differentiating instances for chance-constrained optimization problems. Since most real-world problems involve uncertainty, investigating problems under uncertainty is one of the most critical aspects of the optimization field. Chance-constrained optimization is a method to tackle uncertainty, where constraints that involve stochastic components can be violated with a probability less than a small value of α [5,22].

In our experiments, we consider the maximum coverage problem with a budget chance constraint as an example of chance-constrained problems, where the objective function is submodular. One of the characteristics of submodular functions that makes them relatable to many real-world applications, is diminishing returns. This means the benefit of adding elements to a solution set, does not increase with the growth of this solution set [10,11].

1.1 Related Work

Submodular functions characterize subset selection problems where the marginal gain from adding an element to a solution decreases as more elements are included into this solution. Submodular functions have been widely studied in the literature [3,9,11,12,21]. Recently, the optimization of monotone submodular functions under deterministic and chance-constrained knapsack constraints has been studied [13,16,17,19]. Chance-constrained optimization is a method to deal with uncertainty. Evolutionary algorithms have been shown to be effective in solving chance-constrained optimization problems [6,15,18,23].

Benchmarking is an important aspect of optimization. It helps with the basic assessment of the performance of new algorithms, with assessing algorithms' behavior, and also with algorithm selection and configuration [2]. Analyzing the behavior of different algorithms on specific problems helps to identify the best algorithm to solve them. Diverse sets of differentiating instances are essential for evaluating algorithms' performance in solving a problem [4]. In [8] the authors proposed a new framework to evolve diverse travel salesperson problem (TSP) instances that are easy or hard to solve for a chosen heuristic. They studied the diversity of the features of these instances in multidimensional feature space. Moreover, this approach was adjusted to generate variations of a given image that are similar to it but differ in terms of selected image features [1]. The experimental investigations in [14] demonstrate the performance of popular multi-objective indicators in evolving diverse sets of TSP instances and images according to various features. In [4] Bossek, et al. introduced new creative mutation operators for evolving diverse sets of instances for TSP with multifaceted topologies. In [20] the authors evolved reliable instances for chance-constrained submodular problems that are easy to solve for one algorithm and hard to solve for another by introducing a new method to calculate the performance ratio.

To the best of our knowledge, no study provides methods to generate diverse sets of discriminating instances for chance-constrained optimization problems in the literature. To address this gap, we introduce new feature-based evolutionary diversity algorithms to evolve such instances.

1.2 Our Contribution

We propose two feature-based evolutionary diversity optimization approaches to evolve diverse discriminating instances for chance-constrained optimization problems. We consider the average and standard deviation of both the expected values and the variances of stochastic elements in the chance constraint as the features (that is, four different features). In our experimental investigations, we use the chance-constrained maximum coverage problem with stochastic weights on the nodes. However, our proposed method can be used for any chance-constrained optimization problem where expected values and variances can quantify the stochastic components. We first propose a new mutation operator designed to get diverse discriminating instances based on the average of expected values and variances as features. This method significantly improves the diversity of these features in the evolved instance sets compared to the initial population. Then, we proposed another mutation operator to achieve greater diversity in the standard deviations of the expected values and variances. This operator focuses on increasing the diversity of these features within the set while maintaining the average unchanged in each instance. Our experimental investigations show that this operator has a great ability to improve the diversity of these features in the set. Our method can generate new benchmark sets which can be used as an excellent foundation for feature-based algorithm selection.

In Sect. 2, we introduce the chance-constrained maximum coverage problem, the discounting performance ratio used to indicate how differentiating an instance is, and the diversity measure. In Sect. 3, we introduce our feature-based evolutionary diversity optimization and features. In Sects. 4 and 5 we propose two novel mutation operators for independent and dependent features, respectively, each followed by experimental investigations on the chance-constrained maximum coverage problem. The results are then compared with those of the conventional method to evolve differentiating instances (with no diversity optimization).

2 Preliminaries

In this paper, we evolve diverse sets of discriminating instances for chance-constrained optimization problems using feature-based diversity optimization. We use the chance-constrained maximum coverage problem as an example of chance-constrained optimization problems for our experimental investigations. The objective of this problem is to cover as many vertices as possible from the graph, by picking a subset of the nodes, where each node covers itself and all adjacent nodes. In our setting a stochastic cost is associated with each node in the graph and the problem has a budget chance constraint. Formally, the problem is defined as follows.

$$\text{Maximize} \quad F(x) = N(V'(x)), \tag{1}$$
$$\text{Subject to} \quad Pr(C(x) > B) \leq \alpha, \tag{2}$$

where $V'(x)$ is the subset of nodes in a solution, $N(V'(x))$ is the number of the nodes that are covered in that solution, which are, the nodes in $V'(x)$ and all their neighbors, $C(x)$ is the cost function, and B is the constraint budget. Here α is a small value that bounds the probability of going over the budget in the constraint. In this problem, each node i, $1 \leq i \leq n$ has a stochastic cost, characterized by expected value μ_i and variance σ_i^2. Since the costs of these nodes are independent, the expected cost and variance of a solution respectively are formulated as $\mu(x) = \sum_{i=1}^{n} \mu_i x_i$ and $\sigma^2(x) = \sum_{i=1}^{n} \sigma_i^2 x_i$.

To address the chance constraint we use Chebyshev's inequality to estimate an upper bound for the probability of exceeding the constraint budget and reformulate constraint (2) as

$$\mu(x) + \sqrt{\sigma^2(x) \cdot \frac{(1-\alpha)}{\alpha}} \leq B. \tag{3}$$

According to [22], solutions that satisfy the surrogate constraint 3 also will satisfy the chance constraint $Pr(C(x) > B) \leq \alpha$.

We aim to evolve diverse instances that are easy to solve by one algorithm and hard to solve by another. To make sure the instances are differentiating, we need a measure that reflects the performance ratio of two algorithms for each instance. For this purpose, we use discounting performance ratio (see Eq. (4)) proposed by [20] which allows us to calculate the performance ratio while controlling the reliability of the instances in the chance-constrained setting. We are comparing the performance of heuristic algorithms, which are inherently stochastic. Since these algorithms produce different solutions with each run, to get more reliable instances we perform each pair of algorithms for r independent runs on each instance and use $R'(I)$ as the performance ratio.

$$\begin{aligned} R(I) &= P(A_1^I)/P(A_2^I), \\ R'(I) &= Exp[R(I)] - K_\theta \cdot std[R(I)], \end{aligned} \tag{4}$$

where $P(A_1^I)$ and $P(A_2^I)$ are the best objective values that each algorithm A_1 and A_2 respectively can reach, and $R(I)$ determines the performance ratio of the pair for instance I. If any of the algorithms cannot find a feasible value before the stopping criteria is met, $P(A^I)$ would be set to a small value ϵ.

$Exp[R(I)]$ and $std[R(I)]$ are consequently the expected value and standard deviation of the performance ratios of instance I in these r runs, and K_θ is a constant value specified by the confidence level θ. We use $R'(I)$ to measure the performance difference for our pair of algorithms to ensure all the offspring are differentiating. In this method by penalizing the expected value of the performance ratios by their standard deviation, we ensure the reliability of these chance-constrained instances.

To ensure the diversity of the features in a set, we use the feature-based diversity measure proposed by [8]. This measure indicates the contribution of

Algorithm 1: The $(\mu+1)$ EA_D.

1 Population $P \leftarrow \mu$ individuals with performance ratio of at least T;
2 while *stopping criteria is not met* **do**
3   Calculate the feature value ft for each individual;
4   Sort population $P = \{I_1, \ldots, I_\mu\}$ in ascending order of ft, so that $ft(I_1) \leq ft(I_2) \leq \ldots \leq ft(I_\mu)$;
5   Choose I_j from $\{I_2, \ldots, I_\mu\}$: with probability $\frac{1}{3}$ choose I_1, with probability $\frac{1}{3}$ chose I_μ and with probability $\frac{1}{3}$ choose any number from $\{I_2, \ldots, I_{\mu-1}\}$ u.a.r.;
6 $I'_j \leftarrow$ MUTATE(I_j);
7   **if** $R'(I_j) \geq T$ **then**
8   Add I to P, preserving the ascending order of ft;
9   Recalculate contributions of individuals $d(I_i, P)$;
10   Remove from P an individual with the minimum $d(I_i, P)$, breaking ties u.a.r.;
11 **end**
12 end

each instance I_i to the diversity of a population $P = \{I_1, I_2, \ldots, I_\mu\}$, which is related to the distance between the feature value of an individual and the feature values of the closest individuals in the population. Formally, if we enumerate the population in ascending order of the feature values ft, that is, for all $i \in [2..\mu-1]$ we have $ft(I_1) \leq ft(I_i) \leq ft(I_{i+1}) \leq ft(I_\mu)$, then for these i we define

$$d(I_i, P) = (ft(I_i) - ft(I_{i-1})) \cdot (ft(I_{i+1}) - ft(I_i)). \quad (5)$$

For $ft(I_1)$ and $ft(I_\mu)$ we set the value of $d(I_i, P)$ to $+\infty$ so they are always kept in the population. If any individual has the same feature value as any other individual in the population we set their value of $d(I_i, X)$ to 0.

Furthermore, we calculate the diversity of a population by summing the diversity measures of all instances in the population, except for the instances with minimum and maximum feature values, since their diversity contribution is set to $+\infty$. Thus, the diversity of the population is formulated as follows

$$D_s(P) = \sum_{i=2}^{\mu-1} d(I_i, P). \quad (6)$$

3 Approach

We develop a $(\mu+1)$ evolutionary algorithm to generate diverse discriminating instances for the maximum coverage problem which we call the $(\mu+1)$ EA_D (see Algorithm 1). This algorithm starts with a population containing μ discriminating instances that meet a threshold T of minimum performance ratio. As mentioned before, we are using the discounting performance measure to determine the difference in performance between a pair of algorithms for solving the problem at hand. In each iteration, we choose one individual and alter the costs' expectation or variation of the nodes for that individual to get new offspring. By doing this, we are evolving instances with new constraints while preserving the graph's topology. If this offspring meets the minimum threshold of performance ratio we keep it in the population and remove an individual with the smallest contribution to the diversity. Otherwise, we continue to the next iteration. The features we investigate in this study are the average and standard deviation of expected costs and variances in the graph. We aim to achieve population diversity by focusing on each feature separately. By optimizing diversity across these individual dimensions, we can better explore the solution space and the range of these features. These four features are formulated as follows

$$ft_1(I) = \mu_{ave}^I = \frac{1}{n} \cdot \sum_{i=1}^{n} \mu_i^I, \tag{7}$$

$$ft_2(I) = \sigma_{ave}^I = \frac{1}{n} \cdot \sum_{i=1}^{n} \sigma_i^I, \tag{8}$$

$$ft_3(I) = \sqrt{\frac{1}{n} \cdot \sum_{i=1}^{n} (\mu_i^I - \mu_{ave}^I)^2}, \tag{9}$$

$$ft_4(I) = \sqrt{\frac{1}{n} \cdot \sum_{i=1}^{n} (\sigma_i^I - \sigma_{ave}^I)^2}, \tag{10}$$

where μ_i^I and σ_i^I are respectively the expected cost and variance of node i in individual I.

In Sects. 4 and 5 we introduce two novel mutation operators to improve the diversity of a set with respect to these features. These operators show a great ability to develop offspring with a significant contribution to the population's diversity while maintaining a minimum quality of performance difference.

The idea behind the mutation operators is to reduce the minimum values and increase the maximum values of the features in the set of instances and try to make the values of $(ft(I_i) - ft(I_{i-1}))$ and $(ft(I_{i+1}) - ft(I_i))$ equal in the other individuals. According to the diversity measure (see Eq. 5), this maximizes the value of $(ft(I_i) - ft(I_{i-1})) \cdot (ft(I_{i+1}) - ft(I_i))$ which is the value that shows the contribution of each solution to the set. This approach evolves differentiating instances with increasing contribution to the diversity, resulting in a final population with a diverse range of features.

The fitness function of this algorithm is the contribution of each instance to the diversity of the population $d(I_i, P)$. Note that the fitness of individuals in the population depends on the whole population, and for the same individual, it might change from iteration to iteration. We use the elitist method for selection, which means that in every iteration the individual with the smallest contribution to the diversity is eliminated, with ties broken uniformly at random. The following sections provide a detailed explanation of these new mutation operators and experimental investigations.

Algorithm 2: Mutation for an independent feature, demonstrated for expected values for instance I_j (this works similarly for variances).

1 **for** each node $i \in \{1,\ldots,n\}$ **do**
2   Choose δ_1^i according to normal distribution $N(0,\sigma_1)$;
3      $\mu_i \leftarrow \min\{\mu_{max}, \max\{0, \mu_i + Ind(j) \cdot |\delta_1^i|\}\}$;
4 **end**

4 Evolving Discriminating Instances for Chance Constrained Problems with Diverse Features

In the first mutation operator that we develop (see Algorithm 2), we aim to expand the range of the average values for expected value and variances of costs, as the features (see Eqs. (7) and (8)). In the rest of this study, we refer to these features as feature 1 (ft_1) and feature 2 (ft_1).

In this algorithm, first, we sort the population by their feature value. Then we choose one individual as a parent. The parent is either the individual with minimum or maximum feature value or something in between, each with a probability $P_m = 1/3$. All instances are defined by two vectors representing the expected values (μ_i) and variances (σ_i^2) of their costs, where $i = \{1,\ldots,n\}$ and n is the number of nodes in that instance's graph. There is a value associated with the selected parent, which indicates whether the feature of this instance should be increased or decreased to have the maximum contribution to the diversity of the set. To decide the value of this indicator we have $Ind(j) = 1$ if $(j = \mu \vee (j \neq 1 \wedge (ft(I_j) - (ft(I_{j-1})) \leq (ft(I_{j+1}) - ft(I_j)))$, and $Ind(j) = -1$ otherwise. Using this operator to mutate the individuals leads to an increase in the range of the feature within the population. For individuals I_i, $i \notin \{1,\mu\}$ we change their weights or variances so that the difference in the feature value to I_{i+1} becomes closer to the difference in feature value to I_{i-1}. This implies that feature values get balanced out among individuals that do not take on the maximum or minimum feature value.

For each node i of the mutated individual I_j, we chose a random number $\delta_i \sim N(0,\sigma_1)$, where σ_1 is a parameter of the algorithm which can be used to tune the mutation strength, and then we add $(Ind(j) \cdot |\delta_1^i|)$ to the expected weight or variance of the node. This novel mutation demonstrates a strong capability to evolve discriminating instances with diverse expected costs.

4.1 Experimental Setting

In our experimental investigations, we generate diverse discriminating instances for the chance-constrained maximum coverage problem. We start our $(\mu+1)$ EA$_D$ with a population of 20 discriminating instances ($\mu = 20$), with expected costs $\mu_i \in (0,\mu_{\max})$ and variances $\sigma_i^2 \in (0,\mu_i^2/3)$ in the chance constraint where $\mu_{\max} = 1000$. We calculate the cost budget as $B = n/30 \cdot \mu_{\max}/2$, where n is the number of vertices of the graph, and we set the probability of exceeding the

Table 1. Feature values of the diverse differentiating sets before (conventional) and after $((\mu + 1)\ EA_D)$ using the feature-based diversity optimization method based on the independent features.

Graph	Graph Size	instance set	feature	EA/FGA				EA/GHC			
				Average	Std	Min	Max	Average	Std	Min	Max
lp-recipe	205	conventional	ft_1	497.40	19.77	455.55	530.09	503.43	23.40	467.80	563.94
			ft_2	286.48	10.01	263.73	303.36	288.16	11.90	263.99	314.90
			ft_3	82170.22	5673.14	69551.80	92029.10	83176.23	6846.99	69693.20	99160.90
			ft_4	92502.18	4414.19	82550.00	100310.00	92920.87	3455.68	86684.80	98544.90
		$(\mu + 1)\ EA_D$	ft_1	590.63	109.38	438.70	851.40	499.29	37.29	439.01	563.94
			ft_2	293.35	12.38	262.03	313.74	285.21	15.51	264.11	314.90
			ft_3	81601.93	6792.91	68578.50	92469.40	73517.31	21195.66	5689.40	99245.70
			ft_4	91725.23	4862.35	82557.00	100315.00	86562.02	16676.04	19999.10	98544.60
ca-netscience	379	conventional	ft_1	501.39	12.10	477.04	526.29	497.32	16.77	463.79	532.13
			ft_2	288.66	8.25	276.34	303.10	285.42	9.23	270.86	302.42
			ft_3	83389.89	4781.34	76363.10	91867.80	81552.06	5266.77	73363.50	91455.60
			ft_4	93441.65	3877.30	88127.50	100296.00	92303.41	3329.82	86861.60	97733.80
		$(\mu + 1)\ EA_D$	ft_1	616.71	128.63	428.75	901.78	648.59	142.41	461.43	1000.00
			ft_2	304.57	15.26	270.63	330.67	307.97	18.85	270.76	360.58
			ft_3	83525.52	7197.99	70266.70	97774.60	87193.34	11755.86	73142.40	117415.00
			ft_4	94248.12	4112.99	88128.90	100295.00	93093.14	3434.42	87321.90	98506.00
impcol-d	435	conventional	ft_1	503.75	11.98	476.54	533.78	502.45	11.22	481.40	519.73
			ft_2	289.10	6.73	275.74	305.03	288.21	7.62	273.52	300.09
			ft_3	83623.91	3901.93	76033.60	93041.70	83122.98	4379.62	74815.40	90054.50
			ft_4	93316.19	2310.33	88283.70	98653.50	92923.09	3343.28	87396.60	99784.90
		$(\mu + 1)\ EA_D$	ft_1	636.11	138.97	436.56	951.66	703.55	156.71	469.30	1000.00
			ft_2	306.77	17.75	270.50	339.62	320.29	51.85	272.58	537.77
			ft_3	82269.11	7477.26	66030.00	97221.00	86922.05	12842.93	67947.10	117684.00
			ft_4	93306.45	2625.77	89068.30	98670.40	94427.28	5461.48	85711.20	105159.00
random graph	500	conventional	ft_1	500.06	13.59	481.50	526.34	502.09	15.22	477.09	534.52
			ft_2	286.68	7.68	274.03	300.67	287.05	8.78	269.59	303.60
			ft_3	82245.14	4419.39	75090.00	90401.00	82474.72	5035.59	72676.40	92174.70
			ft_4	92791.03	2756.71	87544.40	97241.50	91929.44	3343.14	84396.70	96554.60
		$(\mu + 1)\ EA_D$	ft_1	500.06	13.59	481.50	526.34	542.59	44.97	470.41	625.91
			ft_2	286.68	7.68	274.03	300.67	295.74	10.91	272.54	307.11
			ft_3	82730.06	4751.16	75090.00	90474.00	75357.12	21022.61	763.78	105918.00
			ft_4	92848.15	2808.63	87544.40	97241.50	85536.89	19544.34	4383.95	97952.90
lp-agg	615	conventional	ft_1	495.50	9.25	471.88	510.73	498.33	10.87	475.78	514.13
			ft_2	285.21	5.89	271.38	294.81	285.69	5.17	277.44	295.03
			ft_3	81379.80	3337.23	73649.40	86913.40	81646.60	2952.12	76975.50	87040.50
			ft_4	92602.10	2839.49	85020.60	98301.60	91886.41	2031.73	88176.10	95144.80
		$(\mu + 1)\ EA_D$	ft_1	608.89	107.78	457.84	884.89	704.10	159.52	475.78	1000.00
			ft_2	297.14	11.48	271.38	320.96	284.56	5.09	278.88	295.03
			ft_3	81466.54	4164.77	72998.30	87869.50	85433.61	7245.04	76756.10	102854.00
			ft_4	92793.41	3539.64	84953.50	98301.60	93224.26	2536.96	88173.20	96939.10
can-715	715	conventional	ft_1	501.79	10.58	487.51	528.17	498.78	10.11	484.43	520.83
			ft_2	288.17	5.45	279.27	300.66	286.10	6.17	277.77	300.41
			ft_3	83069.60	3148.01	77990.60	90398.50	81893.84	3559.84	77158.70	90244.00
			ft_4	93329.39	1562.15	90158.20	96657.70	91737.04	2663.46	85955.40	96008.10
		$(\mu + 1)\ EA_D$	ft_1	610.07	92.73	478.04	795.74	632.91	127.79	467.97	923.97
			ft_2	303.27	10.04	279.40	318.50	301.80	12.91	276.38	328.61
			ft_3	84137.51	5801.29	72738.00	97244.00	80107.78	21471.62	9216.42	118738.00
			ft_4	93312.95	1729.50	90158.80	95554.90	88606.75	15626.40	24665.10	102811.00

chance constraint budget α, to 0.05. We use discriminating instances generated by the discounting performance measure with a 0.9 confidence level ($\theta = 0.9$) from [20]. We generate all 20 instances independently from each other. For each graph, we use a percentage of $R'(I)$ values for that instance as threshold T for

Table 2. D_S values (Equation (6)) for discriminating instance sets evolved by conventional and $(\mu+1)$ EA_D methods regarding average of weights (ft_1) and average of variances (ft_2).

Graph	feature	EA/FGA		EA/GHC	
		conventional	$(\mu+1)$ EA_D	conventional	$(\mu+1)$ EA_D
lp-recipe	ft_1	323.68	10,009.30	268.5	767.32
ca-netscience		115.05	13,778.22	141.85	20651.44
impcol-d		80.68	11,710.95	69.11	14168.5
random graph		111.04	111.04	199.73	1224.66
lp-agg		86.68	10,915.55	98.51	14344.46
can-715		71.11	5,722.73	81.64	10734.53
lp-recipe	ft_2	29,778,773.08	33,665,433.00	30,856,754.00	472,852,600.00
ca-netscience		12,397,990.24	32,682,513.00	15,100,157.88	139,137,998.00
impcol-d		12,368,648.03	55,450,328.00	13,859,104.77	161,954,086.00
random graph		6,323,541.05	7,301,225.92	20,240,043.20	673,454,960.00
lp-agg		7,516,936.70	10,111,003.00	4,727,498.00	50,201,773.00
can-715		6,774,002.20	28,190,477.00	4,526,037.80	772,828,690.00

the performance ratio and make sure that all offspring added to the population throughout the optimization process meet this threshold. We calculate the threshold T as $((0.80 \cdot (R'(I) - 1)) + 1)$, which keeps all thresholds above 1. By doing this, we keep the difference in performance between the two algorithms within a sufficient level while making the set more diverse in each iteration.

We are comparing the performance of the $(1 + 1)$ Evolutionary Algorithm (EA), with the Fast Genetic Algorithm (FGA), and the Greedy Hill Climber (GHC) in our experiments. Detailed explanations of these algorithms can be found in [7]. We evolve instances that are easy to solve by the $(1 + 1)$ EA and hard to solve by the FGA, and easy for the $(1 + 1)$ EA and hard for the GHC, which we denote respectively by EA/FGA and EA/GHC in our experimental results. We run each algorithm for 10000 iterations and perform 10 independent runs for each instance and use the expected value and standard deviation of performance ratios in these runs to calculate the discounting performance ratio. If any of the algorithms can not reach a feasible solution in any of the runs, we set the value of $P(A)$ for that run to a small value $\epsilon = 10^{-2}$.

As mentioned earlier, when mutating an individual, depending on the feature, we change the expected value or variances of the costs by adding or subtracting a small value from all of its values. This value is chosen randomly from a normal distribution $N(0, 3)$ for expected costs and $N(0, 100)$ for costs' variances. We chose these parameters by trying different parameter values in experiments and checking the results to see which value works best for our setting. We run the $(\mu+1)$ EA_D for 10000 iterations and perform 10 runs per each feature and each combination of tested algorithms. This gives us 10 sets of discriminating instances with diverse expected costs, each containing 20 individuals for each graph.

Algorithm 3: Mutation for a dependent feature demonstrated for expected values of instance I_j (this works similarly for variances).

1   $W_l = \{i \in V \mid \mu_i \leq \mu_{ave}\}$, $W_g = V \setminus W_l$;
2   Choose $m \sim Pois(\lambda)$;
3   $K = \min(m + 1, \text{size}(W_l), \text{size}(W_g))$;
4   **for** $v \in \{0, \ldots, K\}$ **do**
5   choose $\delta_2^v \sim N(0, \sigma_2)$;
6   choose nodes $s \in W_l$ and $t \in W_g$ randomly;
7      $\mu_s \leftarrow \min\{\mu_{max}, \max\{0, \mu_s + Ind(j) \cdot |\delta_2^v|\}\}$;
8      $\mu_t \leftarrow \min\{\mu_{max}, \max\{0, \mu_t - Ind(j) \cdot |\delta_2^v|\}\}$;
9   **end**

4.2 Experimental Results

In this section, we show the results of our experiments. They demonstrate the effectiveness of the proposed mutation operator in increasing the diversity of evolved instances. We compare a set of 20 instances, randomly selected from the 10 sets generated using our diverse optimization approach, with 20 instances evolved individually using the $(1 + 1)$ EA introduced by [20] (which we refer to as the conventional method) to ensure a fair comparison. We tested our algorithm on six graphs of varying sizes, ranging from 200 to 715 nodes. As shown in Table 1, in the instances evolved by the feature-based diversity optimization, the range of both the average of expected costs (feature 1) and the average of costs' variances (feature 2), are significantly larger than instances generated by the conventional method. However, the range of standard deviation of the expected value and variances of the costs do not improve noticeably. This is natural as we are shifting all cost values in one direction in each iteration, and this does not significantly change the standard deviations of those values. Additionally, as demonstrated in Table 1, for all features the increased standard deviation of the feature values indicates that the diversity of the first feature has been successfully improved after using our method. Figures 1 and 2 show the coverage of both instance sets evolved by the conventional and the $(\mu + 1)$ EA_D method over the feature space for feature 1 and feature 2 respectively. The instances generated by the conventional method are very clustered in the feature space for both features, despite evolving separately with different random initial solutions. However, the instances evolved using the proposed feature-based diversity optimization provide significantly greater coverage of the feature space. As Fig. 1 shows, evolved instances tend to have larger averages of expected costs and cover the feature space more extensively in that direction. This is because if the costs of the vertices in an instance are too small, then the problem gets too easy to solve by any algorithm. In some instances with very small expected costs, you can even include all of the graph nodes in a solution without exceeding the budget, leaving no opportunity to find discriminating instances.

The values of D_s which represent a set's diversity, are shown in Table 2. As illustrated, these values are significantly larger for almost all graphs for the

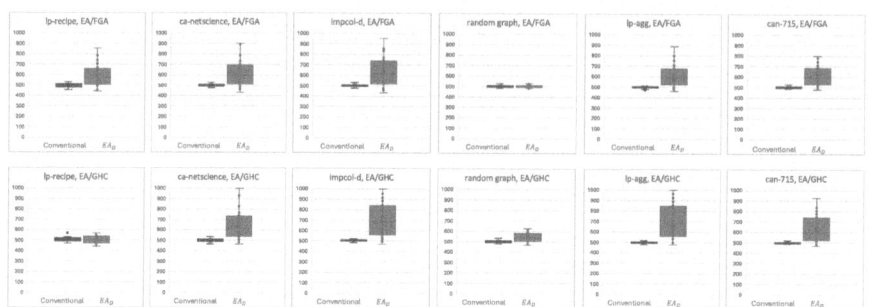

Fig. 1. Distribution of average of weights (ft_1) for 20 instances. The left box plots show the features of the initial population and the right ones show them after using feature-based diversity optimization.

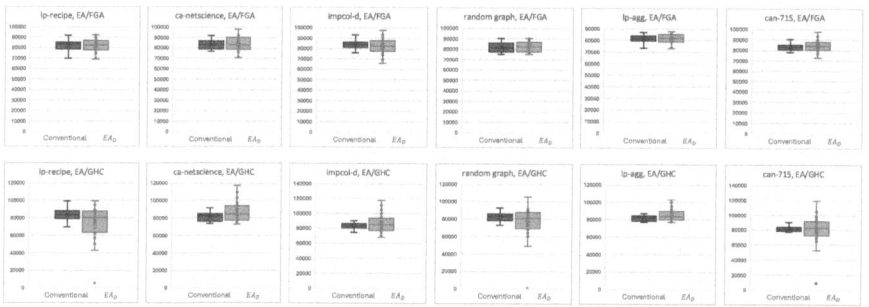

Fig. 2. Distribution of average of variances (ft_2) for 20 instances. The left box plots show the features of the initial population and the right ones show them after using feature-based diversity optimization.

instance sets generated by $(\mu+1)\ EA_D$, which is another indicator of high-quality instance sets in terms of diversity. As stated earlier, the diversity in the standard deviation of expected values and variances of the costs (features 3 and 4), also do not provide sufficient coverage of the feature space (see Table 1). To address this, we propose a new mutation operator to diversify these features. We then proceed to demonstrate the strong performance of the proposed mutation operator through experimental investigations.

5 Evolving Discriminating Chance Constraints with Diverse Dependent Features

The mutation operator introduced in this section is tailored to evolve discriminating instances with diverse dependent features. We aim to increase the range of standard deviations of the expected values and variances of the costs as the features (respectively ft_3 and ft_4) while keeping the average of these values

Table 3. D_S values (Eq. (6)) for discriminating instance sets evolved by conventional and $(\mu+1)$ EA_D methods regarding standard deviation of weights (ft_3) and standard deviation of variances (ft_4).

Graph	feature	EA/FGA		EA/GHC	
		conventional	$(\mu+1)$ EA_D	conventional	$(\mu+1)$ EA_D
lp-recipe	ft_3	79.55	1855.10	38.36	3752.07
		32.78	495.67	15.66	4581.20
		17.06	2287.05	47.27	3898.64
		14.61	15.43	20.64	1474.18
		14.77	229.46	17.43	833.07
		9.69	289.87	31.47	885.66
lp-recipe	ft_4	14,000,313.97	22,296,731.00	6,467,790.40	22,605,636.00
		8,872,710.77	14,888,695.00	5,515,740.80	57,315,042.00
		3,146,046.53	20,687,811.00	9,499,351.87	66,693,355.00
		5,174,800.50	5,152,784.60	4,436,745.78	61,228,746.00
		8,444,281.95	7,957,325.00	3,069,856.78	40,431,827.00
		1,348,041.36	2,395,728.50	3,944,670.89	61,146,158.00

(ft_1, ft_2) unaffected. The core principle of this strategy is similar to the mutation proposed earlier. To indicate if we need to make these standard deviations smaller or larger, for the selected parent we use $Ind(j) \in \{-1, 1\}$. We decide on the value of $Ind(j)$ by considering the position of the instance compared to the other individuals in the population. If it has the maximum feature value we try to make the standard deviation larger by setting the value of $Ind(j)$ to 1, if it has the minimum feature, we will do the opposite. For all other instances I_i with $i \notin \{1, \mu\}$, we determine the value of $Ind(j)$ in a way that brings the values of $(ft(I_i) - (ft(I_{i-1}))$ and $(ft(I_{i+1}) - ft(I_i))$ closer together.

We alter the standard deviation of costs' expected values and variances without changing the averages. To do so we divide the nodes into two groups by looking into the respective averages for each feature. One group with values smaller, and the other with values larger than the average. By adjusting the expected costs or variances of these nodes to be closer to or further from the average by the same value, we can reduce or increase the standard deviation in that instance, without making any changes to the average. To modify the constraint in the mutation, we choose two nodes randomly, one from each category, and select a random normal value δ_2^i, from $N(0, \sigma_2)$. We then add the value of $Ind(j) \cdot |\delta_2^i|$ to the expected cost of the individual chosen from the group with expected costs higher than average and discount the same value from the other one. By doing so, we are changing the standard deviation of the parent while keeping the average fixed. We do this K times in the mutation operator for each iteration. To choose K, we first choose m according to a Poisson distribution with parameter λ and set $K = m + 1$, to avoid iterations without mutating any nodes. The pseudocode of this mutation operator is shown in Algorithm 3.

This new mutation operator significantly enhances the diversity of standard deviations of the expected costs.

5.1 Experimental Setting

This section illustrates the experimental investigations for evolving discriminating chance-constrained instances with diverse dependent features. Here features are the standard deviation of the expected values and variances of the costs in the constraint. We are performing feature-based diversity optimization based on these features while ensuring the average values of the expected values and variances of the costs do not change in the process. The problem setting remains the same as in the previous section. In each iteration, we change costs' expected values or variances (depending on the feature) of $2K$ items from the parent, where $K = m+1$, and m is chosen randomly from the Poisson distribution $Pois(\lambda = 5)$. We are changing the expected costs of these selected nodes by random values which are chosen from the normal distribution $N(0, 100)$ for expected costs and $N(0, 3000)$ for variances. We run the algorithm for 10,000 function evaluations and perform 10 independent runs.

5.2 Experimental Results

This section presents the experimental results for instances generated by our feature-based diversity evolutionary algorithm regarding independent features. We start our algorithm with the same 20 individuals as in the previous section. We run the algorithm 10 independent times and use one randomly to compare it to the initial set. The box plots showing the range of standard deviations of the expected values and variances of the costs before and after using our diversity optimization algorithm to evolve instances are represented in Figs. 3 and 4 respectively. As the plot shows, the proposed $(\mu + 1)$ EA_D is superior to the conventional method in terms of the diversity of the second feature. Obtaining

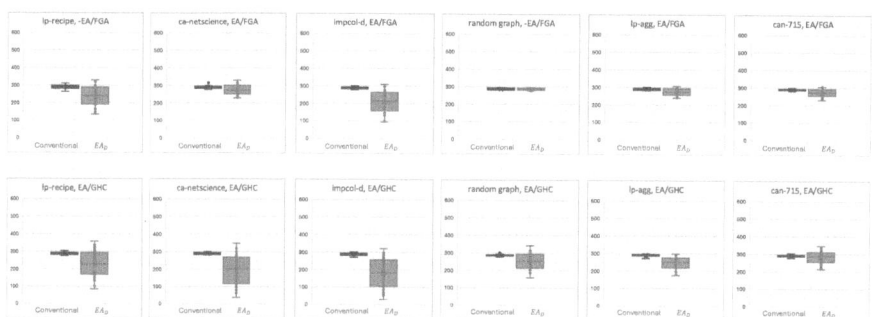

Fig. 3. Distribution of standard deviation of weights (ft_3) for 20 instances. The left box plots show the features of the initial population and the right ones show them after using feature-based diversity optimization.

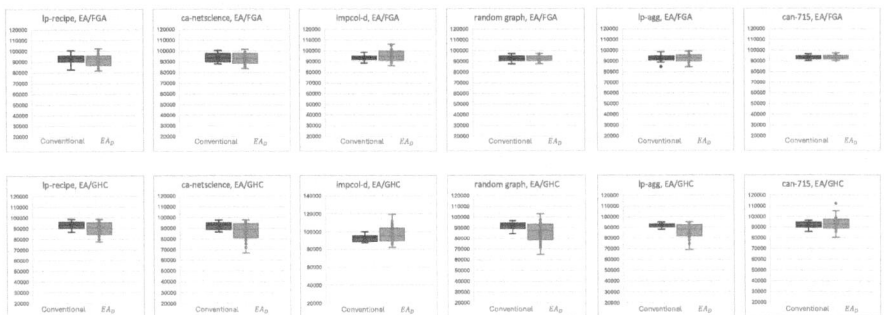

Fig. 4. Distribution of standard deviation of variances (ft_4) for 20 instances. The left box plots show the features of the initial population and the right ones show them after using feature-based diversity optimization.

diverse instances is more challenging for the last two problems with larger graph sizes. It is because we are doing the mutation with the same λ value for all the problems despite the size of the problem. As a result, the number of elements altered during the mutation constitutes a smaller proportion of the total elements in the problem, leading to a smaller impact on the standard deviation. In all six graph problems, the range of ft_3 increases, mostly via decreasing the minimum values while the maximum values slightly improve. This is because raising the standard deviation of the expected costs too high creates an imbalance in the problem, with some nodes having very high costs and others very low. Consequently, the problem becomes too easy for both algorithms, as nodes with smaller expected costs are the obvious choices for covering the graph. This makes it more difficult to find discriminating instances with a performance ratio that meets the threshold, and consequently, it becomes harder to evolve diverse, discriminating instance sets.

The values of D_s for the instance sets generated by the conventional and feature-based diversity algorithms are presented in Table 3. These values, which reflect a set's diversity, are noticeably higher for all graphs after using our method except for "random graph" and "lp-agg" for ft_4. This can happen since D_s is not the fitness function in our algorithm, thus we might be non-elitist in the optimization process regarding D_s. One notable point is that achieving diversity in all of these features is challenging for the problem with the "random graph" when comparing the performance of EA and FGA (see Figs. 1, 2, 3 and 4). This is because of the structure of this random graph which makes it an easier problem to solve and the high similarity of EA and FGA. In this random graph, the probability of an edge appearing is the same for all possible edges. This characteristic results in most nodes having very similar degrees, which makes the problem easier for many algorithms to solve.

6 Conclusions

In this paper, we proposed two feature-based diversity optimization approaches to increase the diversity of the average and standard deviation of expected values and variances of the costs for chance-constrained optimization problems. Focusing on single features in diversity optimization is important as it provides a clear perspective on the potential range of those features. Our methods demonstrate significant improvement in the range of these four features in the experimental investigation conducted on various graph instances. In future studies, it would be interesting to focus on increasing diversity by incorporating multiple features and expanding the range of features over multi-dimensional feature spaces.

Acknowledgments. This work has been supported by the Australian Research Council (ARC) through grant FT200100536.

References

1. Alexander, B., Kortman, J., Neumann, A.: Evolution of artistic image variants through feature based diversity optimisation. In: Proceedings of the Genetic and Evolutionary Computation Conference, GECCO 2017, pp. 171–178. ACM (2017)
2. Bartz-Beielstein, T., et al.: Benchmarking in optimization: Best practice and open issues. CoRR **abs/200** (2020)
3. Bian, A.A., Buhmann, J.M., Krause, A., Tschiatschek, S.: Guarantees for greedy maximization of non-submodular functions with applications. In: Proceedings of the 34th International Conference on Machine Learning, ICML 2017. Proceedings of Machine Learning Research, vol. 70, pp. 498–507. PMLR (2017)
4. Bossek, J., Kerschke, P., Neumann, A., Wagner, M., Neumann, F., Trautmann, H.: Evolving diverse TSP instances by means of novel and creative mutation operators. In: Proceedings of the 15th ACM/SIGEVO Conference on Foundations of Genetic Algorithms, FOGA 2019, pp. 58–71. ACM (2019)
5. Charnes, A., Cooper, W.W.: Chance-constrained programming. Manage. Sci. **6**(1), 73–79 (1959)
6. Doerr, B., Doerr, C., Neumann, A., Neumann, F., Sutton, A.M.: Optimization of chance-constrained submodular functions. In: The Thirty-Fourth AAAI Conference on Artificial Intelligence, AAAI 2020, pp. 1460–1467. AAAI Press (2020)
7. Doerr, C., Ye, F., Horesh, N., Wang, H., Shir, O.M., Bäck, T.: Benchmarking discrete optimization heuristics with IOHprofiler. Appl. Soft Comput. **88**, 106027 (2020)
8. Gao, W., Nallaperuma, S., Neumann, F.: Feature-based diversity optimization for problem instance classification. Evol. Comput. **29**(1), 107–128 (2021)
9. Krause, A., Golovin, D.: Submodular function maximization. In: Tractability, pp. 71–104. Cambridge University Press (2014)
10. Lovász, L.: Submodular functions and convexity. In: Mathematical Programming The State of the Art, XIth International Symposium on Mathematical Programming, ISMP 1982, pp. 235–257. Springer (1982)
11. Nemhauser, G.L., Wolsey, L.A.: Best algorithms for approximating the maximum of a submodular set function. Math. Oper. Res. **3**(3), 177–188 (1978)

12. Nemhauser, G.L., Wolsey, L.A., Fisher, M.L.: An analysis of approximations for maximizing submodular set functions - I. Math. Program. **14**(1), 265–294 (1978)
13. Neumann, A., Bossek, J., Neumann, F.: Diversifying greedy sampling and evolutionary diversity optimisation for constrained monotone submodular functions. In: Genetic and Evolutionary Computation Conference, GECCO 2021, pp. 261–269. ACM (2021)
14. Neumann, A., Gao, W., Wagner, M., Neumann, F.: Evolutionary diversity optimization using multi-objective indicators. In: Proceedings of the Genetic and Evolutionary Computation Conference, GECCO 2019, pp. 837–845. ACM (2019)
15. Neumann, A., Neumann, F.: Optimizing monotone chance-constrained submodular functions using evolutionary multi-objective algorithms. Evol. Comput. 1–35 (2024)
16. Neumann, A., Xie, Y., Neumann, F.: Evolutionary algorithms for limiting the effect of uncertainty for the knapsack problem with stochastic profits. In: PPSN (1). Lecture Notes in Computer Science, vol. 13398, pp. 294–307. Springer (2022)
17. Pathiranage, I.H., Neumann, F., Antipov, D., Neumann, A.: Using 3-objective evolutionary algorithms for the dynamic chance constrained knapsack problem. In: Proceedings of the Genetic and Evolutionary Computation Conference, GECCO 2024. ACM (2024)
18. Perera, K., Neumann, A., Neumann, F.: Multi-objective evolutionary approaches for the knapsack problem with stochastic profits. In: Parallel Problem Solving from Nature - PPSN XVIII - 18th International Conference, PPSN 2024, Proceedings, Part I. Lecture Notes in Computer Science, vol. 15148, pp. 116–132. Springer (2024)
19. Perera, K.K., Neumann, A.: Multi-objective evolutionary algorithms with sliding window selection for the dynamic chance-constrained knapsack problem. In: Proceedings of the Genetic and Evolutionary Computation Conference, GECCO 2024. ACM (2024)
20. Sadeghi Ahouei, S., De Nobel, J., Neumann, A., Bäck, T., Neumann, F.: Evolving reliable differentiating constraints for the chance-constrained maximum coverage problem. In: GECCO 2024, Proceedings of the Genetic and Evolutionary Computation Conference, pp. 1036–1044. ACM (2024)
21. Vondrak, J.: Submodularity and curvature: the optimal algorithm. Ann. Discrete Math **2**, 65–74 (1978)
22. Xie, Y., Harper, O., Assimi, H., Neumann, A., Neumann, F.: Evolutionary algorithms for the chance-constrained knapsack problem. In: Proceedings of the Genetic and Evolutionary Computation Conference, pp. 338–346. ACM (2019)
23. Yan, X., Neumann, A., Neumann, F.: Sliding window bi-objective evolutionary algorithms for optimizing chance-constrained monotone submodular functions. In: Parallel Problem Solving from Nature - PPSN XVIII - 18th International Conference, PPSN 2024, Proceedings, Part I. LNCS, vol. 15148, pp. 20–35. Springer (2024)

Adaptive Neighborhood Search Based on Landscape Learning: A TSP Study

Malek Sarhani[1](✉) and Stefan Voß[2,3]

[1] School of Business Administration, Al Akhawayn University in Ifrane, Ifrane, Morocco
m.sarhani@aui.ma
[2] Institute of Information Systems, University of Hamburg, Hamburg, Germany
stefan.voss@uni-hamburg.de
[3] Escuela de Ingenieria Industrial, Pontificia Universidad Catolica de Valparaiso, Valparaiso, Chile
stefan.voss@pucv.cl

Abstract. Variable Neighborhood Search (VNS) is a prominent metaheuristic for solving combinatorial optimization problems. While adaptive strategies have been explored to enhance VNS, identifying effective criteria for adaptive decision making remains a challenge. This paper introduces an adaptive VNS variant that leverages Fitness Landscape (FL) Analysis to guide neighborhood selection. Our method uses machine learning to determine the best neighborhood structure for each iteration, analyzing key features of the fitness landscape. It evaluates the importance of these features using Shapley values, a technique from explainable artificial intelligence. The effectiveness of the proposed method is validated through experiments on benchmark instances of the Traveling Salesman Problem (TSP), demonstrating superior solution quality compared to traditional VNS and competitive performance relative to other metaheuristics. This study underscores the potential of landscape-aware optimization to enhance the metaheuristic performance for complex combinatorial problems while providing valuable insights into the influence of key landscape features on adaptive search strategies.

Keywords: Metaheuristics · Fitness landscape analysis · Traveling salesman problem · Variable neighborhood search

1 Introduction

Traditionally, optimization algorithms have emphasized algorithmic enhancements, often overlooking the intrinsic characteristics of the problems they address. This limited focus on problem analysis has restricted the understanding of why certain algorithms excel under specific conditions. Fitness landscape (FL) analysis addresses this gap by characterizing the search space of optimization problems, offering insights into algorithm performance and guiding the development of more effective strategies. By linking problem properties to algorithm behavior, FL analysis facilitates informed decision making in algorithm selection and configuration, increasing the likelihood of achieving high-quality solutions.

Variable neighborhood search (VNS) is a widely used metaheuristic for combinatorial optimization problems. By systematically exploring multiple neighborhoods, VNS balances intensification and diversification, enabling it to escape local optima and improve solution quality. Despite its adaptability, challenges remain in defining robust criteria for guiding neighborhood selection dynamically. Addressing these challenges is critical for enhancing the performance of VNS, particularly for large and complex instances where efficient exploration of the search space is paramount.

This paper introduces an adaptive VNS framework informed by FL analysis. The proposed approach computes FL features at regular intervals, leveraging a machine learning model trained on these features to predict the best neighborhood structure to be chosen. Additionally, the importance of these features is assessed using Shapley values, offering explainable insights into the adaptive process. This integration of VNS and FL analysis improves adaptability, bridging the gap between theoretical insights and practical algorithm design. Numerical experiments are provided for the well-known traveling salesman problem (TSP).

The contributions of the paper can be summarized as follows:

- Predicting the best VNS strategy using machine learning (ML) and FL analysis (FLA).
- Identifying the most important features through explainable artificial intelligence techniques, such as Shapley values [17].
- Proposing an adaptive VNS scheme informed by FL analysis and ML.

The rest of the paper is organized as follows: Sect. 2 provides a brief literature review. Section 3 details the proposed approach. Section 4 presents the experiments. Finally, Sect. 5 concludes the paper with key findings and future research directions.

2 Literature Review

Variable neighborhood search (VNS) is a well established metaheuristic proposed by Mladenović and Hansen [6,21] known for about 30 years. In this section we review works related to ours, focusing on methods for addressing the TSP, particularly adaptive approaches and VNS-related strategies.

Various adaptive methods have been applied to the TSP. Zhang et al. [38] addressed premature convergence and low solution accuracy in genetic algorithms (GAs) for the TSP by integrating the 2-opt operator and introducing a self-adaptive ant colony optimization (ACO) approach to enhance convergence and decision consistency. Similarly, Stodola et al. [32] proposed a dynamically controlled pheromone evaporation mechanism in ACO, utilizing information entropy for adjustment. Liu et al. [16] tackled a variant of the vehicle routing problem with time windows using an Adaptive Large Neighborhood Search heuristic.

To the best of our knowledge, no prior work has integrated FL feature information into VNS design or proposed an adaptive metaheuristic approach for the TSP based on these features. Related ideas have been explored in Differential Evolution [7,15], focusing on specific FL characteristics, namely, ruggedness and fitness-distance correlation, respectively. However, as noted in the survey by Zou et al. [39], a single feature is often insufficient for comprehensive landscape analysis. This gap raises two fundamental questions: first, which features are most relevant for adaptive strategies, and second, how to effectively incorporate FL analysis into algorithm design.

Studies of TSP landscapes provide insights into these questions. Boese et al. [3] proposed the "big-valley hypothesis," suggesting a global structure in TSP landscapes, with later studies [24] revisiting this claim. Mapping the global structure of TSP FLs [25] underscores the complexity of these landscapes, highlighting the need for adaptive strategies informed by FL characteristics.

ML has been widely used to enhance metaheuristics. Kalatzantonakis et al. [11] developed a reinforcement-learning-based VNS inspired by the Multi-Armed Bandit model, though it does not integrate FL analysis. Similarly, supervised learning has been applied to leverage computed TSP features for algorithm selection [13,33]. While these approaches demonstrate the potential of ML, our work advances the field by incorporating FL features into an adaptive VNS framework, employing a Random Forest model for strategy prediction, and using Shapley values for feature importance analysis.

Regarding FL features, tools like the `Flacco` package [14] have been proposed for continuous optimization problems. For combinatorial optimization, studies such as Mersmann et al. [19] and Hutter et al. [8] have defined numerous features to characterize problem difficulty. However, computational cost remains a practical challenge [13]. Our approach focuses on representative, computationally efficient features, including ruggedness [35], modality, and fitness-distance correlation, as further detailed in Sect. 3.2.

In summary, understanding the TSP landscape and integrating FL analysis into adaptive strategies remain challenging. This paper addresses these gaps by leveraging the most impactful and computationally efficient FL features to design an adaptive VNS framework, providing a novel approach for solving TSP instances.

3 The Proposed Approach

In this section we present the proposed approach. We start by describing the FLA process, then the adopted features, next the VNS approach and then the learning process as well as the complete approach.

3.1 Fitness Landscape Analysis

In this section we explain the components of a fitness landscape and describe the process of analyzing it to guide the adaptive neighborhood selection in VNS. A fitness landscape is formally defined as a triplet $(X, \mathcal{N}, \phi)$, where:

- X represents the set of candidate solutions, or all possible tours in the TSP.
- $\mathcal{N}$ denotes the neighborhood structure that defines how solutions are connected and – accessible from each other – controlled in VNS by neighborhood operators (e.g., 2-opt, 3-opt).
- $\phi : X \to \mathbb{R}$ is the fitness function, mapping each solution to a numerical fitness value, typically the tour length for the TSP.

To leverage FLA within VNS, we compute specific features of the fitness landscape that relate to neighborhood transitions and landscape complexity. The primary goal of FLA here is to determine the neighborhood structure most likely to yield improvements in each iteration of VNS, based on landscape characteristics like ruggedness, modality, and correlation length. By evaluating these features at regular intervals, the algorithm can adaptively switch between different neighborhood strategies to improve convergence.

3.2 Fitness Landscape Features

In this part, we briefly describe the adopted features. We note that for the choice of the features, we have considered their wide adoption and their significant impact as well as the CPU time. Most of them have either been included in [13], as it incorporates the main previous work extracting features for the TSP, and [34] as it provides a deepened analysis of the significancy of the features. We note that for ruggedness, there are multiple measure, as indicated in [36]. Here, we adopt the entropic approach proposed in [35]. The autocorrelation measure used in the two papers is considered as the autocorrelation feature. Another feature which is not studied in those papers is the Fitness-Distance Correlation (FDC). These two features measuring ruggedness based on entropy and FDC are widely used and are applied as a learning mechanism in [7,15], respectively.

In Table 1 we describe the features and provide references for further information about them. (We note that some features, like FDC, require knowledge of the global optimum.)

3.3 Variable Neighborhood Search for the TSP

The VNS algorithm consists of three primary components: a neighborhood change step, an intensification phase, and a shaking phase [21]. The neighborhood change step drives the search process, deciding the specific neighborhood structure to explore and the acceptance criteria for any solutions obtained. The shaking phase introduces diversity into the search, aiming to escape local optima by applying controlled perturbations to the current solution. Lastly, the intensification phase serves as the main mechanism for improving the incumbent solution, focusing the search on promising regions to find better solutions.

For the TSP, VNS typically applies multiple neighborhood structures, often in the form of increasing k-opt moves, where k represents the size of the neighborhood (e.g., 2-opt, 3-opt, 4-opt). This allows VNS to adaptively switch between neighborhoods, balancing between intensification (exploiting the current solution's vicinity) and diversification (exploring new areas in the solution space).

Table 1. Summary of Fitness Landscape Features

Feature	Description
Ruggedness	Measures the entropy of fitness values across the landscape, representing how rough or smooth the landscape is [26,35].
FDC	Correlation between fitness values and distances to the global optimum, indicating the structure of the landscape [9,10].
Modality	The number of distinct peaks or local optima in the fitness landscape, reflecting the level of complexity [12,13].
Autocorrelation Length	Measures the correlation between fitness values of nearby solutions, indicating how much neighboring solutions are related [18,37].
Density of States	The distribution of fitness values across the landscape, providing insight into the number of solutions at different fitness levels [20,27].
Correlation Length	The extent to which the fitness landscape maintains correlation across solutions at increasing distances, showing landscape continuity [10,29].
Average Basin of Attraction Size	The average size of the regions around local optima where solutions converge, indicating landscape segmentation [30,35].
Distance to Global Optimum	The average distance between local optima and the global optimum, representing how far local solutions are from the best possible one [25,34].
Local Minima Density	The frequency of encountering local minima, indicating the difficulty of escaping local optima [20,29].
Correlation Structure	The overall correlation of the landscape, indicating whether solutions are globally related or fragmented [31,34].
Time to Local Optimum	The time or number of iterations taken to reach a local optimum, representing the efficiency of convergence [25,34].
Number of Nodes	The total number of nodes (or locations) in the problem instance, representing the problem size [1,8].
Mean Cost Between Nodes	The average cost (or distance) between nodes in the problem instance, representing connectivity and travel cost [1,22].

By dynamically adjusting neighborhood structures, VNS efficiently navigates complex landscapes, making it well-suited for TSP instances of varying sizes and structures. Note that we may distinguish between nested (a consecutive set of neighborhoods like 2-opt, 3-opt, etc., with an inherent subset structure) and non-nested neighborhoods [5], where the current approach focuses on nested structures.

In this paper, we enhance the VNS approach with an adaptive strategy that incorporates FL features, aiming to select between 2-opt, 3-opt, 4-opt, or 5-opt neighborhoods based on the characteristics of the landscape. Specifically, our approach computes the FL features to inform the VNS neighborhood selection, with the neighborhood structure adjusted dynamically at each iteration batch to reflect the complexity of the problem landscape. This enables a more refined, context-sensitive search strategy for solving TSP instances.

3.4 The Learning Approach

To apply supervised learning within this optimization framework, a training phase is necessary to build a reliable dataset, which will later support an optimization-based learning approach. During this phase, we train a supervised ML model to identify the most suitable neighborhood strategy. Using multiple TSP instances, we compute feature vectors that represent the problem landscape, test various neighborhoods (e.g., 2-opt, 3-opt, 4-opt, and 5-opt), and select the best-performing neighborhood for each instance. This process creates a training set of feature vectors and corresponding neighborhood labels, allowing the model to learn associations between features and optimal strategies.

The trained model is then used to predict the best neighborhood strategy for new instances. For each new TSP instance, we compute the FL features and use the ML model to suggest the most suitable/best neighborhood. The training and testing algorithms, detailed in Algorithms 1 and 2, outline this approach, where the training phase is conducted once, and the testing phase is repeated for each new instance or update in the neighborhood.

Algorithm 1: Training ML Model for Strategy Determination

Data: Several TSP instances
Result: Trained ML model for neighborhood determination
Initialize: Create empty lists for feature vectors X and labels y;
for *each TSP instance* **do**
　Load instance and create distance matrix;
　Generate initial solution;
　for *each iteration up to num_iterations* **do**
　　Run 2-opt, 3-opt, 4-opt, and 5-opt searches separately;
　　Compute distances and weighted scores considering both distance and time;
　　Select neighborhood with minimum weighted score;
　　Update current solution;
　　if *iteration % decision_interval == 0* **then**
　　　| Compute FL features and append to X and y;
　　end
　end
end
Train a RandomForestClassifier on X and y and save the model;

Another contribution of the paper is to compute the feature importance. Algorithm 3 illustrates the approach for that.

Algorithm 2: ML-Based Strategy Determination for VNS

Data: Trained ML model, new TSP instance
Result: Selected neighborhood for each iteration

Generate an initial solution;
for *each iteration* **do**
  if *iteration % decision_interval == 0* **then**
    Compute FL features for the current solution;
    Use the trained ML model to predict the best neighborhood strategy;
  end
  Perform the local search using the selected neighborhood strategy;
  Move to the best solution found within the selected neighborhood;
  Update the current solution with the best solution from this iteration;
end

Algorithm 3: Feature Importance Calculation

Data: Trained ML model, all new TSP instances
Result: Prediction accuracy and ranked feature importance

Initialize: Set lists for predictions, actuals, and FLA feature vectors;

for *each TSP instance* **do**
  Generate an initial solution;
  for *each iteration* **do**
    if *iteration % decision_interval == 0* **then**
      Compute FL features for the current solution;
      Create feature vector from FL features;
      Predict the best neighborhood using the trained ML model;
      Run 2-opt, 3-opt, 4-opt, and 5-opt local searches separately;
      Select the neighborhood with the shortest tour length;
      Compare the predicted and actual best neighborhood, storing results;
    end
    if *iteration % perturbation_interval == 0* **then**
      Apply a random perturbation to the current solution;
    end
  end
end
Compute Results: Calculate the accuracy of the ML model's predictions;
Use SHAP values to compute feature importance and generate a bar plot;

---

4 Experiments

In this section, we first define the experimental setup. Next, we compare the result of the algorithm with other state-of-the-art methods. Finally, we depict the most important features based on Shapley values.

4.1 Experimental Setup

All experiments were carried out on a computer equipped with an Intel i7-9750H processor and 16GB of RAM. Our algorithm was implemented in Python. For other algorithms, we utilized the package available at https://github.com/Valdecy/pyCombinatorial, and we modified the VNS code to suit our approach. Our approach employs a Random Forest model [4], configured with the default parameters from the Scikit-learn library to maintain a balance between computational efficiency and predictive accuracy. As discussed in Sect. 2, Random Forest was selected for its balanced performance, offering greater accuracy than linear regression models while maintaining shorter execution times compared to more complex methods such as deep neural networks and support vector machines. Table 2 summarizes the Random Forest parameters. The neighborhood selection is updated in batches of ten iterations, with a total of 100 iterations. The training dataset is composed of feature vectors generated from TSP instances, with each vector encapsulating the calculated fitness landscape metrics across each batch of ten iterations.

Table 2. Random Forest Default Parameters

Parameter	Value (Scikit-learn Default)
n_estimators	100
max_depth	None
min_samples_split	2

To manage computational cost, a maximum-time constraint per iteration is added: 0.7 s for training instances and 1.5 s for testing instances. This time limit helps prevent excessive runtime on specific instances. Typically, the algorithm defaults to the 2-opt neighborhood but switches to alternative neighborhoods if 2-opt is likely to be trapped in a local optimum.[1]

Table 3 provides an overview of the parameters used for each algorithm. These settings were selected based on their effectiveness in previous studies and fine-tuned for the TSP context to balance solution quality and computational efficiency.

The algorithm was tested on 26 TSPLib instances, with a 50%-50% split between training and testing: the first 13 instances were used for training, while the remaining 13 instances were used for testing. After training, the model was first validated on the training instances (to assess performance on instances with known best neighborhoods) and subsequently evaluated on the test instances.

---

[1] Further information, including code, is available upon reasonable request.

Table 3. Parameters for Different Algorithms in Solving TSP Instances

Algorithm	Parameter	Values/Settings
VNS	Neighborhood size	5
	Max Iterations	100
	Max Attempts	20
GA	Population Size	50
	Mutation Rate	0.01
	Crossover Rate	0.9
	Max Generations	100
ACO	Ants	30
	Pheromone Decay	0.1
	Alpha	1
	Beta	2
	Max Iterations	50

Each algorithm was executed ten times per instance, with the minimum, mean, and runtime recorded for comparison.

4.2 Results on the Trained Instances

Table 4 shows the obtained results for 13 TSP instances on which the algorithm is trained. For this part, the generalization capacity is not examined but just how choosing the best neighborhood will compare to other algorithms.

In the evaluation of trained instances, VNS-FLA demonstrates improvements over traditional VNS strategies in certain conditions, particularly where the adaptive features of the landscape analysis provide a discernible advantage. However, the performance of VNS-FLA, while showing an improvement over traditional VNS in several trained instances, highlights the need for further tuning to consistently match or surpass the efficacy of ACO in larger datasets. Despite these findings, the results indicated in Table 4 show that VNS-FLA can achieve better or comparable minimum and mean tour lengths across multiple benchmark instances, suggesting its potential as a viable alternative in the landscape of combinatorial optimization strategies.

4.3 Results on the New Tested Instances

We can note from Table 5 across the majority of the instances, that VNS-FLA demonstrates a better performance than traditional VNS in terms of both solution quality (mean and minimum tour lengths) and computational efficiency. While ACO consistently achieves the best solution quality, with the closest values to the optimal in both the minimum and mean measures, it comes at the cost of higher execution times. VNS-FLA, on the other hand, balances competitive

Table 4. Comparison of the performance on the trained benchmark TSP instances

Instance	Opt	Measure	VNS-FLA	VNS	3-opt	GA	ACO
Bays29	2020.0	Min	**2020.0**	2020.0	2020.0	2020.0	2020.0
		Mean	2034.1	2102.4	2099.3	2026.9	**2025.4**
		Time (s)	**12**	15	12	25	22
att48	10628.0	Min	10745.0	11000.0	10950.0	**10700**	10730.0
		Mean	10800.0	11300.0	11250.0	**10750.0**	10790.0
		Time (s)	18	20	**17**	28	25
eil51	426.0	Min	429.0	447.0	438.0	428.0	**427.0**
		Mean	435.7	503.6	466.9	438.3	**433.6**
		Time (s)	11	10	**8**	15	13
berlin52	7542.0	Min	**7542.0**	7963.0	7797.0	**7542.0**	7545.0
		Mean	7786.88	8449.83	8509.86	7830.3	**7600.7**
		Time (s)	22	23	**20**	32	30
st70	675.0	Min	710.0	680.0	680.0	**675.0**	678.0
		Mean	720	725	720	**680**	685
		Time (s)	27	30	**25**	38	35
pr76	108159.0	Min	110000.0	113000.0	112500.0	**109000.0.0**	108500.0
		Mean	110800	115000	114000	**110500**	109800
		Time (s)	**43**	50	42	65	55
eil76	538.0	Min	**540.0**	613.0	591.0	546.0	548.0
		Mean	564.1	653.3	655.1	562.3	**555.5**
		Time (s)	26	**23**	20	36	30
rat99	1211.0	Min	1230.0	1260.0	1255.0	**1225.0**	1230.0
		Mean	1240.0	1280.0	1270.0	**1235.0**	1229.0
		Time (s)	40	**30**	25	35	33
kroA100	21282.0	Min	21500.0	22500.0	22000.0	**21400.0**	21380.0
		Mean	21550.0	23000.0	22750.0	21550.0	**21450.0**
		Time (s)	60	**40**	32	50	45
kroB100	22141.0	Min	22900.0	22300.0	22350.0	**22100.0**	22200.0
		Mean	23250.0	22450.0	22500.0	22250.0	**22200.0**
		Time (s)	62	**43**	36	51	48
eil101	629.0	Min	645.0	650.0	650.0	**634.0**	638.0
		Mean	653.1	673.2	670.0	**638.0**	640.5
		Time (s)	41	27	**23**	42	36
lin105	14379.0	Min	14650.0	15000.0	14750.0	**14520.0**	14580.0
		Mean	14805.7	15302.5	14903.2	**14622.3**	14681.8
		Time (s)	56	46	**35**	54	51
pr107	44303.0	Min	44500.0	45500.0	45000.0	**44350.0**	44380.0
		Mean	45100.0	46000.0	45700.0	44750.0	**44500.0**
		Time (s)	65	45	**36**	55	50

Table 5. Comparison of the Performance on New Larger Benchmark Instances

Instance	Opt	Measure	VNS-FLA	VNS	3-opt	GA	ACO
ch150	6528.0	Min	**6650.0**	6800.0	6850.0	6700.0	6680.0
		Mean	**6700.0**	6900.0	7000.0	6750.0	6720.0
		Time (s)	**35**	40	38	60	55
rat195	2323.0	Min	2400.0	2450.0	2500.0	2370.0	**2355.0**
		Mean	2430	2550	2600	2400	**2380**
		Time (s)	**38**	45	43	65	60
kroA200	29368.0	Min	**29500.0**	30500.0	31000.0	29800.0	29750.0
		Mean	**29780.3**	31502.1	32498.6	30120.5	30010.7
		Time (s)	55	50	**48**	70	65
kroB200	29437.0	Min	**29650.0**	30650.0	31250.0	29900.0	29800.0
		Mean	**29900.7**	31602.5	32598.3	30310.8	30155.2
		Time (s)	42	50	**48**	70	65
tsp225	3916.0	Min	4000.0	4100.0	4200.0	3960.0	**3945.0**
		Mean	4030.5	4180.7	4300.3	3990.2	**3980.1**
		Time (s)	62	**40**	46	70	65
a280	2579.0	Min	2600.0	2650.0	2700.0	2590.0	**2585.0**
		Mean	2630.7	2702.3	2750.8	2610.2	**2600.4**
		Time (s)	45	**42**	52	72	68
pr299	48191.0	Min	48600.0	49500.0	50500.0	48400.0	**48300.0**
		Mean	49000.7	50001.2	51000.5	48600.3	**48550.8**
		Time (s)	50	**45**	58	80	75
lin318	42029.0	Min	42600.0	43600.0	44500.0	42300.0	**42200.0**
		Mean	42900.7	44501.2	45500.6	42600.3	**42450.5**
		Time (s)	52	**50**	60	85	80
rd400	15281.0	Min	15500.0	16000.0	16500.0	15400.0	**15360.0**
		Mean	15650.8	16300.2	16800.5	15500.3	**15420.7**
		Time (s)	60	**58**	70	95	85
fl417	11861.0	Min	12100.0	12600.0	12750.0	12000.0	**11950**
		Mean	12250.3	12850.5	12900.8	12150.2	**12080.4**
		Time (s)	65	**60**	75	100	90
rat575	6773.0	Min	6850.0	7000.0	7100.0	6820.0	**6805**
		Mean	6900.2	7100.5	7200.8	6850.4	**6830.7**
		Time (s)	80	**78**	90	120	110
p654	34643.0	Min	35000.0	36000.0	37000.0	34800.0	**34720.0**
		Mean	35500.4	36500.7	37500.6	35200.3	**35050.8**
		Time (s)	90	**88**	100	130	115
rat783	8806.0	Min	8900.0	9100.0	9250.0	8850.0	**8820.0**
		Mean	9000.4	9200.6	9350.3	8930.2	**8895.7**
		Time (s)	110	**108**	120	150	135

solution quality with relatively low runtime, often outperforming fixed neighborhood strategies like 3-opt and remaining competitive with GA and ACO. This highlights the adaptability and efficiency of VNS-FLA, particularly for larger and more complex TSP instances. The results confirm that VNS-FLA is a robust and effective alternative, offering a strong trade-off between solution quality and computational cost.

In summary, although the implemented approach does not outperform the best-known methods for the TSP, it demonstrates significant improvements over traditional VNS and shows competitiveness with other well-known metaheuristics.

4.4 TSP Feature Importance

The aim of this part is to show the impact regarding feature importance. We note here that the overall prediction accuracy across all the 13 test instances is 86.15%. In a multi-class setup (as in [23]), each neighborhood strategy (e.g., 2-opt, 3-opt) is a distinct class, and a correct prediction is when the model selects the strategy that minimizes the distance for a batch. SHAP values reveal how each fitness landscape feature affects the likelihood of each strategy, with contributions calculated separately for each class. The SHAP bar plot, which typically averages these contributions, highlights features that consistently guide the model toward an optimal strategy, enhancing interpretability across scenarios. Figure 1 shows the most important features.[2]

The SHAP bar plot shows the primary landscape features influencing neighborhood strategy selection in VNS. Among these, ruggedness and FDC emerge as top predictors, indicating the importance of landscape complexity and proximity to optimal solutions in guiding neighborhood choice (we note again that FDC requires knowledge of the local optimum). Autocorrelation length and modality also contribute, suggesting that smoother, more structured landscapes favor certain strategies. These insights confirm that features capturing structural nuances of the landscape provide reliable indicators for strategy adaptation, enhancing the VNS model's performance across diverse TSP instances. Further, these insights pave the way for targeted enhancements to the adaptive mechanisms within the VNS framework, particularly in refining feature weighting and selection processes.

---

[2] This plot quantifies the impact of various features on the model's predictive accuracy. The y-axis lists significant features, and the x-axis shows their corresponding SHAP values.

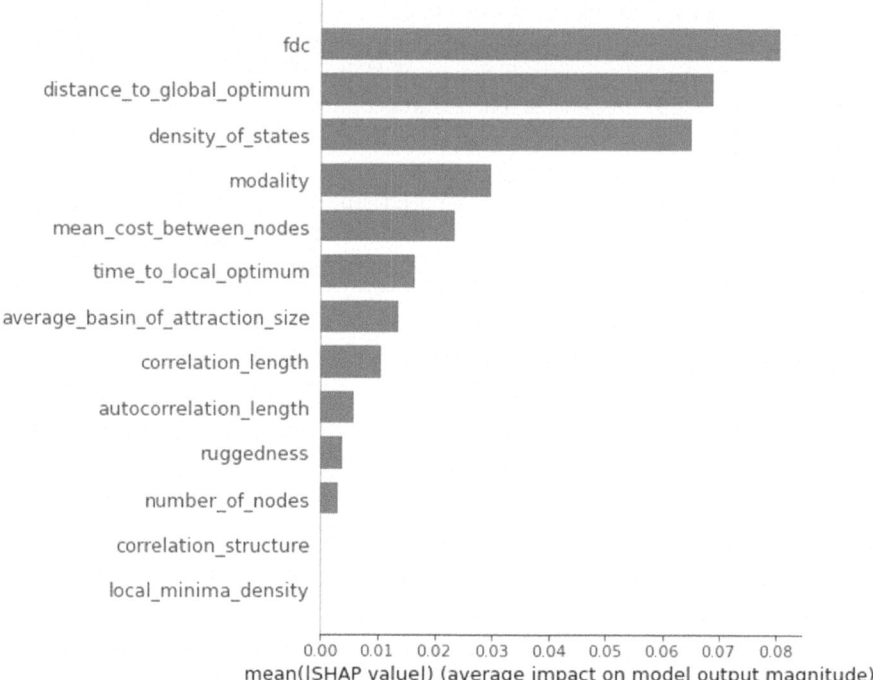

Fig. 1. Significant Features Impacting Neighborhood Prediction: SHAP Value Bar Plot.

5 Conclusion

This study introduced an adaptive variant of variable neighborhood search based on fitness landscape analysis and demonstrated its effectiveness in solving the traveling salesman problem. By leveraging features from fitness landscape analysis and a supervised machine learning model, the proposed adaptive approach dynamically selects the most suitable neighborhood strategy during the search process. Experimental results on both trained and new larger instances confirm the superiority of the proposed method over traditional variable neighborhood search and fixed neighborhood strategies such as the 3-opt heuristic, achieving better or comparable solution quality while maintaining competitive computational efficiency. Although an ant colony optimization algorithm achieved the best solution quality in certain instances, the proposed method offers a strong balance between accuracy and runtime, outperforming ant colony optimization and genetic algorithms in terms of efficiency across multiple benchmarks. To more robustly validate these conclusions, additional testing with a wider range of instances and increased trials is essential to enhance both statistical power and generalizability.

The study highlights the potential of landscape-aware adaptive strategies to enhance the performance of metaheuristics, particularly for large and complex problem instances. The integration of explainable artificial intelligence through Shapley values provided insights into the influence of fitness landscape analysis features on strategy selection, offering interpretability alongside performance gains. While the proposed approach does not outperform state-of-the-art methods such as the Lin-Kernighan heuristic, it demonstrates significant improvements over traditional variable neighborhood search and comparable heuristics, validating the effectiveness of landscape-guided adaptations. Moreover, it provides new insights into existing metaheuristic approaches and their components rather than just developing new ones, fulfilling requests of recent and related literature [2,28].

The adaptive variable neighborhood search framework proposed in this study has significant potential for real-world applications and extensions beyond the TSP. By leveraging fitness landscape analysis and machine learning to guide adaptive decision making, the framework can tackle complex optimization tasks across various domains, including Vehicle Routing Problems in logistics, scheduling problems in workforce and resource allocation, and network optimization in telecommunications.

Future research could focus on exploring additional fitness landscape analysis features to enhance decision making and integrating the framework with other metaheuristics such as Ant Colony Optimization or Genetic Algorithms. Incorporating a wider range of neighborhood operators tailored for permutation-based problems will provide deeper insights into the framework's adaptability and performance. This also asks for the exploration of an integration with the pilot method (extending ideas from [5]). Testing its scalability on more diverse and larger problem sets will further validate the robustness of our approach, while investigating non-nested neighborhoods offers an exciting avenue for enhancing the flexibility and efficiency of the adaptive VNS framework.

References

1. Applegate, D.L., et al.: Certification of an optimal TSP tour through 85,900 cities. Oper. Res. Lett. **37**, 11–15 (2009). https://doi.org/10.1016/j.orl.2008.09.006
2. de Armas, J., Lalla-Ruiz, E., Tilahun, S.L., Voß, S.: Similarity in metaheuristics: a gentle step towards a comparison methodology. Nat. Comput., 1–23 (2021). https://doi.org/10.1007/s11047-020-09837-9
3. Boese, K.D., Kahng, A.B., Muddu, S.: A new adaptive multi-start technique for combinatorial global optimizations. Oper. Res. Lett. **16**(2), 101–113 (1994). https://doi.org/10.1016/0167-6377(94)90065-5
4. Breiman, L.: Random forests. Mach. Learn. **45**(1), 5–32 (2001)
5. Geiger, M., Sevaux, M., Voß, S.: Neigborhood selection in variable neighborhood search. In: Proceedings of the IX Metaheuristics International Conference. pp. 571–573. Udine (2011)
6. Hansen, P., Mladenović, N.: Variable neighborhood search: Principles and applications. Eur. J. Oper. Res. **130**, 449–467 (2001). https://doi.org/10.1016/S0377-2217(00)00100-4

7. Huang, Y., Li, W., Tian, F., Meng, X.: A fitness landscape ruggedness multiobjective differential evolution algorithm with a reinforcement learning strategy. Appl. Soft Comput. **96**, 106693 (2020). https://doi.org/10.1016/j.asoc.2020.106693
8. Hutter, F., Xu, L., Hoos, H.H., Leyton-Brown, K.: Algorithm runtime prediction: methods & evaluation. Artif. Intell. **206**, 79–111 (2014). https://doi.org/10.1016/j.artint.2013.10.003
9. Jones, T., Forrest, S.: Fitness distance correlation as a measure of problem difficulty. In: Proceedings of the 6th International Conference on Genetic Algorithms, pp. 184–192 (1995)
10. Jones, T.: Evolutionary algorithms, fitness landscapes and search. Ph.D. thesis, The University of New Mexico (1995)
11. Kalatzantonakis, P., Sifaleras, A., Samaras, N.: A reinforcement learning-variable neighborhood search method for the capacitated vehicle routing problem. Expert Syst. Appl. **213**, 118812 (2023). https://doi.org/10.1016/j.eswa.2022.118812
12. Kerschke, P., Hoos, H.H., Neumann, F., Trautmann, H.: Automated algorithm selection: survey and perspectives. Evol. Comput. **27**(1), 3–45 (2019). https://doi.org/10.1162/evco_a_00242
13. Kerschke, P., Kotthoff, L., Bossek, J., Hoos, H.H., Trautmann, H.: Leveraging TSP solver complementarity through machine learning. Evol. Comput. **26**(4), 597–620 (2018). https://doi.org/10.1162/evco_a_00215
14. Kerschke, P., Trautmann, H.: Comprehensive feature-based landscape analysis of continuous and constrained optimization problems using the R-package Flacco. In: Bauer, N., Ickstadt, K., Lübke, K., Szepannek, G., Trautmann, H., Vichi, M. (eds.) Applications in Statistical Computing. SCDAKO, pp. 93–123. Springer, Cham (2019). https://doi.org/10.1007/978-3-030-25147-5_7
15. Li, W., Meng, X., Huang, Y.: Fitness distance correlation and mixed search strategy for differential evolution. Neurocomputing **458**, 514–525 (2021). https://doi.org/10.1016/j.neucom.2019.12.141
16. Liu, R., Tao, Y., Xie, X.: An adaptive large neighborhood search heuristic for the vehicle routing problem with time windows and synchronized visits. Comput. Operat. Res. **101**, 250–262 (2019). https://doi.org/10.1016/j.cor.2018.08.002
17. Lundberg, S.M., Lee, S.I.: A unified approach to interpreting model predictions. In: Guyon, I., Luxburg, U.V., Bengio, S., Wallach, H., Fergus, R., Vishwanathan, S., Garnett, R. (eds.) Advances in Neural Information Processing Systems, vol. 30. Curran Associates, Inc. (2017), https://proceedings.neurips.cc/paper_files/paper/2017/file/8a20a8621978632d76c43dfd28b67767-Paper.pdf
18. Malan, K.M., Engelbrecht, A.P.: Fitness landscape analysis for metaheuristic performance prediction. In: Richter, H., Engelbrecht, A. (eds.) Recent Advances in the Theory and Application of Fitness Landscapes. ECC, vol. 6, pp. 103–132. Springer, Heidelberg (2014). https://doi.org/10.1007/978-3-642-41888-4_4
19. Mersmann, O., Bischl, B., Trautmann, H., Wagner, M., Bossek, J., Neumann, F.: A novel feature-based approach to characterize algorithm performance for the traveling salesperson problem. Ann. Math. Artif. Intell. **69**(2), 151–182 (2013). https://doi.org/10.1007/s10472-013-9341-2
20. Merz, P., Freisleben, B.: Fitness landscape analysis and memetic algorithms for the quadratic assignment problem. IEEE Trans. Evol. Comput. **4**(4), 337–352 (2000). https://doi.org/10.1109/4235.887234
21. Mladenović, N., Hansen, P.: Variable neighborhood search. Comput. Operat. Res. **24**(11), 1097–1100 (1997). https://doi.org/10.1016/s0305-0548(97)00031-2

22. Mladenović, N., Urošević, D., Hanafi, S., Ilić, A.: A general variable neighborhood search for the one-commodity pickup-and-delivery travelling salesman problem. Eur. J. Oper. Res. **220**(1), 270–285 (2012). https://doi.org/10.1016/j.ejor.2012.01.036
23. Nahiduzzaman, M., Faisal Abdulrazak, L., Arselene Ayari, M., Khandakar, A., Islam, S.R.: A novel framework for lung cancer classification using lightweight convolutional neural networks and ridge extreme learning machine model with SHapley Additive exPlanations (SHAP). Expert Syst. Appl. **248**, 123392 (2024). https://doi.org/10.1016/j.eswa.2024.123392
24. Ochoa, G., Veerapen, N.: Deconstructing the big valley search space hypothesis. In: Evolutionary Computation in Combinatorial Optimization, pp. 58–73. Springer (2016). https://doi.org/10.1007/978-3-319-30698-8_5
25. Ochoa, G., Veerapen, N.: Mapping the global structure of TSP fitness landscapes. J. Heuristics **24**(3), 265–294 (2017). https://doi.org/10.1007/s10732-017-9334-0
26. Poursoltan, S., Neumann, F.: Ruggedness quantifying for constrained continuous fitness landscapes. In: Datta, R., Deb, K. (eds.) Evolutionary Constrained Optimization. ISFS, pp. 29–50. Springer, New Delhi (2015). https://doi.org/10.1007/978-81-322-2184-5_2
27. Reidys, C.M., Stadler, P.F.: Neutrality in fitness landscapes. Appl. Math. Comput. **117**(2–3), 321–350 (2001). https://doi.org/10.1016/s0096-3003(99)00166-6
28. Sörensen, K.: Metaheuristics - the metaphor exposed. Int. Trans. Oper. Res. **22**(1), 3–18 (2013). https://doi.org/10.1111/itor.12001
29. Stadler, P.F.: Landscapes and their correlation functions. J. Math. Chem. **20**(1), 1–45 (1996). https://doi.org/10.1007/bf01165154
30. Stadler, P.F., Schnabl, W.: The landscape of the traveling salesman problem. Phys. Lett. A **161**(4), 337–344 (1992). https://doi.org/10.1016/0375-9601(92)90557-3
31. Stadler, P.F., Stephens, C.R.: Landscapes and effective fitness. Comments Theoret. Biol. **8**(4–5), 389–431 (2003)
32. Stodola, P., Otřísal, P., Hasilová, K.: Adaptive ant colony optimization with node clustering applied to the travelling salesman problem. Swarm Evol. Comput. **70**, 101056 (2022). https://doi.org/10.1016/j.swevo.2022.101056
33. Sun, Y., Ernst, A., Li, X., Weiner, J.: Generalization of machine learning for problem reduction: a case study on travelling salesman problems. OR Spectrum **43**(3), 607–633 (2020). https://doi.org/10.1007/s00291-020-00604-x
34. Tayarani-N., M.H., Prügel-Bennett, A.: An analysis of the fitness landscape of travelling salesman problem. Evolutionary Comput. **24**, 347–384 (2016). https://doi.org/10.1162/EVCO_a_00154
35. Vassilev, V.K., Fogarty, T.C., Miller, J.F.: Smoothness, ruggedness and neutrality of fitness landscapes: from theory to application. In: Advances in Evolutionary Computing, pp. 3–44. Springer (2003). https://doi.org/10.1007/978-3-642-18965-4_1
36. Wang, W., Shi, J., Sun, J., Liefooghe, A., Zhang, Q.: A new parallel cooperative landscape smoothing algorithm and its applications on TSP and UBQP. Expert Syst. Appl. **263**, 125611 (2025). https://doi.org/10.1016/j.eswa.2024.125611
37. Weinberger, E.: Correlated and uncorrelated fitness landscapes and how to tell the difference. Biol. Cybern. **63**(5), 325–336 (1990). https://doi.org/10.1007/bf00202749
38. Zhang, P., Wang, J., Tian, Z., Sun, S., Li, J., Yang, J.: A genetic algorithm with jumping gene and heuristic operators for traveling salesman problem. Appl. Soft Comput. **127**, 109339 (2022). https://doi.org/10.1016/j.asoc.2022.109339

39. Zou, F., Chen, D., Liu, H., Cao, S., Ji, X., Zhang, Y.: A survey of fitness landscape analysis for optimization. Neurocomputing **503**, 129–139 (2022). https://doi.org/10.1016/j.neucom.2022.06.084

Healthcare Facility Location Problem and Fitness Landscape Analysis

Justin Scouarnec[1(✉)], Corinne Lucet[1], Sara Tari[2], and Laure Brisoux Devendeville[1]

[1] Laboratoire MIS UR 4290, Université de Picardie Jules Verne, Amiens, France
{justin.scouarnec,corinne.lucet,laure.devendeville}@u-picardie.fr
[2] LISIC, Université du Littoral Côte d'Opale, Calais, France
sara.tari@univ-littoral.fr

Abstract. This paper presents an initial study of a Healthcare Facility Location Problem, the Two-Level P-Median Location Problem (TLPMLP). The main goal is to analyze problem instances by studying their associated fitness landscape characteristics, with a focus on ruggedness, neutrality and number of local optima. For each instance, we consider two fitness landscapes defined by the following two functions: the direct neighborhood function ($\mathcal{O}(n)$) complexity) and the two-step Neighborhood function ($\mathcal{O}(n^2)$ complexity). We also conduct a preliminary exploration on the correlation between the fitness landscape characteristics and the solving difficulty by using an exact method (binary linear program) and two straightforward local search methods: a hill-climber and sampled walk. Experiments are conducted on 40 instances from the literature and 27 randomly generated ones. We discuss our results and point out several future research directions on these initial approaches to analyzing TLPMLP.

Keywords: Two Level P Median Location Problem · Fitness Landscape Analysis · Local Search

1 Introduction

When it comes to the location of healthcare facilities various challenges arise, depending on the type of centers under consideration [1]. Usually, addressing those challenges requires to optimize the territory coverage of healthcare facilities, leading to three main categories of optimization problems. For coverage problems, the goal is to ensure the coverage of all demand points in the territory while using the minimum number of centers. P-center problems aim to minimize the maximum distance between any demand point and its nearest healthcare facility while placing a fixed number P of centers. In P-median problems, the objective is to minimize the average distance between demand points and their nearest facility, using a fixed number P of centers. For each of these NP-hard problems, additional constraints can be introduced, such as maximum capacities for centers and non-uniform demand across the territory. This flexibility allows for a wide range of more complex and realistic location problems to be addressed.

Here, we focus on the Multi-level P-Median Location Problem (MLPMLP), a variant of the standard P-median problem that considers several levels of centers, with different functionalities. Variants of this problem have been considered in the literature, for example, the three levels variant has been used to address needs in relation to maternity centers [2] or to optimize blood sample collection [3]. While these studies are efficient to tackle the variant under consideration for specific instances, there is a lack of global unified knowledge on this problem, especially when it comes to the impact of some instances features on the solving difficulty. A more unified view of this problem could help to better understand the impact of instances and of some variants on the optimization process, giving useful information for future studies on the subject. In the context of optimization, fitness landscape analysis is one of the available tools to bridge this gap, as it often provides insights into the solution space structure, helping to understand how solutions are distributed and how easily local search methods can navigate towards optimal or near-optimal solutions. Fitness landscapes can be used to investigate the impact of instances characteristics, but also to identify challenges that can cause heuristics to struggle. In this work, we propose a first step in this direction by providing a first fitness landscape analysis for several instances of the Two Level P-Median Location Problem (TLPMLP) with unitary demands and unlimited capacities for centers [4–6]. The main goal of this preliminary work is to identify the links between some instance features and some landscape characteristics in this context, as well as to observe the impact of two neighborhoods relations on landscape characteristics. We also present an initial experimental study to evaluate the ability of two straightforward local search algorithms to reach high-quality solutions. This evaluation is conducted on small instances where the global optimum can be reached in a reasonable time, as well as on larger instances where the global optimum is not available. We investigate whether there is a correlation between the characteristics of an instance's fitness landscape and the difficulty of solving that instance.

This paper is structured as follows. In Sect. 2, we describe TLPMLP and introduce the concept of fitness landscape analysis as well as the two neighborhood relations we used. Section 3 presents the experimental analysis related to fitness landscapes and local search performance. Finally, the last section discusses this work and provides a wide range of perspectives.

2 Fitness Landscape for TLPMLP

2.1 Problem Definition

The territory to cover is represented by a graph $G = (V, E)$ where V is the set of vertices and E is the set of edges. The neighborhood of a vertex $v \in V$ is $\Gamma(v) = \{u \in V, |(u,v) \in E\}$. Vertices represent points and/or candidate locations for centers, while edges represent a direct path existing between locations. We define $I \subseteq V$ the set of demand points, $J \subseteq V$ the set of candidate locations for level 1 centers and $K \subseteq V$ the set of candidate locations for level 2 centers. The shortest distance between two vertices $u \in V$ and $v \in V$ is noted d_{uv}. The goal of this problem is to locate a fixed number p of level 1 centers and a fixed number q of level 2 centers ($q \leq p$) in order to minimize the sum of the distances from the demand points to their nearest level 1 center, added to the sum of the distances from level 1 centers to their nearest level 2 center.

A solution to this problem can be represented by two sets $P \subseteq J, Q \subseteq K$ where P is the set of placed level 1 centers and Q is the set of placed level 2 centers. We denote such a solution $Sol = (P, Q)$. Since in our model the candidate locations for level 1 and level 2 facilities are not exclusive, P and Q can share common vertices. The score of solution Sol is computed by Eq. 1.

$$f(Sol) = \sum_{i \in I} \min_{j \in P}(d_{ij}) + \sum_{j \in P} \min_{k \in Q}(d_{jk}) \qquad (1)$$

2.2 Fitness Landscapes Analysis

Fitness landscapes were first introduced by Sewall Wright in 1932 as part of his work on population genetics [7]. Nowadays, this concept is also used in evolutionary computation to provide a structured representation of the search space according to a specified neighborhood relation. More formally, a fitness landscape is defined by a triplet $(\mathcal{X}, \mathcal{N}, f)$, with $\mathcal{X}$ the set of possible solutions, $\mathcal{N}$ the neighborhood relation between solutions, and f the fitness function. As an example, the fitness function for TLPMLP is represented by Eq. 1.

Once a fitness landscape is defined, several characteristics can be used to characterize the overall shape of the landscape. Fitness Landscape Analysis (FLA) consists of measuring and estimating some of these characteristics to draw knowledge from optimization problems. For example, one can use FLA to characterize problem instances and thus comparing different instance families for a given problem, but also comparing instances across various optimization problems. FLA is also commonly used to study the behavior of neighborhood-based algorithms, for example by observing which characteristics influence the performance of a given algorithm. Recent works are devoted to using fitness landscape characteristics to predict which algorithm is the most efficient to optimize a given problem [8]. For a comprehensive overview of FLA, the reader is referred to [9–11].

Here, we consider four main characteristics widely used in FLA: the landscape size (graph size, number of centers), the ruggedness, the neutrality rate, and the proportion of local optima.

Ruggedness. The ruggedness of a fitness landscape refers to the number and distribution of local optima. A smooth landscape has few local optima with large basins of attraction. A rugged landscape, on the other hand, has many local optima with small basins of attraction. Local search algorithms tend to be less efficient on rugged landscapes as they are more easily trapped in the basins of attraction of low-quality local optima. Conversely, a smooth landscape is easier to explore by local search.

Several indicators exist to estimate ruggedness, the most widely used being the function [12], $\rho(n) \in [0, 1]$, described in Eq. 2. Usually, the autocorrelation function is computed by using the sequence of fitness value in a random walk. $\rho(n)$ indicates the extent to which a value in a sequence is correlated with the value that follows it after a given number of steps n.

$$\rho(n) = \frac{E\left[(f(x_i) - \bar{f})(f(x_{i+n}) - \bar{f})\right]}{\text{Var}(f(x_i))} \quad (2)$$

Only using the first autocorrelation step $\rho(1)$ is usually sufficient. A low value indicates a rugged landscape, while a high value indicates a smooth landscape.

Neutrality. The neutrality rate ν of a landscape refers to the presence of neighboring solutions sharing the same score (see Eq. 3).

$$\nu = \frac{\#\{(Sol, Sol'), Sol' \in \mathcal{N}(Sol), f(Sol) = f(Sol')\}}{\#\{(Sol, Sol'), Sol' \in \mathcal{N}(Sol)\}} \quad (3)$$

High neutrality rates usually induce plateaus in the landscape, making its exploration potentially difficult for local search algorithms as they tend to get stuck into these plateaus preventing them from finding better solutions. Neutrality rates can be estimated by using several indicators. Here, we randomly generate several pairs of neighboring solutions and the ratio of pairs of neutral neighbors over the whole set.

Proportion of Local Optimal. The proportion of local optima can be estimated using the length of adaptive walks, as it gives an idea of the size of the attraction basins and therefore of the local optima proportion. Longer walks imply a smaller number of local optima. A high proportion of local optima makes the landscape difficult to explore by local search, while a low proportion makes it easier to explore by local search.

2.3 Neighborhood Functions

In this work, we use FLA to investigate the impact of instance features on fitness landscapes derived from TLPMLP, but also to determine a suitable neighborhood relation for this problem. As such, we define the two following neighborhood functions.

Direct Neighborhood Function. This function considers two solutions $Sol \in \mathcal{X}$ and $Sol' \in \mathcal{X}$ to be neighbors (i.e. $Sol' \in \mathcal{N}(Sol)$) when Sol' can be obtained by moving a single center (any level) from Sol to one of its adjacent vertices.

Let $Sol = (P, Q)$ and $v_1 \in P$ (resp. $v_1 \in Q$). $Sol' \in \mathcal{N}(Sol)$ iff $\exists v_2 \in \Gamma(v_1) \setminus P$ (resp. $\exists v_2 \in \Gamma(v_1) \setminus Q$) s.t. $Sol' = (P \setminus \{v_1\} \cup \{v_2\}, Q)$ (resp. $Sol' = (P, Q \setminus \{v_1\} \cup \{v_2\})$).

With $\delta(G)$ being the density of graph G, the complexity of generating all neighbor solutions of a given solution using this function is $\mathcal{O}((p+q) \cdot \delta(G) \cdot |V|)$.

Two-Step Neighborhood Function. This function considers two solutions $Sol \in \mathcal{X}$ and $Sol' \in \mathcal{X}$ to be neighbors (i.e. $Sol' \in \mathcal{N}(Sol)$) when Sol' can be obtained by moving a single center (any level) from Sol to one of its adjacent vertices or to one of the adjacent vertices to its adjacent vertices.

Let $Sol = (P, Q)$ and $v_1 \in P$ (resp. $v_1 \in Q$). $Sol' \in \mathcal{N}(Sol)$ iff $\exists v_2 \in \{\Gamma(v_1) \bigcup_{u \in \Gamma(v_1)} \Gamma(u)\} \setminus P$ (resp. $\exists v_2 \in \{\Gamma(v_1) \bigcup_{u \in \Gamma(v_1)} \Gamma(u)\} \setminus Q$) s.t. $Sol' = (P \setminus \{v_1\} \cup \{v_2\}, Q)$ (resp. $Sol' = (P, Q \setminus \{v_1\} \cup \{v_2\})$).

The complexity of generating all neighbor solution of a given solution using this function is $\mathcal{O}((p+q) \cdot (\delta(G) \cdot |V|)^2)$.

2.4 Local Search Methods

We use two local search algorithms to sample the solution space and estimate the landscape characteristics: a simple random walk and a first improvement hill-climbing algorithm referred to as adaptive walk. For both, the two neighborhood functions mentioned above are used to generate neighboring solutions and compared in experiments Sect. 3.

Random Walk: The algorithm starts from a random initial solution Sol and moves to randomly selected solution $Sol' \in \mathcal{N}(Sol)$ for a fixed number of steps $nbSteps$.

Adaptive Walk: The algorithm starts from a random initial solution. Neighboring solutions $Sol' \in \mathcal{N}(Sol)$ are examined in a random order, the first improving solution is selected as the new current solution. The process is repeated until a local optimum is reached.

3 Experiments and Results

Our first objective is to extract landscape characteristics from various instances. To observe the impact of the number of centers to be placed on the landscape characteristics derived from the instance, we arbitrarily vary the number of level 1 centers to be placed from 1 to 10% of the total number of vertices in the instance. The number of level 2 centers to be placed is arbitrarily set to 10% of the number of level 1 centers (raised to the next higher integer).

3.1 Instances

Literature Instances: As a first step, we decided to study pmed instances [13], which are commonly used to study the p-median problem. This allows us to examine the impact of graph size on landscape characteristics, for instances with a similar structure. Each pmed/x references 3 to 5 instances with x vertices. We also use DIMACS instances [14] to study differently structured graphs, specifically queen, myciel, miles, and DSJC instances, which do not have weighted edges. To address this, random weights ranging from 1 to 20 were assigned to the edges. The characteristics of these instances are summarized in Table 1.

Table 1. "Literature instance characteristics. "Instance" refers to the name of the instance or the group of instances (for pmed), "N" is the number of vertices, "M" the number of edges, "D" the graph's density, and "$\Delta(G)$" its maximum or average maximum degree (for groups of instances).

Instance	N	M	D	$\Delta(G)$	Instance	N	M	D	$\Delta(G)$
DSJC-125/0.1	125	736	0.10	23	miles/1000	128	3216	0.39	86
DSJC-125/0.5	125	3891	0.50	75	miles/1500	128	5198	0.64	106
DSJC-125/0.9	125	6961	0.90	120	myciel/3	11	20	0.36	5
DSJC-250/0.1	250	3218	0.10	38	myciel/4	23	71	0.28	11
DSJC-250/0.5	250	15668	0.50	147	myciel/5	47	236	0.21	23
DSJC-250/0.9	250	27897	0.90	234	myciel/6	95	755	0.17	47
DSJC-500/0.1	500	12458	0.10	68	myciel/7	191	2360	0.13	95
DSJC-500/0.5	500	62624	0.50	286	queen-5/5	25	160	0.53	16
DSJC-500/0.9	500	112437	0.90	471	queen-6/6	36	290	0.46	19
pmed/100	100	199.8	0.04	8.40	queen-7/7	49	476	0.40	24
pmed/200	200	800	0.04	16.20	queen-8/12	96	1368	0.30	32
pmed/300	300	1800	0.04	21.40	queen-8/8	64	728	0.36	27
pmed/400	400	3200	0.04	28.00	queen-9/9	81	1056	0.32	32
pmed/500	500	5000	0.04	32.00	queen-10/10	100	1470	0.30	35
pmed/600	600	7200	0.04	39.00	queen-11/11	121	1980	0.28	40
pmed/700	700	9800	0.04	47.25	queen-12/12	144	2596	0.25	43
pmed/800	800	12800	0.04	46.66	queen-13/13	169	3328	0.23	48
pmed/900	900	16200	0.04	57.33	queen-14/14	196	4186	0.22	51
miles/500	128	1170	0.15	38	queen-15/15	225	5180	0.21	56
miles/750	128	2113	0.26	64	queen-16/16	256	6320	0.19	59

Generated Instances: To study the impact of characteristics other than graph size (density in particular), we studied randomly generated instances with a fixed density (0.1, 0.5, 0.9) for a graph size ranging from 100 to 900 vertices (10 instances per size-density pair). These instances are noted (number of vertices)-(density) in the following figures.

3.2 Fitness Landscapes Analysis

For the sake of space and readability, generated instances with 200, 400, 600 and 800 vertices as well as pmed instances with 200, 300, 400, 600, 700 and 800 vertices were omitted in the following FLA results tables since the characteristics evolve linearly with the number of vertices on these types of instances. Additionally, results are provided only for the direct neighborhood relation, as despite the difference of their complexity, tendencies for the two neighborhood relations are similar. We still mention the minor differences when relevant.

The autocorrelation $\rho(1)$ is computed from the scores met during a single 1000 step random walk for each landscape. Results are reported in Fig. 1(a) for generated instances, and Fig. 2(a) for literature instances.

The neutrality is estimated by using 1000 random couples of neighboring solutions for each landscape. Results are reported in Fig. 1(b) for generated instances, and Fig. 2(b) for literature instances.

For each landscape, the local optima proportion is estimated using the average length of 30 adaptive walks. Results are reported in Fig. 3.

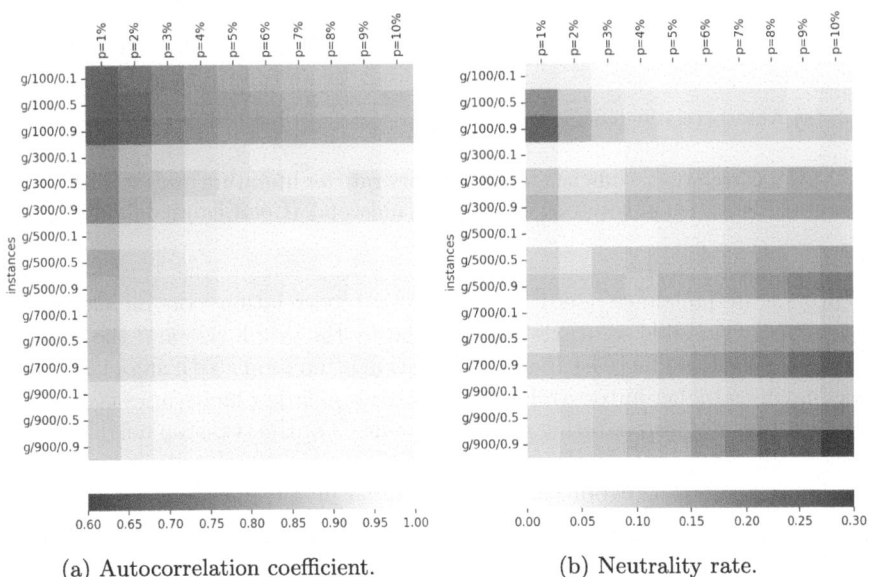

(a) Autocorrelation coefficient. (b) Neutrality rate.

Fig. 1. Autocorrelation coefficient and neutrality rate for generated instances. A darker shade of red indicates a more rugged/neutral landscape. (Color figure online)

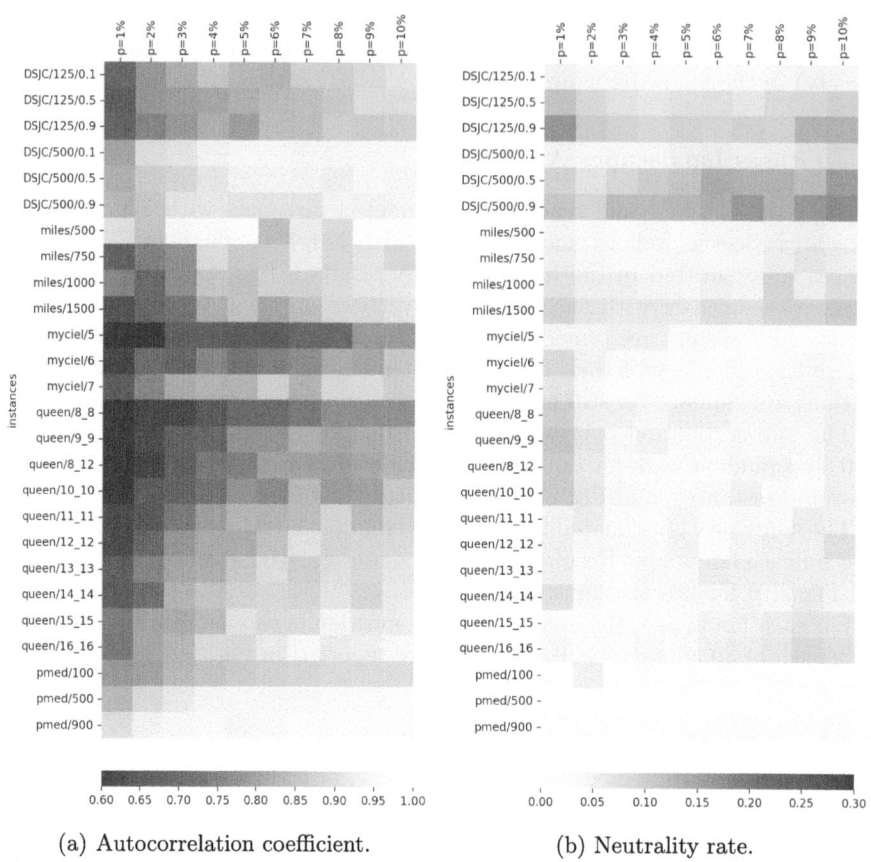

(a) Autocorrelation coefficient.   (b) Neutrality rate.

Fig. 2. Autocorrelation coefficient and neutrality rate for literature instances. A darker shade of red indicates a more rugged/neutral landscape. (Color figure online)

On generated instances (see Fig. 1(a)), the autocorrelation score increases with the number of possible solutions (determined by the graph size and the number of centers). While an increase in autocorrelation often indicates a smoother landscape, having a higher autocorrelation coefficient on larger landscapes is common since the impact of a step during a walk is smaller. For the two-step neighborhood function, the ruggedness levels are slightly higher when the density is low and on small landscape but its evolution is rather similar on all generated instances.

Results reported for literature instances (see Fig. 2(a)), show similar ruggedness levels and evolution between previously seen generated instances and pmed instances. To a lesser extent, this is also the case for DSJC, queen and myciel instances. The ruggedness evolution according to the number of centers is slightly different on miles landscapes. A partial explanation could be that these instances are based on real-world data, but queen and myciel instances are not randomly generated.

Measured neutrality rates on generated instances (see Fig. 1(b)) highlight that density has a major influence on neutrality rates. With a high density, neutrality rates are particularly high (up to 30%). Considering the nature of TLPMLP, this result suggests a high number of symmetrical solutions. Neutrality levels also increase with the number of centers, except on smaller landscapes with few centers. This is a bias induced by our protocol where $p \times 0.1$ level 2 centers must be placed, rounded up to the next integer, resulting in 1 level 2 center for $[1-10]$ level 1 centers. Neutrality-wise, there is no significant difference for the other neighborhood relation.

The neutrality rates of literature instances are reported in Fig. 2(b). Neutrality levels and evolution are similar on generated and DSJC instances. Neutrality levels are very low and do not evolve on pmed instances, which can be explained by their low density. Despite varying densities, queen, myciel and miles instances neutrality levels differences are lower than on randomly generated instances of similar densities. This suggests a lower symmetry than for randomly generated instances. Results are similar between the landscapes defined by the two neighboring solutions.

Figure 3(a) reports the mean adaptive walk length on landscapes derived from generated instances. As expected, on generated instances, the length of adaptive walks increases with the landscape size (graph size, and number of centers). Their length decrease with density, which is likely due to the increase on neutrality levels. Indeed, our adaptive walk does not accept neutral moves making the search likely to get trapped early on a local optimum when neutrality levels are high. While the walks lengths are slightly higher for the two-level neighborhood relation due to the higher connectivity between solutions, results are similar.

Observations are mostly similar between generated instances and instances from the literature (see Fig. 3(b)). The way instances were generated does not seem to significantly affect the number of local optima for similar landscapes size.

3.3 Performance of both Sampled and Adaptive Walks

We propose a first performance analysis on our landscapes by using optimal solutions on small instances, as well as two straightforward local search methods on all instances. The aim is not to achieve the best results possible but rather to grasp the varying levels of difficulty according to our instances. The first local search is a first improvement hill-climbing algorithm (HC), described in Sect. 2.4. The second one, called Sampled Walk (SW) [15], is a straightforward partial neighborhood local search which requires a single parameter λ. At each step, SW evaluates λ neighbors of the current solution, and selects the best one. Unlike the hill-climber, SW accepts deteriorating neighbors.

To find the optimal solutions on the smallest instances, we used a branch & cut algorithm running on CPLEX 22.1.0 for 6 h for each instance, using an AMD EPYC 7513 32-Core Processor. Given HC, SW and the two neighborhood relations, we have 4 local search variants. For each couple (instance, variant), 30 runs were performed, starting from the same initial set of 30 randomly generated

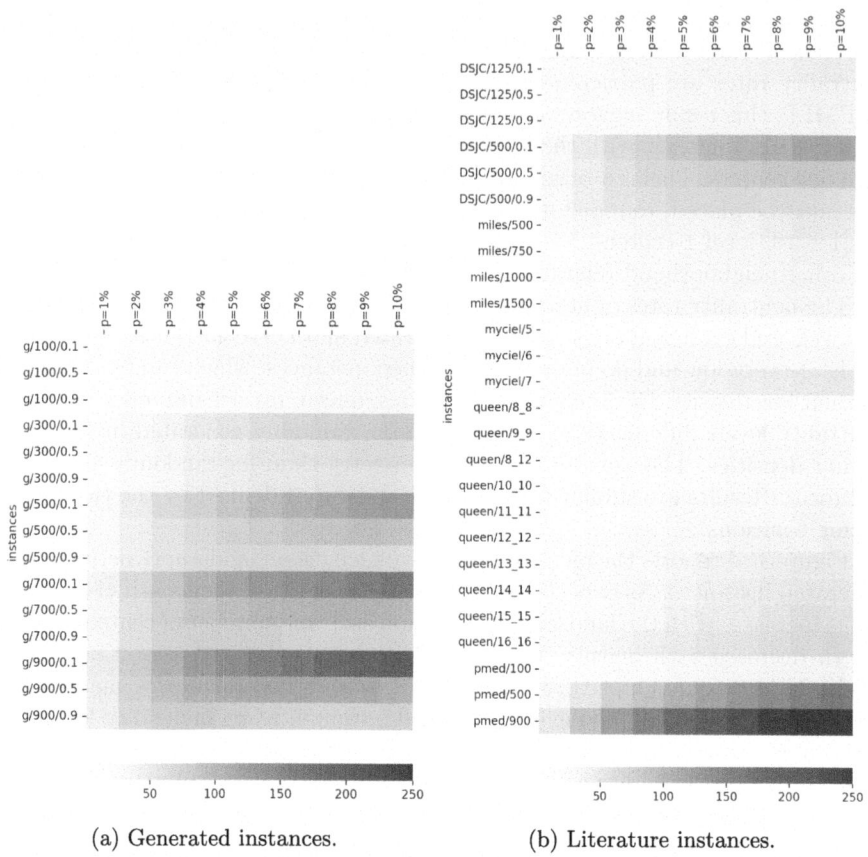

Fig. 3. Mean adaptive walk length for generated and literature instances. A darker shade of red indicates longer walks. (Color figure online)

solutions. For SW, we set $\lambda = (|V| \cdot (p+q) \cdot Density) \cdot 0.2$ (squared for the two-step Neighborhood Function), and a stopping criterion of $(|V| \cdot (p+q) \cdot Density) \cdot 0.2 \cdot |V|$ evaluations. Note that these values were determined through preliminary tests but are by no means optimal for all instances, as the goal is not to optimize our instances for best results.

To save space, we only present the results using the most efficient neighborhood relation for HC (two-step) and SW (direct). We still consider the results across the 120 runs to determine the best-known bk on a given instance. All figures display when an instance is optimally solved by using shades of red. In this case, for each instance the color gradient corresponds to the gap between the average local optimum and the global optimum, and the value to the number of runs where this optimum is reached. Instances with values in shades of blue correspond to instances that were not optimally solved. In this case, for each instance, we report the gap between bk and the average values across all runs with the color gradient and number of runs where bk was found.

Results on generated instances are reported in Fig. 4. Instances solved exactly by CPLEX are the ones where $N <= 300$. Except for $N = 100$, not all instances are optimally solved. HC finds the global optimum more often on smaller instances (given the number of vertices and the value of p). Note that using the two-step neighborhood leads more often to global optimum, especially for $p = 1\%$ when N increases, and for $N = 100$ when the number of centers increases. Except on the smallest instance, SW reaches the global optimum more often than HC, especially when p increases. As λ corresponds to a neighborhood proportion, this suggests that a smaller proportion could be required on smaller landscapes. On instances that are not optimally solved, for a given N, bk is found more often by HC when $p = 1\%$ for both neighborhoods. Except on a few instances with the two-step neighborhood, HC fails to find bk. The gap to bk increases with the number of centers up to $N = 500$ when the density is 0.9 and up to $N = 700$ when the density is 0.5. On the largest graphs, the gap to bk tends to decrease with the instance size when the density is high. On these instances, bk is sometimes found for higher values of p when $N \geq 700$, which is not the case on smaller graphs. This change correlates well with the higher neutrality rates observed in the previous section. The bk value is probably affected by these high neutrality rates, and there may be several solutions in the search space with the same fitness value. SW reaches bk more often than HC, except for $p = 1\%$ with small densities. Overall, the gap between the average solutions and bk is significantly lower than for HC, which is not surprising as the search process accepts deteriorating neighbors. For SW, the gap follows a similar evolution than for HC on larger landscapes with high density. Note that on instances with $N = 900$, while HC regularly reaches bk at least once, this is not the case for SW. This difference is probably due to the high number of evaluations of HC with the two-step neighborhood for these large instances.

Results on instances from the literature are reported in Fig. 5. Except on DSJC instances where $N = 500$ and pmed instances where $N = 400$, all graphs are optimally solved for at least up to $p = 3\%$. The global optimum is always found with HC on 26 instances, up to $p = 4\%$ SW reaches the global optimum more often than HC on miles instances, as well as on most queen instances, except the smaller ones. However, SW performs poorly on myciel instances. Since these instances are particularly small, SW lack of efficiency could be a consequence of a too-high value of λ preventing a sufficient convergence. SW also performs worse than HC on pmed instances. An explanation could be the inadequation between the value of λ and the low density of these instances.

On instances that are not optimally solved, HC finds bk at least once on each pmed instance. Its behavior on DSJC is similar to the one observed on generated instances, which is not surprising considering their similarities in landscape structure. HC finds bk more often on pmed with high values of N than on DSJC instances of similar size and displays a lower gap between its average results and bk. SW is more efficient on DSJC instances, except when their density is low and $p = 1\%$, confirming the similarity in results to our generated instances. Note that SW does not reach bk on larger pmed instances, highlighting its different requirements in parameterization for these low-density instances.

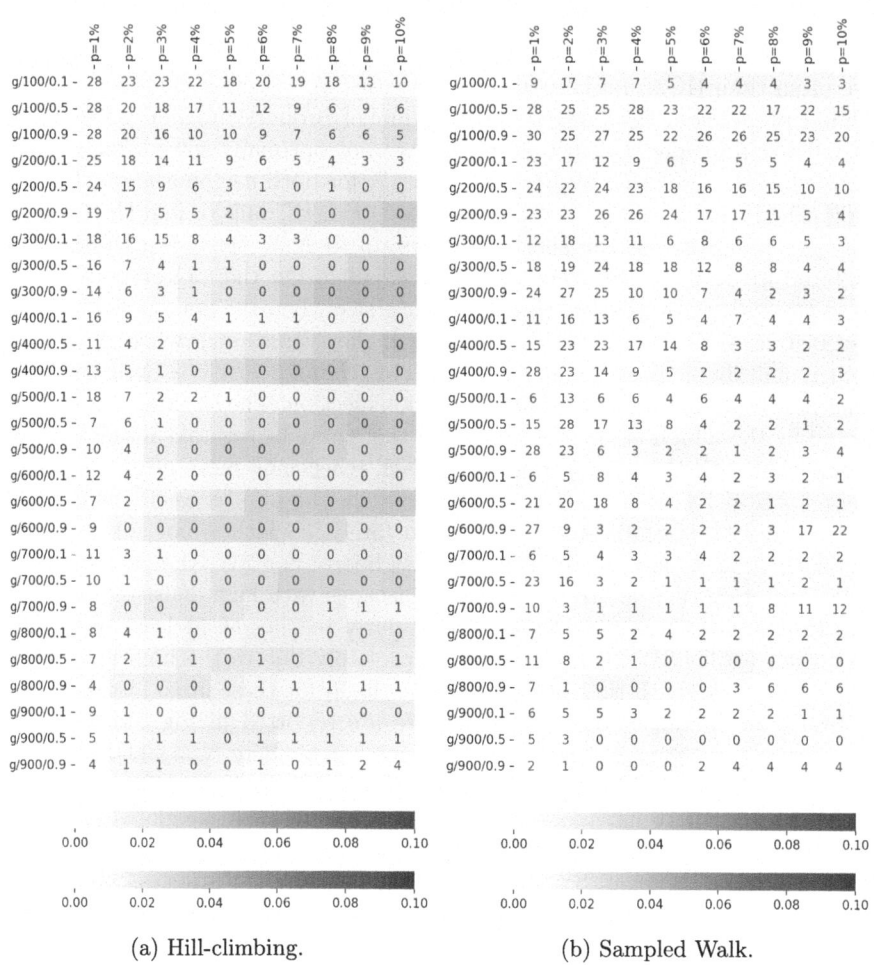

Fig. 4. Mean fitness distance to the best-encountered solution and (average) number of runs where it is found for hill-climbing algorithm and sampled walk algorithm on generated instances. Red corresponds to landscapes where the global optimum was found by using CPLEX. (Color figure online)

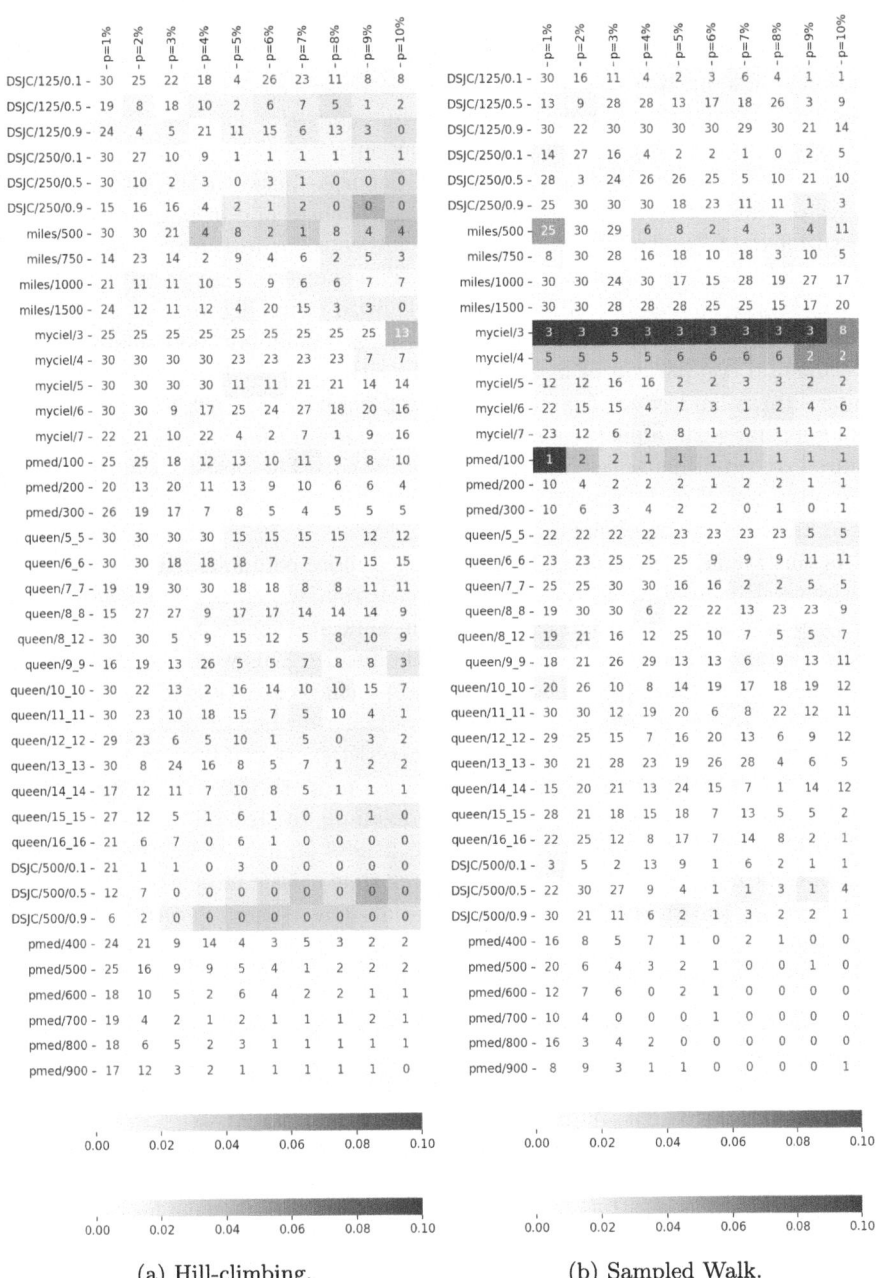

(a) Hill-climbing.   (b) Sampled Walk.

Fig. 5. Mean fitness distance to the best-encountered solution and (average) number of runs where it is found for sampled-walk algorithms. Red corresponds to landscapes where the global optimum was found by using CPLEX. (Color figure online)

4 Conclusion and Perspectives

This work was devoted to a first landscape analysis of the two-level P-median location problem. To this aim, we estimated widely-used landscape characteristics on several instances with varying characteristics for two neighborhood relations and provided a first investigation on the solving difficulty for these instances. The results highlighted some differences between the ruggedness of landscapes derived from randomly generated instances and from instances based on real-life data. Neutrality levels are mostly affected by instance densities, especially when they are randomly generated. This suggests that the symmetry of such instances should be reduced. Overall, little difference has been observed in the landscape structure for the two considered neighborhood relations. The experimental comparison of local search algorithms showed similar behavior on similar fitness landscapes (namely generated instances and DSJC). Our results highlight that a different proportion of neighborhood for the sampled walk is required on small instances and with a low density. The best-known solution is more often found on instances with high levels of neutrality, which is likely due to the high symmetry of such instances.

As this study is a first step in providing a more comprehensive view of the multi-level P-median location problem, much remains to be done. Since neutrality results suggest a high symmetry on some instances, a preprocessing to reduce symmetry should be considered in the future. This would imply to study the impact of various approaches on landscape characteristics. Since most variants of this problem require to use constraints, we plan to study the impact of capacity constraints on landscape characteristics, depending on their distribution. This will also require work on how to properly take those constraints into account in local search algorithms, for example by using appropriate neighborhood relations. We plan to expand the fitness landscape analysis by using local optima network to highlight differences that classic indicators fail to capture, both on the current problem and the aforementioned variations. Finally, improving the overall knowledge of this problem also implies proposing and experimentally studying more advanced local search algorithms.

References

1. Ahmadi-Javid, A., Seyedi, P., Syam, S.S.: A survey of healthcare facility location. Comput. Operat. Res. **79**, 223–263 (2017). https://doi.org/10.1016/j.cor.2016.05.018, https://www.sciencedirect.com/science/article/pii/S0305054816301253,
2. Galvão, R.D., Espejo, L.G.A., Boffey, Brian: A hierarchical model for the location of perinatal facilities in the municipality of rio de janeiro. Euro. J. Operat. Res. **138**(3), 495–517 (2002). https://doi.org/10.1016/S0377-2217(01)00172-2, https://www.sciencedirect.com/science/article/pii/S0377221701001722,
3. Elalouf, A., Hovav, S., Tsadikovich, D., Yedidsion, L.: Minimizing operational costs by restructuring the blood sample collection chain. Operat. Res. Health Care **7**, 81–93 (2015); ORAHS 2014 - The 40th international conference of the EURO working group on Operational Research Applied to Health

Services. https://doi.org/10.1016/j.orhc.2015.08.004, https://www.sciencedirect.com/science/article/pii/S2211692314200610
4. Daskin, M.: What you should know about location modeling. Naval Res. Logist. (NRL) **55**, 283–294 (2008). https://doi.org/10.1002/nav.20284
5. Kariv, O., Hakimi, S.L.: An algorithmic approach to network location problems ii. The p-medians. SIAM J. Appli. Math. **37**(3), 539–560 (1979). http://www.jstor.org/stable/2100911
6. Megiddo, N., Supowit, K.J.: On the complexity of some common geometric location problems. SIAM J. Comput. **13**(1), 182–196 (1984), arXiv:https://doi.org/10.1137/0213014, https://doi.org/10.1137/0213014
7. Wright, S., et al.: The roles of mutation, inbreeding, crossbreeding, and selection in evolution. In: Proceedings of the VI International Congress of Genetics, pp. 356–366 (1932)
8. Liefooghe, A., Daolio, F., Verel, S., Derbel, B., Aguirre, H., Tanaka, K.: Landscape-aware performance prediction for evolutionary multiobjective optimization. IEEE Trans. Evolutionary Comput. **24**(6), 1063–1077 (2020). https://doi.org/10.1109/TEVC.2019.2940828
9. Malan, K.M., Engelbrecht, A.P.: A survey of techniques for characterising fitness landscapes and some possible ways forward. Inform. Sci. **241**, 148–163 (2013). https://doi.org/10.1016/j.ins.2013.04.015, https://www.sciencedirect.com/science/article/pii/S0020025513003125
10. Malan, K.M.: A survey of advances in landscape analysis for optimisation. Algorithms **14**(2) (2021). https://doi.org/10.3390/a14020040, https://www.mdpi.com/1999-4893/14/2/40
11. Pitzer, E., Affenzeller, M.: A comprehensive survey on fitness landscape analysis. Recent Adv. Intell. Eng. Syst., 161–191 (2012)
12. Weinberger, E.: Correlated and uncorrelated fitness landscapes and how to tell the difference. Biol. Cybern. **63**(5), 325–336 (1990)
13. Beasley, J.E.: A note on solving large p-median problems. Eur. J. Oper. Res. **21**(2), 270–273 (1985)
14. https://mat.tepper.cmu.edu/COLOR02
15. Tari, S., Basseur, M., Goëffon, A.: Sampled walk and binary fitness landscapes exploration. In: Lutton, E., Legrand, P., Parrend, P., Monmarché, N., Schoenauer, M. (eds.) EA 2017. LNCS, vol. 10764, pp. 47–57. Springer, Cham (2018). https://doi.org/10.1007/978-3-319-78133-4_4

Generating (Semi-)active Schedules for Dynamic Multi-mode Project Scheduling Using Genetic Programming Hyper-heuristics

Yuan Tian[✉], Yi Mei, and Mengjie Zhang

Centre for Data Science and Artificial Intelligence and School of Engineering and Computer Science, Victoria University of Wellington, Wellington 6012, New Zealand
{yuan.tian,yi.mei,mengjie.zhang}@ecs.vuw.ac.nz

Abstract. Dynamic multi-mode resource-constrained project scheduling problem (DMRCPSP) is crucial for effectively managing complex projects where activities have multiple options of resource demand and durations are uncertain. Efficient solutions to this problem are vital for industrial applications, where optimal scheduling can significantly impact project costs and timelines. Genetic programming hyper-heuristic (GPHH) has been applied to evolve effective scheduling heuristics to make real-time decisions. However, current decision-making procedures for solving DMRCPSP primarily generate non-delay schedules, which are often suboptimal. This paper proposes two new decision-making procedures to generate (semi-)active schedules. The core idea is to expand the decision set and allow activities that cannot immediately start to reserve resources for future execution. We further developed GPHH to evolve scheduling heuristics based on these procedures and a comparison was conducted with existing decision-making procedures. The results demonstrated that the proposed GP approaches managed to evolve rules that generate more effective (semi-)active schedules than the best-known non-delayed schedules in real-time.

Keywords: Genetic Programming · Dynamic project scheduling · Multiple modes · Uncertain duration · Decision-making procedure

1 Introduction

Multi-mode resource-constrained project scheduling problem (MRCPSP) [1] is a well-known NP-hard optimization problem in project management. It extends the classic RCPSP [17] by allowing each activity to be executed in one of several modes, with each mode representing a different combination of resource demand and duration. The challenge of MRCPSP lies in determining the optimal sequence and mode of activities while satisfying resource availability constraints and precedence relations between activities. In practice, project scheduling is rarely static. Real-world projects frequently encounter uncertainty [19] such as resource availability and demand, or activity durations. This introduces the need for dynamic project scheduling that considers uncertainties that may arise during project execution. In this paper, we focus on dynamic MRCPSP with uncertain activity durations.

In the DMRCPSP, heuristic rules [1,8,24] are applied at each decision point to compute a priority value for all executable activities, after which the highest-priority activity and its mode are selected for execution. However, the design of effective heuristic rules requires substantial domain expertise and extensive computational experimentation, as they must be tailored to the specific characteristics of the problem.

Genetic programming hyper-heuristics (GPHH) have emerged as a powerful method for automating the generation of heuristics, particularly for solving complex combinatorial problems, such as job shop scheduling [36] and routing problems [27]. There has been some research into applying GP to either static [3,9,14,25] or dynamic [4,7] single-mode RCPSP, few studies [2,16] have addressed its application to the dynamic MRCPSP. This paper aims to fill this gap by applying GPHH to a dynamic multi-mode scheduling environment.

Project scheduling is driven by a series of decisions made at key decision points, typically triggered upon the completion of an activity. At each decision point, a heuristic rule is applied to the set of eligible activities to compute their priorities, determining the order in which activities are scheduled for execution. The quality of the generated schedule depends on the decision-making procedure, the definition of the eligible activity set (known as the decision set), and the heuristic rule used to prioritize activities. Most existing approaches [4,7] adapted from parallel schedule generation scheme [21] which only focus on scheduling activities that can start immediately and inherently produce non-delay schedules. However, the scheduling theory [30,31] indicates that the set of non-delay schedules is only a subset of active schedules and it might not contain optimal schedules for a regular performance measure. More importantly, this greedy strategy seems to enhance resource utilization, but it often leads to longer project completion times. This is because crucial activities, such as those on the critical path that cannot start immediately, may be delayed, ultimately extending the overall project duration. This limitation becomes even more pronounced in the MRCPSP context [32], where at any decision point, faster resource-intensive modes are often excluded from the decision set, leaving only resource-efficient but slower modes.

To address this limitation, we propose a novel decision-making approach that enables activities which cannot start immediately due to resource limitations and allows certain activities to be executed in faster modes when resources become available. This approach can improve the quality of the schedule and generate (semi-)active schedules [30]. We then develop a GPHH approach that evolves effective scheduling heuristics for generating high-quality schedules. This paper has the following research objectives:

- Propose two decision-making procedures that generate (semi-) active schedules by enabling the delayed start of activities.

– Develop GPHH based on the newly proposed decision-making procedures to learn scheduling heuristics that generate effective (semi-)active schedules.
– Validate the effectiveness of GPHH with the proposed procedures and compare the performance of GPHH with existing non-delay procedures.

2 Background

2.1 Problem Description

In the DMRCPSP, a project includes a set of activities J to complete and a set of resources R available for allocation. Precedence constraints exist among activities, i.e., an activity i can only start after all of its predecessors $j \in P_i$ have finished. Each resource $r \in R$ has availability K_r in each time unit throughout the project horizon. Each activity i can be executed in one of several modes $m \in M_i$. The mode m determines the demand for resources $k_{i,m,r}$ and the expected duration $\hat{d}_{i,m}$ of the activity. In a dynamic setting, the actual duration d_i of an activity i is not known in advance. Instead, it is drawn from an uniform distribution $d_i \in [d^{min}_{i,m}, d^{max}_{i,m}]$, where $d^{min}_{i,m}$ and $d^{max}_{i,m}$ represent the optimistic and pessimistic durations, respectively. During project execution, the actual duration d_i is revealed only when the activity starts. At any time during the execution of the project, the total resource demand cannot exceed K_r. Scheduling decisions (activity selection and mode assignment) are made dynamically as the project progresses, and the objective is to minimize the project makespan while respecting both resource and precedence constraints under uncertainty.

Figure 1 gives an example of DMRCPSP. The project information is given on the left table, seven activities (activities 1 and 7 are dummy start and end, respectively), two modes for each activity and one resource type. A schedule in resource view is illustrated on the right side. The dashed line is the resource availability. Each activity is executed in a mode and represented as a rectangle with its width as the duration and the height as the demand of R1. All activities are completed on time according to the expected duration, except for activity 3 which is a one time unit late compared to the expected duration.

Act.i	P_i	Mode	$\hat{d}_{i,m}$	$d^{min}_{i,m}$	$d^{max}_{i,m}$	R1
1	-	1	0	0	0	0
2	1	1	5	4	6	7
		2	7	6	10	5
3	1	1	6	4	7	8
		2	9	8	11	5
4	3	1	4	3	5	9
		2	8	7	10	8
5	2, 4	1	3	2	4	5
		2	5	4	8	4
6	4	1	4	3	5	9
		2	6	4	7	7
7	5, 6	1	0	0	0	0
Resource availabilities						12

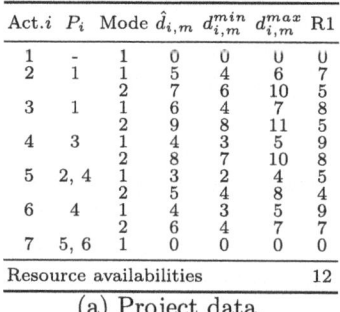

(a) Project data.

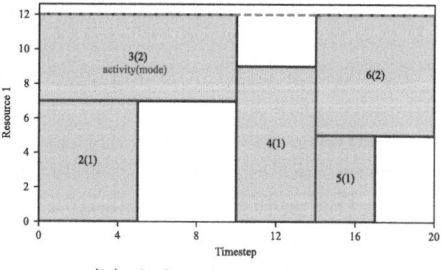

(b) A feasible schedule.

Fig. 1. A DMRCPSP example.

2.2 Genetic Programming Hyper-heuristics for Project Scheduling

Genetic programming (GP), as a hyper-heuristic algorithm, has been successfully applied to project scheduling. From the problem modelling perspective, these studies mainly cover single-mode RCPSP [3,9,14,25,26], multi-project RCPSP [7,10,11], and RCPSP with dynamic resource disruption [4]. From the algorithm design perspective, the research includes representation [3,6,9], ensemble learning [6,13], surrogate models [10,23,25,37], rollout approach [5,14], co-evolution [38] and fitness function design [26], aimed at improving the GP performance.

Research on using GP to solve the MRCPSP is still in its early stages. The main difference between MRCPSP and single-mode RCPSP lies in the fact that two decisions need to be made at each decision point: activity selection and mode assignment. In [32], the impact of these two decision sequences on scheduling quality was studied, and a representation that integrates these two decisions was proposed, which works as follows: (1) transforming activities and their eligible modes in the decision set into individual activity-mode tuples. (2) a heuristic rule is used to select from these tuples. This approach reduces the search space of the heuristic rule and avoids the need to consider the matching between the two types of rules.

3 Proposed Method

3.1 Overall Framework

The GP framework used in this paper is shown in Algorithm 1. The GP system first sets up the training set $\mathcal{S}$ and randomly initialises the population. The GP individual adopts a tree-based representation proposed in [32] and each individual is composed of selected elements from terminals and functions (See Sect. 3.3).

At each generation, each heuristic rule (individual) Δ_i is applied to generate schedules for all instances in the training set $\mathcal{S}$ where multiple combinations of decision-making procedures and decision set variants θ (See Sect. 3.2) can be used. The relative deviation $\delta_{i,j}$ is adopted and the lower bound for each instance comes from the makespan calculated by the critical path method [20] which disregards the resource constraints. Then, the fitness value $fitness_{\Delta_i}$ is calculated by averaging the deviations over instances. Next, the best rule Δ^* in the current generation is identified. After that, the GP system will apply parent selection and genetic operators such as reproduction, subtree crossover and subtree mutation to the current population to generate new individuals for the next generation. When the maximum number of generations is reached, the GP system will stop and return the best-evolved rule Δ^* so far.

3.2 Fitness Evaluation Based on (Semi)-active Schedule Generation

A decision-making procedure manages the execution flow of the project by determining which activities should be executed in which modes at any given time.

```
1  Input: A set of training instances S = {I₁,...,...,Iₜ}
2  Output: the best evolved rule Δ*
3  Initialisation: Randomly initialise the population pop ← {Δ₁,...Δₙ};
4  Δ ← null , gen ← 0 ;
5  while gen <= maxGen do
6      forall the Δᵢ ∈ pop do
7          forall the Iⱼ ∈ S do
8          |   δᵢ,ⱼ ← (makespan(decision_making(Iⱼ, Δᵢ, θ)) − LB_{Iⱼ})/LB_{Iⱼ} ;
9          fitness_{Δᵢ} ← ∑_{Iⱼ∈S} δᵢ,ⱼ/T ;
10     Δ* ← find the best rule in pop ;
11     O ← apply parent selection to pop ;
12     pop ← apply crossover, mutation and reproduction to O ;
13     gen ← gen + 1
14 return Δ*;
```

Algorithm 1: GP framework for evolving heuristics for DMRCPSP.

The behaviour of a procedure is primarily determined by two aspects: (1) how the decision set is defined and (2) how to select a feasible subset of activities from the decision set.

We propose two decision-making procedures that allow for considering activities that cannot start immediately due to resource constraints but have already satisfied their precedence constraints. These procedures reserve resources for such activities to allow them to start in faster modes. Accordingly, we introduce the following variants of the decision set, listed in increasing order of scope:

- **Non-delay Set:** The set of activities and modes that can immediately start at the current time, i.e., those that satisfy both precedence and resource constraints.
- **Expanded Set:** The non-delay set plus activity-mode tuples that cannot immediately start but whose completion time would not be later than the latest finish time of any activity-mode tuple in the non-delay set.
- **Full Precedence Feasible Set:** The set of all activities that satisfy the precedence constraints at the current time, along with all their modes.

Immediate Decision-Making Procedure. The idea of the immediate decision-making procedure is that if an activity-mode tuple that cannot be immediately started is selected, it is placed in a queue set Q, and then the process moves to the next decision time. Algorithm 2 outlines the process of this procedure, with the differences from the conventional parallel scheme highlighted in brown. At each decision time t, the current project status is first updated by checking if any activities have been completed (line 8), updating the set of completed activities C, the ongoing activity set P, and the availability of resources $\mathcal{R}$. Then, the procedure checks if any activities in the queue set Q can start (lines 10–12). If an activity can start, it is added to the ongoing set P and allocated corresponding resources from $\mathcal{R}$; if not, the procedure moves directly to the next decision time $t+1$. This is because the activities in Q cannot start due to resource availability, and only the completion of activities in P can release the necessary resources at the next decision time.

If Q is empty, the decision loop (lines 13–21) begins to schedule activities in the decision set. The decision set D is calculated based on all activities in the project $\mathcal{I}$, the ongoing activities P, the completed activities C, the current resource availability $\mathcal{R}$ and the selected decision set option. The heuristic rule Δ is then applied to D to compute the priority of each activity-mode tuple, and the tuple with the highest priority is selected (line 15). The procedure then checks the earliest possible start time of the selected activity j^* in the current mode m^* under current resource availability $\mathcal{R}$ (line 16). If the activity can start immediately (line 17), it is executed and the loop continues; otherwise, the tuple is added to Q, and the procedure moves to the next decision time at $t + 1$.

---

1 **Input**: instance $\mathcal{I}$; heuristic rule Δ; option for decision set $\theta = N|E|F$
2 **Output**: A feasible schedule
3 **Initialisation**:
4   current time: $t \leftarrow 0$; resource availability: $\mathcal{R} = \{K_r | \forall r \in R, \forall t\}$
5   set of activities already complete: $C \leftarrow \{\}$;
6   set of activities being processed: $P \leftarrow \{\}$;
7   set of activity-mode tuples in queue: $Q \leftarrow \{\}$;
8 **while** $|C| < J$ **do**
9     $P, C, \mathcal{R} \leftarrow$ Update project status at time t;
10   **forall the** (j, m) in Q **do**
11   **if** (j, m) can start right now **then**
12             $P, \mathcal{R} \leftarrow scheduleActivity(j, m, t, \mathcal{R})$
13   **if** Q is $\varnothing$ **then**
14   **while** $D \leftarrow getDecisionSet(\mathcal{I}, P, C, \mathcal{R}, \theta)$ **do**
15             $j^*, m^* \leftarrow choose(D, \Delta)$;
16             $start \leftarrow earliestStartTime(j^*, m^*, \mathcal{R})$;
17   **if** $start == t$ **then**
18                 $P, \mathcal{R} \leftarrow scheduleActivity(j^*, m^*, start, \mathcal{R})$;
19   **else**
20                 $Q \leftarrow Q \cup (j^*, m^*)$;
21   break;
22     $t \leftarrow t + 1$;
23 **return** the schedule;

Algorithm 2: Immediate decision-making procedure.

This procedure allows selection from an enlarged decision set (θ in line 14 can be E or F), and after selecting an activity that cannot start immediately, it refrains from scheduling other activities to free up more resources. One limitation might be it heavily relies on the quality of the heuristic rule–if an activity that cannot start immediately is selected, the decision loop for the current time just breaks and go to the next decision time. However, the availability of resources at that time may still be sufficient for other activities. In that case, this procedure generates semi-active [30] schedules.

Two-Phase Decision-Making Procedure. The two-phase decision-making procedure balances reserving resources for activities that cannot start immediately with maximizing the number of activities executed at the current time. The key idea is that when an activity-mode tuple is selected that cannot start immediately, the procedure does not directly go to the next decision time. Instead, resources are pre-allocated to the selected activity, and the process continues to check if any other activities can start under the updated resource status.

```
1  Input: instance I; heuristic rule Δ; option for decision set θ = N|E|F
2  Output: A feasible schedule
3  Initialisation:
4     current time t ← 0; resource availability R = {K_r|∀r ∈ R, ∀t};
5     set of activities already complete: C ← {};
6     set of activities being processing: P ← {};
7     set of activity-mode-start tuples in queue: Q ← {};
8  while |C| < J do
9     P, C, R ← Update project status at time t;
10    forall the (j, m) in Q do
11       if (j, m) can start right now then
12          P, R ← scheduleActivity(j, m, t, R)
13    Q.removeAll();
14    if Q is ∅ then /* Decision round begins */
15       /* Planning phase */
16       A ← {}; // Activities that can start immediately
17       R*_{r,t} ← copy(R_{r,t});
18       while D ← getDecisionSet(I, P, C, A, Q, R*, θ) do
19          j*, m* ← choose(D, Δ);
20          start ← earliestStartTime(j*, m*, R*);
21          if start == t then
22             A ← A ∪ {(j*, m*)};
23          else
24             if start not greater than any start time in Q then
25                Q ← Q ∪ {(j*, m*, start)};
26             else
27                θ ← N;
28                continue;
29          R* ← scheduleActivity(j*, m*, start, R*);
30       /* Execution phase */
31       forall the (j, m) in A do  P, R ← scheduleActivity(j, m, t, R))
32    t ← t + 1;
33 return the schedule;
```

Algorithm 3: Two-phase decision-making procedure.

As shown in Algorithm 3 (with the differences from Algorithm 2 highlighted in brown), when it comes to decision time t, after scheduling all startable activities in Q (lines 10–12), this set will be cleared (line 13) as the project status changes and the priorities calculated from the previous time are outdated. The decision round (lines 14–31) has two phases: the planning phase and the execution phase. In the planning phase, an empty set A is initialized which will contain all activity-mode tuples that can start immediately at time t (line 16). A copy of resource availability R is created (line 17). Then eligible activity-mode tuples are calculated and then one tuple will be selected from the decision set D given on rule Δ. The decision set D is determined by all activities I, a set of activities in P, C, A, Q, resource availability $\mathcal{R}$ and the variant option θ. All selected activity-mode tuples, regardless of whether they can start immediately or not, are allocated resources in advance (lines 21–29). If the activity can start immediately, it is placed in a temporary set A; if it cannot start, it is added to the queue set Q. Then resources are reserved for this activity and the loop continues. When the start time of the selected activity-mode tuple is later than any start time in Q, this activity will not be included in Q. This way, we can improve the efficiency of the decision-making process without affecting its performance.

In the execution phase, only the activities in set A, which can start immediately, are executed. This two-phase procedure allows for a more strategic allocation of resources, balancing the need to pre-schedule future activities to maximize immediate resource utilization.

3.3 Terminals and Functions

Functions and terminals used are shown in Table 1. Terminals are problem-specific attributes that can be classified into precedence-related, time-related, and resource-related attributes. Two terminals – TTW and FTFN – are newly introduced to indicate the waiting time to start and the expected finish time from now for this activity-mode tuple, respectively. Some attributes are dynamically calculated based on the current project state. For example, AvgRA represents the average resource availability at the current time. All time-related attributes have been transformed to be time-invariant [28], meaning that they are independent of the absolute time, to improve the generalization ability of the heuristic rule.

Table 1. The terminals and functions.

Notation	Description	Notation	Description
EST[1,2] [15]	Earliest Start Time	GRPW [3]	Greatest Rank Positional Weight
EFT[1,2] [15]	Earliest Finish Time	GRPW* [15]	Greatest Rank Positional Weight All
LST[1,2] [15]	Latest Start Time	RR [15]	Resource Required
LFT[1,2] [15]	Latest Finish Time	GRD [26]	Greatest Resource Demand
SLK[1,2]	Slack Time	AvgRA[1]	Average Resource Availability
D	Duration	MaxRA[1]	Maximum Resource Availability
MinD	Optimistic Duration	MinRA[1]	Minimum Resource Availability
MaxD	Pessimistic Duration	AvgRR [3]	Average Resource Requirement
TTW[1,2]	Time to Wait for Start	MaxRR [3]	Maximum Resource Requirement
FTFN[1,2]	Time to Finish From Now	MinRR [3]	Minimum Resource Requirement
DPC [15]	Direct Predecessor Count	DSC [15]	Direct Successor Count
TPC [15]	Total Predecessor Count	TSC [15]	Total Successor Count
Function set: $+, -, *, \div$ (protected division), min, max, neg, abs			

1: Dynamic attributes; 2: Time-invariant attributes

4 Experiment Design

A set of experiments has been conducted to investigate the performance of GP in these proposed decision-making procedures with different decision set variants.

4.1 Simulation Configuration

The simulation model is adopted to examine the performance of rules and decision-making procedures in a dynamic environment. The simulation model can load different project instances and we synthesized some projects with 200 activities and 4 types of resources. Each activity can be executed in one of three modes, ranging from fast to slow, with all modes requiring some amount of each resource type. The expected duration $\hat{d}_{i,m}$ of the fastest mode is uniformly sampled from the range $[5, 10]$ while the optimistic durations $d_{i,m}^{min}$ and pessimistic durations $d_{i,m}^{max}$ fluctuate around this range by $\pm[1, 3]$. The demands of each resource type for the slowest mode range from $[1, 6]$ and then the resource demands increase as the expected duration reduces. The total availability of resources for the project is determined based on a resource strength (RS) [22] value of 0.25, which is used to measure the scarcity of each type of resource.

Table 2. The parameter settings of GP.

Parameter	Value
Population size	1000
Number of generations	50
Method for initialising population	ramped-half-and-half
Initial minimum/maximum depth	2 / 6
Elitism	10
Maximal program depth	8
Crossover rate	0.80
Mutation rate	0.15
Reproduction rate	0.05
Parent selection	Tournament selection with size 7

The precedence relations between activities are generated using RanGen [12], a generator that utilizes the Order Strength (OS) indicator to measure the complexity of precedence relations in the project. As summarized in [18], a higher OS indicates more precedence constraints between activities, resulting in fewer possible activity sequences and a smaller solution space. Conversely, a lower OS means fewer precedence constraints, allowing for more possible sequences and a larger solution space. Inspired by the settings of a static MRCPSP benchmark dataset MMLIB [29] where OS sets to 0.75 (high), 0.5 (medium) and 0.25 (low), we created six scenarios with OS values of 0.8, 0.75, 0.6, 0.5, 0.25, 0.2 to more accurately compare the relationship between algorithm performance and OS. For each scenario, five instances were generated. In the training stage, the actual durations of each instance are resampled per generation to improve the generalisation of evolved rules. In the test stage, the actual durations of each instance are sampled 50 times to improve the accuracy of the best-evolved rule in a single run.

4.2 Compared Methods and Parameter Settings

According to the decision-making procedure and the decision set used by GP, we adopt the format of GP<decision procedure>-<decision set> to name the algorithm. The decision procedure options are parallel scheme (PL), immediate procedure (IP) and two-phase procedure (TP). The decision set has three variants, **N**, **E**, and **F**. The parallel scheme, adapted from [32], is the baseline algorithm. This method can only select an activity mode tuple that can start immediately at any given time, so the only decision set it can select is **N** set. IP and TP give the same results as PL-N when paired with **N** set, so only IP and TP are tested when paired with either **E** or **F** set. It is worth noting that literature [32] has investigated that with the parallel scheme, GP-evolved rules outperform existing manually-designed heuristic rules; therefore, existing heuristic rules are not included for comparison here. Therefore, we compare five algorithms, GPPL-N, GPIP-E/F, GPTP-E/F. All algorithms share the same parameter settings referred from previous related research [33–35], as detailed in Table 2.

5 Results and Discussions

Thirty independent runs are conducted for comparison. The differences between these algorithms are verified by the Wilcoxon rank sum test with a significance level of 0.05. In the following results, "↑", "↓" and "≈" indicate the corresponding result is significantly higher than, lower than or similar to its counterpart.

5.1 Test Performance

Table 3 shows the mean and standard deviation of the objective value of the five algorithms over 30 independent runs for six scenarios. Figure 2 shows the average objective value on test instances of the best-evolved rule over generations.

Table 3. The mean (standard deviation) of the objective values of these algorithms over 30 independent runs for six scenarios.

Scenario	GPPL-N	GPIP-E	GPIP-F	GPTP-E	GPTP-F
0.8	1.58(0.01)	1.60(0.03)(↑)	1.63(0.02)(↑)(↑)	1.54(0.03)(↓)(↓)(↓)	1.54(0.02)(↓)(↓)(↓)(≈)
0.75	1.62(0.02)	1.65(0.03)(↑)	1.68(0.01)(↑)(↑)	1.61(0.03)(≈)(↓)(↓)	1.61(0.02)(↓)(↓)(↓)(≈)
0.6	1.57(0.02)	1.59(0.04)(↑)	1.62(0.02)(↑)(↑)	1.55(0.03)(↓)(↓)(↓)	1.55(0.03)(↓)(↓)(↓)(≈)
0.5	1.60(0.02)	1.58(0.04)(≈)	1.64(0.05)(↑)(↑)	1.58(0.03)(↓)(≈)(↓)	1.58(0.02)(↓)(≈)(↓)(≈)
0.25	1.54(0.02)	1.50(0.04)(↓)	1.56(0.04)(≈)(↑)	1.55(0.03)(≈)(↑)(≈)	1.55(0.02)(≈)(↑)(≈)(≈)
0.2	1.53(0.01)	1.39(0.03)(↓)	1.41(0.03)(↓)(↑)	1.54(0.03)(≈)(↑)(↑)	1.54(0.03)(≈)(↑)(↑)(≈)

The two-phase procedure (GPTP) outperforms the parallel scheme (GPPL) in most scenarios, specifically in 4 out of the 6 tested scenarios. The largest performance gap between these two approaches occurs when OS is 0.8. As

OS decreases, the performance difference between the two algorithms gradually decreases, eventually showing no significant difference at OS = 0.25 and 0.2. This indicates that the two-phase procedure is particularly suited to scenarios with a high OS, where precedence relations between activities are dense. High OS implies that the duration of activities (i.e., mode selection) significantly impacts the scheduling of subsequent activities. Therefore, allowing activities to delay their start in favour of executing in faster modes proves to be a more optimal strategy. In scenarios with sparse precedence relations (low OS), the sequence in which activities are executed has minimal impact on the decision set at the next time step. For example, in the OS = 0.9 scenario, there may be only 4 activities available for selection at time $t = 0$, meaning that choosing different activities to start will unlock different sets of subsequent activities. However, in the OS = 0.25 scenario, nearly 30 activities may be available for selection at $t = 0$, so even if the heuristic rule selects a different order, many activities can still be initiated simultaneously. Even if they cannot start together, the decision set in the next time step remains largely unchanged.

The performance difference between GPTPs using the two types of decision set (expanded and full) is minimal. This is related to the update and clearing mechanism of the queue set Q in the procedure. This phenomenon can be attributed to the following: (1) In the planning phase, any activity-mode tuple with a start time later than the latest start time in the Q set will not be accepted. This update mechanism makes it less likely for activity-mode tuples exclusive to the full set to be added to the Q set. (2) At the new decision time, after scheduling as many activities from the Q set as possible, the set is immediately cleared and then goes to the decision loop instead of advancing to the next time. This clearing mechanism helps maintain flexibility in execution, mitigating the negative impact of selecting long-duration activities that are not ready to start immediately.

The GPIP algorithm shows a significant difference in performance between the two decision sets: using the expanded set performs considerably better than the full set. The full set encompasses a larger range of activities than the expanded set. In cases where the heuristic rule is of low quality, it is more likely to select an activity that meets precedence constraints but cannot start immediately and has a long finish time. Therefore, using a more compact decision set like the expanded set results in better performance for GPIP, as it reduces the likelihood of selecting such delayed activities.

Both GPIP variants tend to underperform compared to the baseline GPPL algorithm in scenarios of high and medium OS. This is due to a key characteristic of the immediate procedure: when it selects an activity that cannot start immediately, it adds the activity to the queue set Q, immediately ends the current decision loop, and advances to the time when that activity can start. At the end of the loop, the procedure does not consider whether other activities can begin with the current available resources. As a result, some activities that could have started at the current time are delayed due to the selection of an activity that cannot start immediately.

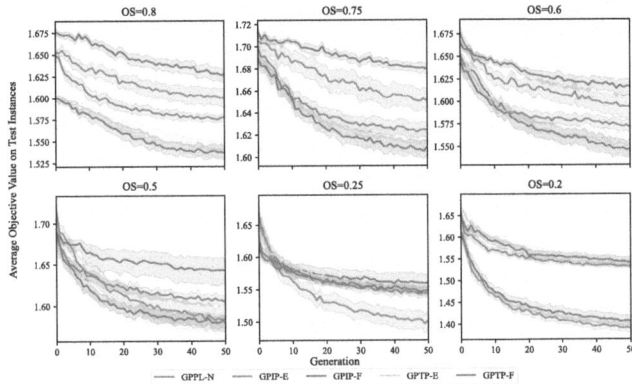

Fig. 2. Convergence curves over 30 independent runs in six scenarios.

In most scenarios, the GPTP algorithm outperforms GPIP. However, in the scenario of low OS, GPIP-E performs better than the other algorithms. This could be because, in scenarios with loose precedence constraints (where most activities can start simultaneously), selecting some activities that start later but execute in a faster mode is effective. GPIP can stop the current decision loop in time after selecting such activities, preserving more resources for the next decision time, and allowing subsequent activities to start in faster modes. On the other hand, after selecting activities that cannot start immediately, GPTP still tries to execute other activities in slower modes with lower resource consumption. This greedy strategy is not always effective when precedence constraints are less strict. As a result, in such situations, the GPIP-E algorithm performs better than GPTP.

5.2 Rule Size

The rule size reflects the complexity of the evolved rules in different scenarios. From Fig. 3, it can be seen that in most scenarios, the rule sizes generated by the same scenario are roughly the same. Combined with the test performance of different algorithms, it can be concluded that GPTP achieves better results while maintaining similar rule sizes, which highlights the advantages of the two-phase decision-making procedure. The rule size of GPIP-E is slightly larger than other algorithms in $OS = 0.5$ and 0.25. At the same time, its performance in these two scenarios is equal to or even slightly better than other algorithms. This suggests that although the immediate decision-making procedure is not the optimal approach, more complex rules can still achieve similar or even superior results.

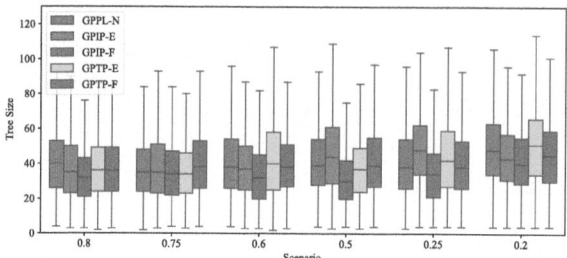

Fig. 3. Tree size of best-evolved rules over 30 independent runs in six scenarios.

5.3 Attribute Occurrence

Different attributes (or terminals) are preferred depending on the scenario and decision-making procedure. Figure 4 shows the proportion of attribute usage in the best-evolved rules from each of the five algorithms when OS = 0.5 over 30 independent runs. It can be observed that LST is a preferred attribute across all algorithms, and it is inherently an effective heuristic rule [21]. Rules from the GPIP algorithms frequently use TTW. Since these algorithms end the decision loop once an activity that cannot start immediately is selected, more consideration is given to whether the current activity-mode tuple can start immediately.

Among resource-related attributes, the rules evolved by GPTP tend to use GRD more often, which considers the resource duration required for the activity-mode tuple. This is likely because the procedure allows activities to be placed in the queue set, reserving resources. GPPL rules, on the other hand, tend to use AvgRA more frequently, which measures the current average resource availability. For GPIP, the GPIP-E rules use RR more frequently, whereas GPIP-F rules tend to use GRD more.

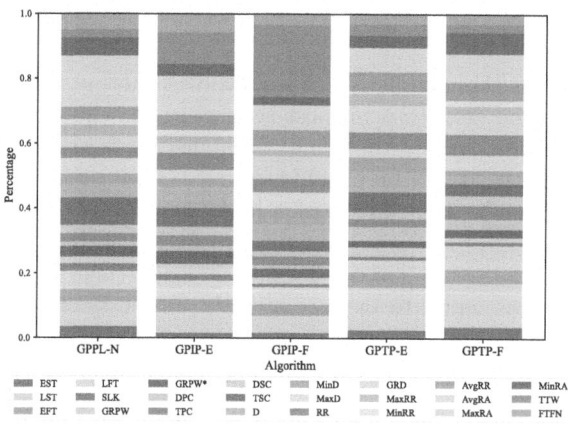

Fig. 4. Attribute distribution of the best-evolved rules over 30 independent runs in scenario OS = 0.5.

5.4 Training Time

Most of the training time for the algorithms is spent on evaluation, which involves applying the heuristic rule in the simulation model to solve five instances within each scenario. Therefore, the evaluation time is related to the number of times priority values are calculated, the size of the heuristic (tree) rule, and the complexity of the procedure. From Fig. 5, it can be seen that the training time for all algorithms increases as OS decreases. This is because, as OS decreases, the size of the decision set at any given moment increases, which results in more calculations of priority values, thus lengthening the training time.

Comparing the algorithms, it is evident that GPTPs have the longest training time in all scenarios and GPTP-F takes more time than GPTP-E. The reason is that the two-phase procedure is more complex; in the planning phase, extensive calculations are required to assess the resource usage for the activities after scheduling them. The full set covers a larger range than the expanded set, leading to more calculations of priority values at each decision point. The GPIP with the full set takes less time than that with the expanded set in most scenarios. In some scenarios, GPIP-E even takes less time than GPPL. The reason might be the rule size evolved in GPIP-E is smaller than in GPIP-F, and in some scenarios, the rule size of GPIP-E is even smaller than that of GPPL.

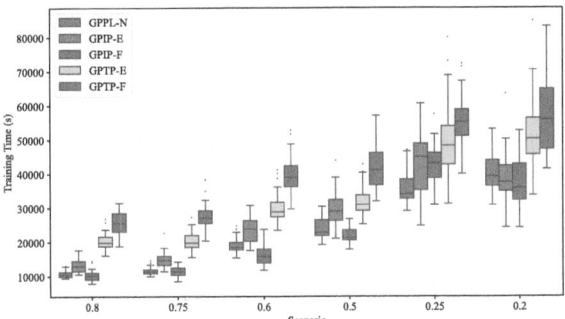

Fig. 5. Training time over 30 independent runs in six scenarios.

6 Conclusions and Future Work

This paper improves the decision-making procedure for solving the DMRCPSP by transforming the generated schedule from a non-delay schedule to a (semi-)active schedule. The main idea is to allow activities to start later but execute faster at the current decision point. Based on how to end the current decision loop, we designed two decision-making procedures derived from the parallel scheme: the immediate procedure, which ends the loop immediately after selecting an activity that cannot start, and the two-phase procedure, which continues scheduling other activities that can start immediately even after selecting a non-startable activity. Accordingly, we designed two decision sets that include activities that will be executed later: the expanded set and the full set.

We develop a GPHH based on the proposed decision-making procedures that can evolve rules to generate high-quality schedules. To verify the effectiveness of the developed GPHH with the proposed decision-making procedures, we compared GP with different procedures and decision sets on multiple scenarios. The results demonstrated that the two-phase procedure (GPTP-E/F) performs better than GPPL and GPIP in projects with complex precedence relations (high/medium OS). Although the immediate procedure has a promising intent, its performance is generally poor across most scenarios due to its characteristic of ending the decision loop prematurely after selecting a non-startable activity. However, this feature allows GPIP to perform better than the other algorithms in scenarios with low OS values. Overall, the two proposed methods are suitable for different scenarios and outperform the existing methods.

One of the challenges of multi-mode RCPSP is selecting appropriate modes for activities and reserving resources to ensure the project is executed efficiently. Future work will focus on further improving the two-phase procedure, addressing its current greedy behaviour. This could involve mechanisms like look-ahead or rollout to better determine when to end the loop and reserve resources for subsequent activities. Future research will also consider using GPHH to solve more constrained and dynamic multi-mode RCPSP problems.

References

1. Boctor, F.F.: A new and efficient heuristic for scheduling projects with resource restrictions and multiple execution modes. Eur. J. Oper. Res. **90**(2), 349–361 (1996)
2. Chakrabortty, R.K., Sarker, R.A., Essam, D.L.: Multi-mode resource constrained project scheduling under resource disruptions. Comput. Chem. Eng. **88**, 13–29 (2016)
3. Chand, S., Huynh, Q., Singh, H., Ray, T., Wagner, M.: On the use of genetic programming to evolve priority rules for resource constrained project scheduling problems. Inf. Sci. **432**, 146–163 (2018)
4. Chand, S., Singh, H., Ray, T.: Evolving heuristics for the resource constrained project scheduling problem with dynamic resource disruptions. Swarm Evol. Comput. **44**, 897–912 (2019)
5. Chand, S., Singh, H., Ray, T.: Evolving rollout-justification based heuristics for resource constrained project scheduling problems. Swarm Evol. Comput. **50**, 100556 (2019)
6. Chen, H., Ding, G., Qin, S., Zhang, J.: A hyper-heuristic based ensemble genetic programming approach for stochastic resource constrained project scheduling problem. Expert Syst. Appl. **167**, 114174 (2021)
7. Chen, H., Ding, G., Zhang, J., Li, R., Jiang, L., Qin, S.: A filtering genetic programming framework for stochastic resource constrained multi-project scheduling problem under new project insertions. Expert Syst. Appl. **198**, 116911 (2022)
8. Chen, H., Ding, G., Zhang, J., Qin, S.: Research on priority rules for the stochastic resource constrained multi-project scheduling problem with new project arrival. Comput. Indust. Eng. **137**, 106060 (2019)
9. Chen, H., Li, X., Gao, L.: A guided genetic programming with attribute node activation encoding for resource constrained project scheduling problem. Swarm Evol. Comput. **83**, 101418 (2023)

10. Chen, H., Li, X., Gao, L.: A surrogate-assisted dual-tree genetic programming framework for dynamic resource constrained multi-project scheduling problem. Inter. J. Product. Res. 1–23 (2023)
11. Chen, H., Zhang, J., Li, R., Ding, G., Qin, S.: A two-stage genetic programming framework for stochastic resource constrained multi-project scheduling problem under new project insertions. Appl. Soft Comput. **124**, 109087 (2022)
12. Demeulemeester, E., Vanhoucke, M., Herroelen, W.: RanGen: a random network generator for activity-on-the-node networks. J. Sched. **6**(1), 17–38 (2003)
13. Đumić, M., Jakobović, D.: Ensembles of priority rules for resource constrained project scheduling problem. Applied Soft Comput. **110**, 107606 (2021)
14. Đumić, M., Jakobović, D.: Using priority rules for resource-constrained project scheduling problem in static environment. Comput. Indust. Eng. **169**, 108239 (2022)
15. Đumić, M., Šišejković, D., Čorić, R., Jakobović, D.: Evolving priority rules for resource constrained project scheduling problem with genetic programming. Future Generation Comput. Syst. **86**, 211–221 (2018)
16. Elloumi, S., Loukil, T., Fortemps, P.: Reactive heuristics for disrupted multi-mode resource-constrained project scheduling problem. Expert Syst. Appl. **167**, 114132 (2021)
17. Hartmann, S., Briskorn, D.: An updated survey of variants and extensions of the resource-constrained project scheduling problem. Eur. J. Oper. Res. **297**(1), 1–14 (2022)
18. Hartmann, S., Kolisch, R.: Experimental evaluation of state-of-the-art heuristics for the resource-constrained project scheduling problem. Eur. J. Oper. Res. **127**(2), 394–407 (2000)
19. Herroelen, W., Leus, R.: Project scheduling under uncertainty: survey and research potentials. Eur. J. Oper. Res. **165**(2), 289–306 (2005)
20. Kelley, J.E.: Critical-path planning and scheduling: mathematical basis. Oper. Res. **9**(3), 296–320 (1961)
21. Kolisch, R.: Serial and parallel resource-constrained project scheduling methods revisited: theory and computation. Eur. J. Oper. Res. **90**(2), 320–333 (1996)
22. Kolisch, R., Sprecher, A.: PSPLIB-a project scheduling problem library: OR software-ORSEP operations research software exchange program. Eur. J. Oper. Res. **96**(1), 205–216 (1997)
23. Li, L., Zhang, H., Bai, S.: A multi-surrogate genetic programming hyper-heuristic algorithm for the manufacturing project scheduling problem with setup times under dynamic and interference environments. Expert Syst. Appli., 123854 (2024)
24. Lova, A., Tormos, P., Barber, F.: Multi-mode resource constrained project scheduling: scheduling schemes, priority rules and mode selection rules. Inteligencia Artificial. Revista Iberoamericana de Inteligencia Artificial **10**(30) (2006)
25. Luo, J., Vanhoucke, M., Coelho, J.: Automated design of priority rules for resource-constrained project scheduling problem using surrogate-assisted genetic programming. Swarm Evol. Comput. **81**, 101339 (2023)
26. Luo, J., Vanhoucke, M., Coelho, J., Guo, W.: An efficient genetic programming approach to design priority rules for resource-constrained project scheduling problem. Expert Syst. Appl. **198**(1), 116753 (2022)
27. MacLachlan, J., Mei, Y., Branke, J., Zhang, M.: Genetic programming hyper-heuristics with vehicle collaboration for uncertain capacitated Arc routing problems. Evol. Comput. **28**(4), 563–593 (2020)

28. Mei, Y., Nguyen, S., Zhang, M.: Evolving time-invariant dispatching rules in job shop scheduling with genetic programming. In: McDermott, J., Castelli, M., Sekanina, L., Haasdijk, E., García-Sánchez, P. (eds.) EuroGP 2017. LNCS, vol. 10196, pp. 147–163. Springer, Cham (2017). https://doi.org/10.1007/978-3-319-55696-3_10
29. Peteghem, V.V., Vanhoucke, M.: An experimental investigation of metaheuristics for the multi-mode resource-constrained project scheduling problem on new dataset instances. Eur. J. Oper. Res. **235**(1), 62–72 (2014)
30. Pinedo, M.L.: Scheduling: Theory, Algorithms, and Systems. Springer International Publishing, Cham (2022)
31. Schwindt, C., Zimmermann, J. (eds.): Handbook on Project Management and Scheduling Vol.1. IHIS, Springer, Cham (2015). https://doi.org/10.1007/978-3-319-05443-8
32. Tian, Y., Mei, Y., Zhang, M.: Learning Heuristics via Genetic Programming for Multi-mode Resource-constrained Project Scheduling. In: 2024 IEEE Congress on Evolutionary Computation (CEC 2024), Yokohama, Japan, pp. 7.99 (Jun 2024)
33. Xu, M., Mei, Y., Zhang, F., Zhang, M.: Genetic programming for dynamic flexible job shop scheduling: evolution with single individuals and ensembles. IEEE Trans. Evolutionary Computat. (2023)
34. Xu, M., Mei, Y., Zhang, F., Zhang, M.: genetic programming with lexicase selection for large-scale dynamic flexible job shop scheduling. IEEE Trans. Evolutionary Comput. (2023)
35. Zhang, F., Mei, Y., Nguyen, S., Zhang, M.: Phenotype based surrogate-assisted multi-objective genetic programming with brood recombination for dynamic flexible job shop scheduling. In: 2022 IEEE Symposium Series on Computational Intelligence (SSCI), pp. 1218–1225 (Dec 2022)
36. Zhang, F., Mei, Y., Nguyen, S., Zhang, M.: Survey on genetic programming and machine learning techniques for heuristic design in job shop scheduling. IEEE Trans. Evolutionary Comput. (2023)
37. Zhang, H., Demeulemeester, E., Li, L., Bai, S.: Surrogate-assisted cooperative learning genetic programming for the resource-constrained project scheduling problem with stochastic activity durations and transfer times. Comput. Operat. Res., 106816 (2024)
38. Zhang, H., Li, L., Bai, S., Zhang, J.: Co-evolution genetic programming-based hyper-heuristics for the stochastic project scheduling problem with resource transfer and idle costs. Swarm Evol. Comput. **90**, 101678 (2024)

Price-and-Branch Heuristic for Vector Bin Packing

Ze Wang[1], Tim Süß[2], Nikolay Popov[3], and Lars Nagel[1](✉)

[1] Loughborough University, Loughborough, LE11 3TU, UK
{z.wang6,l.nagel}@lboro.ac.uk
[2] Fulda University of Applied Science, 36037 Fulda, Germany
tim.suess@cs.hs-fulda.de
[3] iC-Haus GmbH, 55294 Bodenheim, Germany
popov_nikolay@yahoo.de

Abstract. Vector bin packing is an NP-hard problem in which a set of item vectors must be packed into a minimum number of bins such that, in each bin, the sum of the vectors does not exceed the bin's vector capacity. Vector bin packing has many applications such as scheduling virtual machines in cloud computing.

In this paper, we introduce the *Pattern Heuristic*, a price-and-branch heuristic which applies column generation. It creates a large pool of valid packings called patterns by running fast heuristics. Then it optimally solves the integer linear program of finding a minimum set of patterns which covers the complete set of items. Despite its complexity, it finds optimal or near-optimal solutions for benchmark instances with 200 items in a matter of seconds. The pattern pool is created by a *Local Search Heuristic* which, while being very fast, achieves surprisingly good results. As a stand-alone heuristic, it is suitable for scenarios with real-time demands.

The new heuristics are compared with algorithms from the literature in experiments using established benchmarks. They demonstrate a good trade-off between running time and packing quality. The Pattern Heuristic computes new optimal solutions for a 20-dimensional benchmark.

1 Introduction

Vector bin packing (VBP) is the problem of packing a given set of d-dimensional item vectors into d-dimensional bins with the objective to minimise the number of bins required. More formally: Let $\mathcal{J}$ be a set of items where each item $v \in \mathcal{J}$ is a d-dimensional vector $(v_1, \ldots, v_d)$ with $v_i \in [0, 1]$. Let $\mathcal{B}$ be a set of bins where the capacity of each bin $b \in \mathcal{B}$ is a d-dimensional vector $(1, \ldots, 1)$. The goal is to find a minimum partition of $\mathcal{J}$ into disjoint sets $\mathcal{J}_1, \ldots, \mathcal{J}_m$ such that the sum of the vectors in each $\mathcal{J}_k$ is at most the capacity of a bin, i.e. $\forall \mathcal{J}_k, \forall i \in [d], \sum_{v \in \mathcal{J}_k} v_i \leq 1$. As a generalisation of the classical bin packing problem, VBP is NP-hard, and it was even shown to be APX-hard [35].

The VBP problem can serve as a model for many applications. One example is the scheduling of batch jobs on computer clusters, where the resource requirements of the jobs, such as CPU and memory, and storage, can be represented as vectors which must be packed onto servers whose capacity limit is also given as a vector [28].

Due to its fundamental importance in theory and a wide range of real-world use cases, extensive work has been conducted to tackle the VBP problem. This includes lower bounding schemes [1,7,31] that calculate a lower bound for a given instance, approximation algorithms [3,4,9,11,18,19] that provide worst case guarantees for both, packing quality and runtime, exact algorithms [1,5,13,29,31,33] that yield optimal solutions, and heuristics [2,6,10,17,22,23,25,26,28,30,32,34] which tend to run much faster and usually provide acceptable packing solutions. Experiments have demonstrated that exact algorithms are unsuitable for real-time scheduling because their runtime rapidly increases with the number of items and dimensions [25,26]. These algorithms are limited to small-scale or less time-critical scenarios. On the other hand, heuristics are frequently employed in the scheduling of computing resources [24,27,28,30,34,36].

In this work, we propose two heuristics for the VBP problem. The first heuristic is the *Bin-centric Local Search Heuristic* which packs one bin after the other. It first assigns items to that bin using a fast heuristic, and then improves the assignment by making swaps between the items in the bin and unpacked items. This design is shown to produce a good packing quality in a short time.

As our main contribution, we propose the *Pattern Heuristic*, which, to the best of our knowledge, is the first *price-and-branch algorithm* [20] designed for VBP. A *pattern* is a bin packing, i.e. any subset of items that fit into one bin. The Pattern Heuristic leverages solutions created by the Bin-centric Local Search Heuristic to create a pool of patterns. It then solves the set cover problem of finding a minimum set of patterns covering all items by phrasing it as an integer linear programming (ILP) problem and solving it using the CPLEX [14].

Extensive experiments were conducted to evaluate the performance of the proposed heuristics against state-of-the-art algorithms on well-known benchmarks. The experiments demonstrate that the Bin-centric Local Search Heuristic is a strong alternative to fast constructive heuristics as it achieves superior packing results while maintaining a competitive runtime. The Pattern Heuristic achieves the best packing quality among all heuristics. In most cases it computes optimal or near-optimal results within seconds, demonstrating the effectiveness of applying the price-and-branch idea to the VBP problem.

While the Pattern Heuristic is completely new, the Local Search Heuristic is based on a previous version [26] which is extended by adding new powerful variants that improve vector sorting, swapping times and swapping operations.

In the remainder, a summary of the existing algorithms is given in Sect. 2. The new algorithms are introduced in Sect. 3 and evaluated in Sect. 4.

2 Related Work

In this section we survey the literature on algorithms for vector bin packing.

First-Fit and Best-Fit Heuristics. Kou and Markowsky [17] introduced first-fit decreasing (FFD) and best-fit decreasing (BFD) heuristics. The algorithms first sort the vectors in decreasing order and then pack them using first-fit or best-fit. To sort vectors, the authors use the infinity norm (L_∞), the sum of all vector components (L_1) or lexicographical ordering. Panigrahy et al. [28] distinguished item-centric and bin-centric FFD and BFD where a bin-centric approach packs one bin at a time. It greedily packs the item which reduces the remaining capacity of the bin most. The capacity is measured using L_1, L_2 (the Euclidean norm), L_∞, and the dot product. Weight functions are used to prioritise coordinates. Gabay and Zaourar [10] suggested additional norms or cost functions for the bin-centric approach. Aringhieri et al. [2] proposed a variant of the BFD heuristic called BFD-DYNLB, which uses dynamically computed lower bounds to make decisions. It is shown to pack better than BFD in their experiments. Mommessin et al. [25] systematically analysed parameters of item-centric, bin-centric and multi-bin heuristics: sorting methods, weight vectors, and norms (called score types). The multi-bin heuristics are designed to solve vector scheduling. They open all m bins at the beginning. In every iteration, every item-bin pair is assessed using a norm, and the best pair is picked for the assignment. To solve vector bin packing, the algorithm is executed multiple times to determine the best m. A selection of these heuristics (from [10,17,25,28]) is included in the evaluation in Sect. 4.

Local Search, Genetic Algorithms and Meta-heuristics. Wilcox et al. [34] designed the *Reordering Grouping Genetic Algorithm*. It improves a population of packing sequences. The crossover operator uses exon shuffling; the mutation uses swaps or moves items or reassigns a complete bin. Masson et al. [23] proposed a *Multi-start Iterated Local Search* (MS-ILS) scheme. It improves an initial solution using local search in combination with a shaking procedure. Sait and Shahid [30] applied *Simulated Evolution* [16]. Their heuristic improves an initial FFD solution by performing evaluation, selection and allocation steps in a loop. Items are unpacked with a specific probability and reassigned using FFD. Mannai et al. [22] presented a *tabu search* for the 2D problem. Buljubašić and Vasquez [6] proposed a hybrid scheme named *Consistent Neighbourhood Search* (CNS). It improves an initial solution generated by removing one bin per loop iteration using tabu search and an optimal packing scheme for small item sets. Aringhieri et al. [2] applied *tabu search* on initial solutions created by a randomised BFD-DYNLB. Turky et al. [32] developed a *Hyper-heuristic Framework* (HH) for solving optimisation problems which applies local-search heuristics such as simulated annealing.

Exact Algorithms. Brandão and Pedroso [5] generalised the bin packing arc-flow formulation by de Carvalho [8] to solve vector bin packing. The computation is accelerated by applying a lossless compression technique to the arc-flow graph. Heßler et al. [13] systematically studied VBP formulations and proposed a branch-and-price approach (named B&P1 in Sect. 4). The pricing problem is modelled as the shortest path problem with resource constraints [15]. Dual inequalities are applied to stabilise the column generation process. Wei et al.

[33] introduced a branch-and-price algorithm for the 2D problem (named B&P2 in Sect. 4). Unlike the approach of Heßler et al. [13], the pricing problem is modelled as a two-constraint knapsack problem with conflicts. Pessoa et al. [29] proposed VRPSolver, a generic branch-cut-and-price solver that can be applied to vector bin packing. (It is also called VRPSolver in Sect. 4.)

The quality and computational complexity of lower bounds are important in designing branch-and-bound algorithms. Spieksma [31] introduced a lower bound based on identifying the number of disallowed item pairs. Caprara and Toth [7] studied lower bounds adapted from the bin packing problem, and proposed a strong bound by solving the linear relaxation through column generation. Alves et al. [1] developed bounds by extending the principles of dual-feasible functions to multiple dimensions. Their quality is close to that of column generation but with significantly less computational time.

3 New Algorithms

In this section, we present the new vector bin packing strategies: *Bin-centric Local Search Heuristic* and *Pattern Heuristic*.

3.1 Bin-Centric Local Search Heuristic

The *Bin-centric Local Search Heuristic* (LS) differs from other local-search approaches as it applies the technique to only one bin at a time instead of the complete solution. When the current bin cannot be improved further, it is closed, and a new bin is opened.

The main logic is given in Algorithm 1. It takes a set $\mathcal{J}$ of items, a fast heuristic $\mathcal{H}$, a norm N, and a swap type t as input. Starting with an empty set of bins, it runs a nested loop until all items are packed. In every iteration of the outer while loop, it creates a new bin (Line 3-4), which is then packed by the inner do-while loop. The inner loop runs $\mathcal{H}$ to fill the bin (Line 6) before it calls the function Swap which reduces the bin's remaining capacity by swapping unpacked items with items in the current assignment (Line 10). If no items are left, the algorithm stops and returns the bins. If Swap performs at least one swap, the do-while loop will continue. In every iteration it tries to assign additional items, which makes sense because the changed assignment may offer new gaps.

In our implementation, $\mathcal{H}$ is either an FF or an FFD heuristic (see Sect. 2). In the case of FF, the largest unpacked item is packed in the first iteration to prevent (or mitigate) an accumulation of large items to be packed at the end. Then the remaining unpacked items are assigned in a shuffled order using FF. Other than that the assignment follows a random order. In the case of FFD, the unpacked items are sorted in descending order using N (which is L_1, L_2 or L_∞ in our experiments) before they are packed with FF.

The Swap procedure takes as inputs $\mathcal{J}$, the current bin b, the norm N to scalarise the remaining capacity, and the swap type t. Swap checks all possible swaps of a packed item with an unpacked item. If there is at least one that

Algorithm 1: $LS(\mathcal{J}, \mathcal{H}, N, t)$

Input: $\mathcal{J}$: set of items, $\mathcal{H}$: heuristic, N: norm, t: swap type (1, 2 or 3)
Output: $\mathcal{B}$: packed bins

1. $\mathcal{B} \leftarrow \emptyset$
2. **while** $\mathcal{J} \neq \emptyset$ **do**
3.   Create a new bin b
4. $\mathcal{B} \leftarrow \mathcal{B} \cup \{b\}$
5.   **do**
6.   Pack items from $\mathcal{J}$ into b using $\mathcal{H}$
7.   Remove all items packed into b from $\mathcal{J}$
8.   **if** $\mathcal{J} = \emptyset$ **then**
9.   **return** $\mathcal{B}$
10. $flag \leftarrow Swap(\mathcal{J}, b, N, t)$
11.   **while** $flag = true$;
12. **return** $\mathcal{B}$

would reduce the remaining capacity, at least one swap will be performed, and the procedure will return true. The performance of LS mainly depends on Swap. Three different versions are used that swap pairs with different levels of greed. The first one ($t = 1$) swaps every suitable pair immediately. The second one ($t = 2$) considers all potential swaps for an unassigned item and then performs the best one. The third version ($t = 3$) checks all pairs before it performs the best swap. While the last one makes the optimal choice, the motivation behind the first two versions is that swaps are performed more frequently and that a local optimum is reached more quickly. Aside from varying t, LS is run with three different N, namely L_1, L_2 and L_∞, so that there are nine versions in total.

We also implemented a variant of Algorithm 1 in which up to two items can be swapped (in and out). The algorithm checks for every two items in the bin if replacing one or both of them by one or two items from the set of unpacked items would improve the assignment. To prevent packing small items at the beginning and leaving large items unpacked, we disallow swapping more items in than in out. Again, there are nine versions differing in the swapping type ($t = 1, 2, 3$) and the norm ($N = L_1, L_2, L_\infty$). In the experiments, we will denote by LS_I1 and LS_I2 the variants that swap at most one or two items, respectively.

3.2 Pattern Heuristic

Our main contribution is the *Pattern Heuristic* which uses the Bin-centric Local Search Heuristic, i.e. LS, as a subroutine. In our experiments we observed that LS generates solutions with many tightly packed bins. By running LS with the FF heuristic several times and feeding different random seeds to it, a pool of good patterns can be produced. This pool of patterns can then be used to create new solutions which in many cases are better than the ones computed by LS. We use integer linear programming to compute an optimal result based on the pool of patterns. In what follows, we will provide a formal problem description, the ILP model and finally the design of the Pattern Heuristic.

Problem Statement. Given a pool of patterns $\mathcal{P}$, the problem is to select a subset $\mathcal{S} \subseteq \mathcal{P}$ such that all items are packed, and $|\mathcal{S}|$ is minimised. It can be formulated as the following ILP:

$$\min \sum_{p \in \mathcal{P}} x_p \tag{1a}$$

$$\text{s.t.} \sum_{p \in \mathcal{P}} a_i^p x_p = 1, \quad i \in \mathcal{I} \tag{1b}$$

$$x_p \in \{0, 1\}, \quad p \in \mathcal{P} \tag{1c}$$

where $\mathcal{I} = \{1, 2, 3, \ldots, n\}$ is the index set of the item set $\mathcal{J}$. x_p indicates whether pattern p is in the solution $\mathcal{S}$; that is, $p \in \mathcal{S}$ if and only if $x_p = 1$. The $a_i^p \in \{0, 1\}$ indicate whether item i is included in pattern p. Hence, the vector

$$(a_1^p, a_2^p, a_3^p, \ldots, a_n^p) \tag{2}$$

is an encoding of the pattern p. The objective is to minimise the number of patterns, and thus bins, in $\mathcal{S}$ (Eq. (1a)). The constraints ensure that all items are included in $\mathcal{S}$ (Eq. (1b)) and define the domain of x_p (Eq. (1c)).

The above ILP is a well-known set-partitioning formulation for the cutting stock problem introduced by Gilmore and Gomory [12]. If $\mathcal{P}$ was the set $\mathcal{P}_{all}$ of all possible patterns, an optimal solution would also be an optimal solution to the *Master Problem* (MP), i.e. the original Vector Bin Packing problem. However, it is usually not possible or sensible to generate and work with $\mathcal{P}_{all}$ because of its size. For this reason, we only generate a subset $\mathcal{P}_{sub} \subseteq \mathcal{P}_{all}$ and solve the *Restricted Master Problem* (RMP) for $\mathcal{P} = \mathcal{P}_{sub}$.

Pattern Generation. For generating the patterns of $\mathcal{P}_{sub}$, we will run the Local Search Heuristic LS_I1 with different random seeds and parameters. For assessing the quality of $\mathcal{P}_{sub}$, we turn to the linear relaxation of the MP which relaxes the integrality constraint in Eq. (1c) to $x_p \in [0, 1]$, thus allowing x_p to be fractional. The aim is to include enough patterns in $\mathcal{P}_{sub}$ so that the relaxed MP can be optimally solved. That is, if z_{RMP}^{LP} and z_{MP}^{LP} denote the optimal objective values of the RMP and MP relaxations, then the aim is to create a set of patterns such that $z_{RMP}^{LP} = z_{MP}^{LP}$. If this succeeds, then $\mathcal{P}_{sub}$ is likely to be a good pattern set for solving the unrelaxed problem, and the hope is that an optimal solution to the RMP with $\mathcal{P}_{sub}$ is close to the optimal solution to the MP. As a by-product, the optimal solution to the relaxed problem also provides a strong lower bound for the MP which helps to assess the quality of any solution to the MP [7].

The relaxed MP can be solved by *Column Generation* [12,21], which iteratively adds new patterns to $\mathcal{P}$ that reduce z_{MP}^{LP} until there is no such pattern left. This process is also known as finding patterns with negative reduced cost. Let p' be a new pattern and $x_{p'}$ its variable. The reduced cost of the pattern p'

Algorithm 2: *PatternHeuristic(J)*

Input: $\mathcal{J}$: set of items
Output: $\mathcal{B}$: set of packed bins

1    $\mathcal{P} \leftarrow \emptyset$ // create set of patterns
2    $\mathcal{H} \leftarrow FFD$ // use FFD in LS for initialisation of $\mathcal{P}$
3    $LB \leftarrow \lceil \| \sum_{v \in \mathcal{J}} v \|_\infty \rceil$ // initialise lower bound
4   **do**
5       $\mathcal{P}_{new} \leftarrow$ run $LS(\mathcal{J}, \mathcal{H}, N, t)$ once for every combination of N and t to create pattern set
6       $\mathcal{H} \leftarrow FF$ // switch to FF
7       $\mathcal{B} \leftarrow$ best solution from new and old LS runs
8   **if** $|\mathcal{B}| = LB$ **then**
9   **return** $\mathcal{B}$
10   **if** $\mathcal{P} \cup \mathcal{P}_{new} = \mathcal{P}$ **then**
11          $\mathcal{B} \leftarrow$ solve RMP with $\mathcal{P}$ using CPLEX
12   **return** $\mathcal{B}$
13      $\mathcal{P} \leftarrow \mathcal{P} \cup \mathcal{P}_{new}$
14      $z_{RMP}^{LP} \leftarrow$ solve linear relaxation of RMP with $\mathcal{P}$
15      $\pi \leftarrow$ dual solution of the RMP
16      $rc \leftarrow$ solve MKP with π to update the reduced cost
17 **while** $rc < 0$;
18 $LB \leftarrow \lceil z_{RMP}^{LP} \rceil$
19 **if** $|\mathcal{B}| = LB$ **then**
20   **return** $\mathcal{B}$
21 $\mathcal{B} \leftarrow$ solve RMP with $\mathcal{P}$ using CPLEX
22 **return** $\mathcal{B}$

is calculated as $1 - \pi^\mathsf{T} \mathbf{a}^{p'}$, where $\pi = (\pi_1, \ldots, \pi_n)$ is the current dual solution of the relaxed RMP, $\mathbf{a}^{p'}$ is a vector encoding p' as in Eq. (2), and 1 is the cost of $x_{p'}$, i.e. the coefficient of $x_{p'}$ in Eq. (1a). If there is no pattern p' with negative reduced cost, then z_{MP}^{LP} will not decrease further, and thus, the optimal value is reached.

Proving the non-existence of new patterns with negative reduced cost is equivalent to demonstrating that the optimal objective value of the following *Multi-Dimensional Knapsack* (MKP) problem is not greater than 1.

$$\max \quad \sum_{j \in \mathcal{I}} \pi_j a_j^{p'} \tag{3a}$$

$$\text{s.t.} \quad \sum_{j \in \mathcal{I}} a_j^{p'} v_i^j \leq 1, \quad i \in [d] \tag{3b}$$

$$a_j^{p'} \in \{0, 1\} \tag{3c}$$

The objective of the MKP problem, as given in Eq. (3a), is to maximise $\pi^\mathsf{T} \mathbf{a}^{p'}$ while ensuring that the sum of the item vectors v^j of the pattern does not exceed the capacity of the bin in any dimension (Eq. (3b)).

Algorithm. Now we can use the elements described above to write down the Pattern Heuristic which is given as Algorithm 2. It begins by initialising the pattern set $\mathcal{P}$, the heuristic $\mathcal{H}$ and the lower bound LB, which is computed by summing up all item vectors and rounding up the largest component in the sum. It serves as a termination condition in Line 8 and 19. In every iteration of the do-while loop, LS is run with different parameters to create a set $\mathcal{P}_{new}$ of patterns. In the first iteration LS applies the deterministic FFD, but then replaces it by FF which uses different random seeds every time. The best solution of these and all previous LS runs is stored in $\mathcal{B}$. If $|\mathcal{B}|$ matches LB, $\mathcal{B}$ is returned as an optimal solution, and the algorithm terminates (Line 8-9). If $\mathcal{P}_{new}$ does not contain new patterns, the algorithm solves the RMP on $\mathcal{P}$, returns the solution and terminates (Line 10-12). Otherwise, the heuristic solves the relaxed RMP and MKP to compute the reduced cost (Line 14-16). If the reduced cost is negative, the loop enters the next iteration. If they are not negative, an optimal solution to the relaxed RMP is found. This solution rounded up forms a new lower bound LB (Line 18) which is usually better than the one computed at the beginning (Line 3). If the new LB matches the number of bins in the currently best solution $|\mathcal{B}|$, $\mathcal{B}$ is returned as an optimal solution. Otherwise, the final, but not necessarily optimal solution is computed by solving the RMP on $\mathcal{P}$.

It is interesting to note that, in the experiments, new patterns were found in every loop iteration so that it was never necessary to prematurely end the loop returning a suboptimal solution. The lower bound checks, on the other hand, were of great value as they often cut the running time short.

4 Evaluation

The performance of the *Bin-centric Local Search Heuristic* and the *Pattern Heuristic* are assessed by their packing quality, measured by the number of bins used, and their runtime. We start with the *Bin-centric Local Search Heuristic* in Sect. 4.1. It is compared with fast constructive heuristics from the literature including item-centric (IC), bin-centric (BC) and multi-bin (MBP) heuristics from [10,17,25,28]. The *Pattern Heuristic*, which is the main contribution, is evaluated in Sect. 4.2. It is compared with exact algorithms (*VPSolver* [5], *B&P1* [13], *VRPSolver* [29], *B&P2* [33]) and advanced heuristics (*MS-ILS* [23], *CNS* [6], *HH* [32]). Note that B&P1 and B&P2 refer to branch-and-price algorithms that were not explicitly named in their respective publications.

Environment. We implemented our heuristics and the item-centric and bin-centric heuristics [17,28] in C++; they are available on *Bitbucket*[1] The implementations of MBP, VPSolver and VRPSolver were downloaded and run on our test machine. The solvers whose code is not available are compared using the results from their respective papers. All experiments were performed on a single thread of an Intel(R) Core(TM) i5-9400 CPU with a frequency of 2.90 GHz and 16 GiB RAM. The runtime limit was set to one hour. To solve the ILP and LP,

---

[1] https://bitbucket.org/vsvbp/vectorbinpacking/src/master/.

the Pattern Heuristic and the VRPSolver were run with CPLEX 12.10, while the VPSolver was run with the recommended GUROBI solver.

Benchmarks. The primal benchmark we used is a well-known 2-dimensional benchmark [7]. It comprises 10 data classes, C1-C10, which differ in how items are generated. Each data class contains four subclasses with different numbers of items: [25, 50, 100, 200] for C1-C9 and [24, 51, 99, 201] for C10. Within each subclass, ten instances were randomly created so that the benchmark contains 400 instances in total. The items in the first six classes were generated by randomly drawing their components from a given interval $[a, b]$.

C1: $[0.1, 0.4]$ **C2:** $[0.001, 1]$ **C3:** $[0.2, 0.8]$
C4: $[0.05, 0.2]$ **C5:** $[0.025, 0.1]$ **C6:** $[\frac{20}{150}, \frac{100}{150}]$ Class C7 and C8 introduce positive and negative correlations between the two dimensions. For both classes, the first dimension value v_1 is randomly drawn from the C6 interval. The second value is drawn from $[v_1 - 10/150, v_1 + 10/150]$ and $[110/150 - v_1, 130/150 - v_1]$, respectively. C9 generates vectors using the interval of C1, but adapts the bin capacity to create tight optimal solutions. C10 contains items which can be perfectly fit into bins with no remaining capacity.

To test the performance on a larger number of dimensions, we used the 20-dimensional benchmark from [5]. Each instance was created by combining the ten instances of a subclass from the 2-dimensional benchmark. Hence, there is one 20-dimensional instance for each subclass, resulting in a total of 40 instances. For all classes, we scale bins and items so that the bins have capacity $(1, ..., 1)$.

4.1 Performance of the Bin-Centric Local Search Heuristic

The Bin-centric Local Search Heuristic is compared with item-centric, bin-centric and multi-bin heuristics from the literature.

Variants of Heuristics. Almost all item-centric, bin-centric and multi-bin heuristics from the literature establish an order among the item vectors by using a weighted norm or cost function, which allows to sort the vectors and assign larger items first. For the bin-centric and multi-bin heuristics, we selected the 12 best cost functions from the publications in which the heuristics were introduced and applied two weight vectors, resulting in a total of 24 functions. The first-fit and best-fit item-centric heuristics were run with the functions suggested (L_1, L_∞, lexicographical ordering) as well as L_2.

The Bin-centric Local Search Heuristic has two variants: LS_I1 and LS_I2 (see Sect. 3.1). The former allows swapping at most one item at a time, while the latter allows swapping at most two items. For both we created 9 subvariants which differ by when the swap operation is executed and by the cost function used to measure a bin's remaining capacity (L_1, L_2, or L_∞).

For every data class, we determined the variant of each heuristic with the best packing quality and applied that variant to all instances of that data class. (We assume that one would know enough about the data set in a real setting to be able to pick the most suitable cost function.) In cases where multiple variants yielded the least number of bins, we selected the one with the shortest runtime.

Note that, while considering a wide variety of variants, this paper does not focus on discussing the variants, but on the overall performance of a heuristic using its best variant as a representative. Comprehensive studies on the impact of cost function and weight vector can be found in [25,26,28].

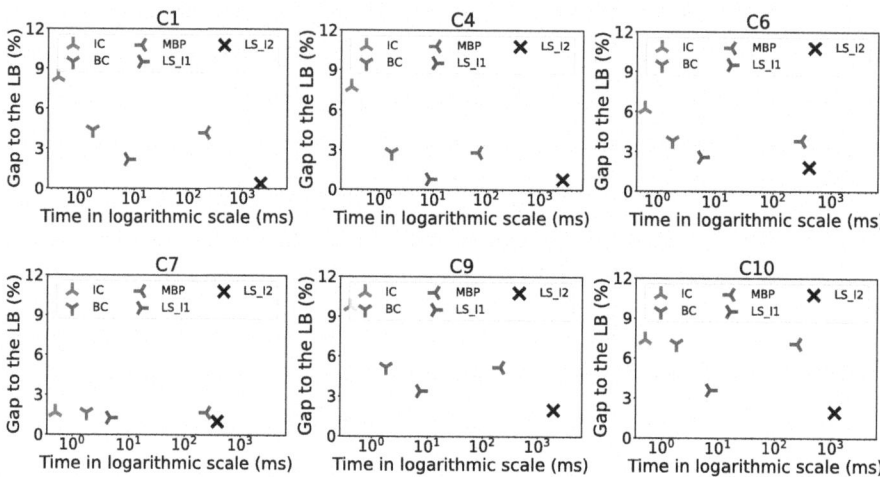

Fig. 1. Results of Bin-centric Local Search Heuristics and constructive heuristics on the 2-dimensional benchmark.

Results of Heuristics. The results for 2D and 20D instances with 200 items are shown in Fig. 1 and Fig. 2. Since all heuristics perform equally well for C2, C3, C5 and C8, these plots are left out. The x-axis shows average runtime in milliseconds (ms) on a logarithmic scale and y-axis represents the packing quality by $\frac{|\mathcal{B}|-LB}{LB} \cdot 100$, i.e. the percental gap between the number $|\mathcal{B}|$ of bins used and a lower bound LB (which is the optimal value for all classes except C9 in the 2D benchmark and C4 and C5 in the 20D benchmark). The values displayed are the averages taken over the ten instances. Figure 3 shows aggregated results for each heuristic, namely the average gap to the lower bound taken over all classes.

From Fig. 1, we can observe a general trade-off between packing quality and runtime. The item-centric heuristics are extremely fast across all data classes usually taking under 1ms, but exhibit an uneven packing quality. They achieve optimal results for C2, C3, and C8 (not shown in plots) while performing much worse than all other heuristics for C1, C4, C6, and C9. The bin-centric heuristics run in 1–2 ms and about ten times slower than the item-centric heuristics, but they exhibit a superior packing quality which is reflected in a smaller lower bound gap (Fig. 3). The MBP are significantly slower (20–200 ms), but do not improve the lower bound gap by much. The LS_I1 achieve an excellent trade-off between packing quality and runtime. While they are only about four times slower than the bin-centric approach, their packing quality is at least as good as the packing quality of the other heuristics and outperforms them for C1, C4, C6, C7, C9, and C10. The LS_I2 tend to pack the bins even tighter, but take up to 400ms. For LS_I1 and LS_I2, the average gap to the lower bound is 1.38% and 0.80%, respectively, while the other heuristics have a gap of 2.49% or more.

Reasons why bin-centric heuristics tend to be better than item-centric heuristics are given in [28]. It is not too surprising that the LS variants are better than simple bin-centric heuristics because they also pack the bins one by one, but improve the initial packing using local search.

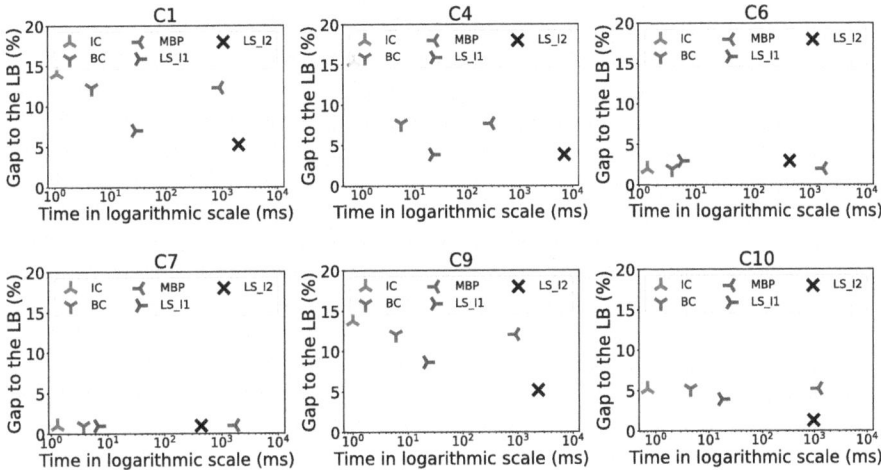

Fig. 2. Results of Bin-centric Local Search Heuristics and constructive heuristics on the 20-dimensional benchmark.

The 20D results are given in Fig. 2. Generally, the observations made from the 2D benchmark remain valid, but unlike the 2D case where LS_I1 and LS_I2 always dominate the others, there is one class, namely C6, where they are slightly worse. The percental gaps being more than doubled for most heuristics (Fig. 3) suggests that packing 20D items is much harder than packing 2D ones.

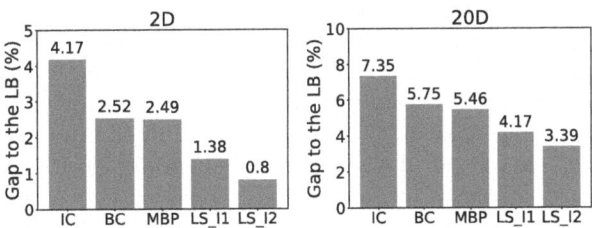

Fig. 3. Average gap to lower bound across all data classes of 2D / 20D benchmark.

Table 1. Comparison of Pattern Heuristic and exact solvers on 2D benchmark.

C	$|\mathcal{J}|$	VPSolver		B&P1		VRPSolver		B&P2		Pattern	
		Opt	T(s)	Opt	T(s)	Opt	T(s)	Opt	T(s)	Opt	T(s)
1	200	0	3600.0	3	2899.4	10	150.6	10	7.6	10	128.9
2	200	10	5.8	10	0.1	10	29.2	10	0.2	10	1.3
3	200	10	0.2	10	0.1	10	16.9	10	0.2	10	1.2
4	25	10	25.6	10	14.1	10	2.3	10	0.04	10	0.0
	50	0	3600.0	10	46.9	10	4.8	10	0.2	10	0.0
	100	0	3600.0	10	305.2	10	14.1	10	0.02	10	0.0
	200	0	3600.0	3	2973.4	10	126.0	10	5.4	10	0.1
5	25	10	12.6	10	16.5	10	2.0	10	0.01	10	0.0
	50	0	3600.0	10	58.5	10	5.6	10	0.3	10	0.0
	100	0	3600.0	10	386.3	10	21.7	10	0.01	10	0.0
	200	0	3600.0	7	2566.5	10	377.5	10	0.03	10	0.0
6	200	10	5.3	10	14.2	10	117.2	10	1.4	10	6.8
7	200	10	34.4	10	23.8	10	58.0	10	0.8	10	9.5
8	200	10	0.3	10	0.3	10	3.9	10	0.1	10	8.7
9	200	0	3600.0	0	3600.0	6	1959.0	1	3542.7	4	2417.1
10	201	10	113.4	7	1675.2	9	774.2	10	16.0	10	19.8
Total		**80**		**130**		**155**		151		154	

4.2 Performance of the Pattern Heuristic

In this section, the Pattern Heuristic is compared with the state-of-the-art exact algorithms and complex heuristics.

Comparison with Exact Algorithms. The Pattern Heuristic is compared with VPSolver [5], B&P1 [13], VRPSolver [29], and B&P2 [33]. As the downloaded code of the VRPSolver is not suitable for the 20D benchmark, we only ran it on the 2D benchmark and take the 20D results from their paper.

The results of the 2D benchmark are presented in Table 1. Columns C specifies the data class of the benchmark and $|\mathcal{J}|$, by the number of items per instance, the subclass. *Opt* gives the number of instances within the subclass optimally solved by the corresponding algorithm, and $T(s)$ the average runtime. If the algorithm fails to terminate within the time limit of 1h (= 3600s), the runtime is set to 3600s. Due to space limitations, the results for instances with 25, 50 and 100 instances are left out if they were solved by all algorithms.

All algorithms solved all instances in C2, C3, C6, C7, and C8 optimally. The instances of classes C1, C4 and C5 are among the most challenging as they contain only small items. This implies a much larger number of patterns, which is a problem for VPSolver in particular because it must enumerate all patterns. Pattern Heuristic, VRPSolver and B&P2 solved all instances, but VPSolver and

B&P1 failed to solve 70 or 17 instances, respectively. The instances of C9 and C10 have tight optimal solutions, and the most challenging subclass overall is C9 with 200 items, which none of the algorithms solved perfectly. VRPSolver found 6 optimal solutions, Pattern Heuristic 4, B&P2 1, and B&P1 and VPSolver 0. For C10, VRPSolver and B&P1 failed to solve 1 or 3 instances, respectively.

Table 2. Comparison of Pattern Heuristic with exact solvers on 20D benchmark

C	$|\mathcal{J}|$	VPSolver		B&P1		VRPSolver		Pattern	
		Opt	T(s)	Opt	T(s)	Opt	T(s)	Opt	T(s)
1	200	1	1257.2	1	2141.8	1	510.0	1	121.3
2	200	1	0.1	1	<0.1	1	<10.0	1	3.3
3	200	1	0.1	1	<0.1	1	<10.0	1	3.3
4	25	1	164.6	1	130.2	1	<10.0	1	0.0
	50	0	3600.0	0	3600.0	1	1278.0	1	0.0
	100	0	3600.0	0	3600.0	0	3600.0	0	3600.0
	200	0	3600.0	0	3600.0	0	3600.0	0	3600.0
5	25	1	2045.4	1	2020.7	1	28.0	1	0.0
	50	0	3600.0	0	3600.0	0	3600.0	1	0.0
	100	0	3600.0	0	3600.0	0	3600.0	1	0.0
	200	0	3600.0	0	3600.0	0	3600.0	0	3600.0
6	200	1	0.5	1	0.2	1	<10.0	1	5.3
7	200	1	0.6	1	0.6	1	<10.0	1	5.5
8	200	1	0.3	1	<0.1	1	<10.0	1	5.1
9	200	0	3600.0	1	3399.9	1	131.0	1	124.8
10	201	1	27.6	1	278.7	1	159.0	1	16.9
Total		9		10		11		**13**	

The results of the 20D benchmark are given in Table 2 where B&P2 is omitted as it is only designed for the 2D case. The Pattern Heuristic optimally solved all instances that were solved by the others and achieved new optimal solutions for two instances of C5. 32 of 37 subclasses were solved in under 10 s, 20 of them within 1 s, and some solutions (including the two new ones) even in less than 50ms. The fast results were achieved when our solver only needed a small pattern pool and the simple lower bound to reach and prove optimality.

In summary, the Pattern Heuristic achieved a packing quality superior to VPSolver, B&P1, and B&P2 and about equal to VRPSolver. The Pattern Heuristic solved one 2D instance less than VRPSolver and two 20D instances more. The computational time of the Pattern Heuristic is generally much faster than VPSolver, B&P1, and VRPSolver, and is comparable to that of B&P2. In the 2D benchmark, the Pattern Heuristic solved 38 out of 40 subclasses within one minute, with 35 of them solved in under ten seconds. For two subclasses, it took longer than one minute due to the difficulty of finding the minimum set of patterns for the RMP. In the 20D benchmark, there are three instances with small items where the Pattern Heuristic fails to finish within the 3600-second limit. This is because the pattern pool for these instances is too large to find all patterns with negative reduced cost.

Comparison with Heuristics. The Pattern Heuristic is compared with three strong heuristics from the literature: MS-ILS [23], CNS [6], and HH [32]. As their implementations were not available, we present the results from the respective papers which only cover the classes C1, C6, C7, C9, and C10. Due to lack of space we only show the subclasses with 200 items. In those MS-ILS was executed 10 times with a time limit of 300 s, CNS 50 times with a limit of 10 s, and HH 31 times with no specified limit. Note that since the algorithms were tested on different machines, the comparison of the running times is inaccurate.

The results of the 2-dimensional benchmark are given in Table 3. The columns C and $|\mathcal{J}|$ specify class and subclass, the column LB the sum of the lower bounds on the number of bins required to pack the ten instances of the subclass. Then for each algorithm, $Best$ sums up the total number of bins required for all subclass instances, and $T(s)$ is the average runtime of all runs in seconds.

Table 3. Comparison of Pattern Heuristic with other heuristics on 2D benchmark

C	$|\mathcal{J}|$	LB	MS-ILS[5]		CNS		HH		Pattern	
			Best	T(s)	Best	T(s)	Best	T(s)	Best	T(s)
1	200	503	503	60.00	503	0.01	503	189.00	503	128.9
6	200	811	811	60.00	811	8.00	823	132.00	811	6.8
7	200	801	802	60.00	801	2.00	812	196.00	801	9.5
9	200	503	513	60.00	513	10.01	513	184.00	509	2417.12
10	201	670	678	60.00	670	0.91	681	117.00	670	19.8

[2]We divided the MS-ILS times by 5 as the authors of [6] argued that the machine used for MS-ILS is approximately five times slower than the one used for CNS.

We observe that the Pattern Heuristic consistently outperforms the other heuristics in terms of packing quality as it uses the fewest number of bins. It is followed by CNS which successfully solved all instances in C1, C6, C7 and C10 and only failed at some of the difficult instances in C9 with 200 items.

The Pattern Heuristic usually runs in seconds and is considerably faster than MS-ILS and HH, especially on small instances (not shown in the table), with the exception of two of the 40 subclasses. For these subclasses it takes an average time of about 2 or 40 min, respectively. The reason, as described above, is the size of the pattern pool required to solve the RMP. However, one should note that the Pattern Heuritic requires only 2 s to output the result achieved by the other heuristics for C9. (It can output intermediate results in Line 7 of Algorithm 2.)

Nevertheless, CNS, which is tailored for 2D instances, tends to be faster than the Pattern Heuristic when packing more than 50 items.

5 Conclusion

With the fast Bin-centric Local Search Heuristics and the near-optimal Pattern Heuristic, we have presented two new types of heuristics for vector bin packing which are highly competitive in that they offer good trade-offs between speed and packing quality. The experiments demonstrate that the Local Search Heuristics are suitable for real-time scheduling and provide better results than other fast heuristics. The Pattern Heuristic can compete with state-of-the-art exact algorithms and heuristics. While it achieves a better packing quality than all other algorithms, the runtime is not as good as those of algorithms tailored for 2D like B&P2 and CNS. For practical applications, the speed of the Pattern Heuristic could be improved by parallelising the pattern collection, by terminating it prematurely or by filtering out bad patterns to speed up the LP solver.

References

1. Alves, C., de Carvalho, J.V., Clautiaux, F., Rietz, J.: Multidimensional Dual-feasible Functions and Fast Lower Bounds for the Vector Packing Problem. Eur. J. Oper. Res. **233**(1), 43–63 (2014)
2. Aringhieri, R., Duma, D., Grosso, A., Hosteins, P.: Simple but effective heuristics for the 2-constraint bin packing problem. J. Heuristics **24**, 345–357 (2018)
3. Bansal, N., Caprara, A., Sviridenko, M.: A new approximation method for set covering problems, with applications to multidimensional bin packing. SIAM J. Comput. **39**(4), 1256–1278 (2010)
4. Bansal, N., Eliáš, M., Khan, A.: improved approximation for vector bin packing. In: Proceedings of the Twenty-Seventh Annual ACM-SIAM Symposium on Discrete Algorithms, pp. 1561–1579. SIAM (2016)
5. Brandão, F., Pedroso, J.P.: Bin packing and related problems: general arc-flow formulation with graph compression. Comput. Oper. Res. **69**, 56–67 (2016)
6. Buljubašić, M., Vasquez, M.: Consistent neighborhood search for one-dimensional bin packing and two-dimensional vector packing. Comput. Oper. Res. **76**, 12–21 (2016)
7. Caprara, A., Toth, P.: Lower bounds and algorithms for the 2-dimensional vector packing problem. Discret. Appl. Math. **111**(3), 231–262 (2001)
8. de Carvalho, J.: Exact solution of bin-packing problems using column generation and branch-and-bound. Ann. Oper. Res. **86**, 629–659 (1999)
9. Chekuri, C., Khanna, S.: On multidimensional packing problems. SIAM J. Comput. **33**(4), 837–851 (2004)
10. Gabay, M., Zaourar, S.: Vector bin packing with heterogeneous bins: application to the machine reassignment problem. Ann. Oper. Res. **242**(1), 161–194 (2016)
11. Garey, M.R., Graham, R.L., Johnson, D.S., Yao, A.C.: Resource constrained scheduling as generalized bin packing. J. Comb. Theory, Ser. A **21**(3), 257–298 (1976)

12. Gilmore, P.C., Gomory, R.E.: A linear programming approach to the cutting-stock problem. Oper. Res. **9**(6), 849–859 (1961)
13. Heßler, K., Gschwind, T., Irnich, S.: Stabilized branch-and-price algorithms for vector packing problems. Eur. J. Oper. Res. **271**(2), 401–419 (2018)
14. IBM: IBM ILOG CPLEX Optimizer (2022)
15. Irnich, S., Desaulniers, G.: Shortest path problems with resource constraints. In: Column Generation, pp. 33–65. Springer (2005)
16. Kling, R.M., Banerjee, P.: ESP: a new standard cell placement package using simulated evolution. In: 24th ACM/IEEE Design Automation Conference, pp. 60–66. IEEE (1987)
17. Kou, L.T., Markowsky, G.: Multidimensional bin packing algorithms. IBM J. Res. Dev. **21**(5), 443–448 (1977)
18. Kulik, A., Mnich, M., Shachnai, H.: Improved approximations for vector bin packing via iterative randomized rounding. In: 2023 IEEE 64th Annual Symposium on Foundations of Computer Science (FOCS), pp. 1366–1376. IEEE Computer Society, Los Alamitos, CA, USA (nov 2023)
19. Fernandez de La Vega, W., Lueker, G.S.: Bin packing can be solved within $1+\varepsilon$ in linear time. Combinatorica **1**(4), 349–355 (1981)
20. Lübbecke, M.E.: Column Generation. Wiley Encycl. Operat. Res. Manag. Sci. **17**, 18–19 (2010)
21. Lübbecke, M.E., Desrosiers, J.: Selected topics in column generation. Oper. Res. **53**(6), 1007–1023 (2005)
22. Mannai, F., Boulehmi, M.: A guided tabu search for the vector bin packing problem. In: Intern. Conf. of the African Fed. of Oper. Res. Soc. (AFROS) (2018)
23. Masson, R., et al.: An iterated local search heuristic for multi-capacity bin packing and machine reassignment problems. Expert Syst. Appl. **40**(13), 5266–5275 (2013)
24. Mommessin, C., et al.: Affinity-aware resource provisioning for long-running applications in shared clusters. J. Parallel Distrib. Comput. **177**, 1–16 (2023)
25. Mommessin, C., Erlebach, T., Shakhlevich, N.V.: Classification and evaluation of the algorithms for vector bin packing. Comput. Operat. Res. **173**, 106860 (2025)
26. Nagel, L., Popov, N., Süß, T., Wang, Z.: Analysis of heuristics for vector scheduling and vector bin packing. In: Sellmann, M., Tierney, K. (eds.) Learning and Intelligent Optimization - 17th International Conference, LION 17, Nice, France, June 4-8, 2023, Revised Selected Papers. LNCS. vol. 14286, pp. 583–598. Springer (2023)
27. Pandit, D., Chattopadhyay, S., Chattopadhyay, M., Chaki, N.: Resource allocation in cloud using simulated annealing. In: 2014 Applications and Innovations in Mobile Computing (AIMoC), pp. 21–27. IEEE (2014)
28. Panigrahy, R., Talwar, K., Uyeda, L., Wieder, U.: Heuristics for Vector Bin Packing. research. microsoft. com (2011)
29. Pessoa, A., Sadykov, R., Uchoa, E., Vanderbeck, F.: A generic exact solver for vehicle routing and related problems. Math. Program. **183**, 483–523 (2020)
30. Sadiq, M.S., Shahid, K.S.: Optimal multi-dimensional vector bin packing using simulated evolution. J. Supercomput. **73**(12), 5516–5538 (2017)
31. Spieksma, F.C.: A branch-and-bound algorithm for the two-dimensional vector packing problem. Comput. Operat. Res. **21**(1), 19–25 (1994)
32. Turky, A., Sabar, N.R., Dunstall, S., Song, A.: Hyper-heuristic local search for combinatorial optimisation problems. Knowl.-Based Syst. **205**, 106264 (2020)
33. Wei, L., Lai, M., Lim, A., Hu, Q.: A branch-and-price algorithm for the two-dimensional vector packing problem. Eur. J. Oper. Res. **281**(1), 25–35 (2020)

34. Wilcox, D., McNabb, A., Seppi, K.: Solving virtual machine packing with a reordering grouping genetic algorithm. In: 2011 IEEE Congress of Evolutionary Computation (CEC), pp. 362–369. IEEE (2011)
35. Woeginger, G.J.: There is no asymptotic PTAS for two-dimensional vector packing. Inf. Process. Lett. **64**(6), 293–297 (1997)
36. Wood, T., Shenoy, P.J., Venkataramani, A., Yousif, M.S.: Black-box and gray-box strategies for virtual machine migration. In: 4th Symposium on Networked Systems Design and Implementation (NSDI). USENIX (2007)

Author Index

A
Adak, Sumit 1
Ahouei, Saba Sadeghi 184
Aishwaryaprajna, 18
Alfandari, Laurent 116
Anderer, Simon 33
Antipov, Denis 184

B
Basseur, Matthieu 50
Boufar, Tarek 50

D
Da Ros, Francesca 66
Darmasaputra, Samuel Alan 101
Devendeville, Laure Brisoux 217
Di Gaspero, Luca 66
Do, Sy Hoang 101
Donne, Diego Delle 116

G
Goëffon, Adrien 84
Goudet, Olivier 84
Gunawan, Aldy 101

J
Justen, Nicolas 33
Juvigny, Corentin 116

K
Kötzing, Timo 167

L
Lackner, Marie-Louise 66
Liefooghe, Arnaud 133
Lucet, Corinne 217

M
Masson, Corentin 150
Mei, Yi 232
Mengshoel, Ole Jakob 150
Mostaghim, Sanaz 33
Musliu, Nysret 66

N
Nagel, Lars 249
Neumann, Aneta 184
Neumann, Frank 184

O
Ochoa, Gabriela 133

P
Popov, Nikolay 249

R
Radhakrishnan, Aishwarya 167
Rifki, Omar 50
Rowe, Jonathan E. 18

S
Sánchez-Díaz, Xavier F. C. 150
Sarhani, Malek 200
Saubion, Frédéric 84
Scheuermann, Bernd 33
Scouarnec, Justin 217
Süß, Tim 249

T
Tari, Sara 217
Tian, Yuan 232

V
Verel, Sébastien 84, 133
Voß, Stefan 200

W
Wang, Ze 249
Witt, Carsten 1

Y
Yu, Vincent F. 101

Z
Zhang, Mengjie 232